Advances in the Applications of Membrane-Mimetic Chemistry

Advances in the Applications of Membrane-Mimetic Chemistry

Edited by

Teh Fu Yen

University of Southern California
Los Angeles, California

Richard D. Gilbert

North Carolina State University
Raleigh, North Carolina

and

Janos H. Fendler

Syracuse University
Syracuse, New York

Springer Science+Business Media, LLC

Library of Congress Cataloging-in-Publication Data

On file

Based on the proceedings of an American Chemical Society Macromolecular Secretariate Symposium
on Advances in Membrane–Mimetic Chemistry and Its Applications, held April 15–16, 1991, in
Atlanta, Georgia

ISBN 978-0-306-44828-7   ISBN 978-1-4615-2580-6 (eBook)
DOI 10.1007/978-1-4615-2580-6

© 1994 Springer Science+Business Media New York

Originally published by Plenum Press in 1994

PREFACE

This volume had its birth from a symposium organized by the Macromolecular Secretariat of the American Chemical Society in Atlanta, GA, 1991. Since Macromolecular Secretariat has five participating divisions—Polymer Chemistry; Polymer Materials: Science and Engineering Division; Colloid and Surface Chemistry Division; Cellulose, Paper and Textile Division; and Rubber Division—the speakers were invited from these disciplinaries and they are truly interdisciplinary in multidisciplinary areas.

A number of papers are from the presentations at this symposium. However, some papers were subsequently invited to be sent in. Therefore, many papers have cited references with dates as late as this current year. This book emphasizes applications, and some of the papers were finished in 1993. Therefore, it is timely for scientists and engineers interested in this area of progress.

For scientists and engineers who are not familiar with this field, since the development is still youthful, this volume will cover some new frontiers, such as electronics, medical devices, fossil fuels, asphaltics, geochemistry, and environmental engineering. With that in mind, this book can be very useful as a reference. We do include a number of review papers in this volume. In summary, this book contains sixteen chapters with twenty-eight authors from various organizations and specialties.

We would like to take this opportunity to acknowledge the American Chemical Society for making it possible for all three editors to collaborate. The editors want to thank all the contributors for their effort in making the present volume possible. We equally appreciate their patience in waiting for this work to be completed.

Lastly, the editors would like to thank Plenum Publishing Corporation for their intent and patience. We especially would like to acknowledge Ms. Patricia M. Vann, editor, for her monumental help and thoughtful understanding, and also other copy editors of Plenum Publishing Corporation. We would like to mention Ms. Garine Gabrielian for processing the manuscripts and completing the painstaking process of checking each paper. Some parts of this book have also been checked and processed by Mr. Michael Lee and Mr. Brian Whitten, for which we are grateful.

Teh Fu Yen
University of Southern California

Richard D. Gilbert
North Carolina State University

Janos H. Fendler
Syracuse University

CONTENTS

MEMBRANE-MIMETIC APPROACH TO NANOTECHNOLOGY

Janos H. Fendler

Department of Chemistry
Syracuse University
Syracuse, New York 13244-4100

INTRODUCTION

Advanced materials in the nanometer dimension are the subject of ever increasing scrutiny, reports, and review articles.[1] Attention is increasingly focused upon the formation of "intelligent" materials which can, to different degrees, self assemble, self diagnose, self repair, and recognize and discriminate physical and/or chemical stimuli, and, at the extreme, have the capability of learning and self replicating.

Nanosized materials are size quantized; that is to say that their dimensions are comparable to the length of the de Broglie electron, the wavelengths of phonons, and the mean free paths of excitons.[2-5] Electron-hole confinement in nanosized spherical particles results in three-dimensional, quantum-size effects, i.e. in the formation of "quantum dots", "quantum crystallites", or "zero-dimensional excitons". In one-dimensional size quantization, the exciton is free to move only in two dimensions with the resultant formation of "quantum wells" or "two-dimensional excitons". In quantum wells, size quantization manifests in the growth direction, while bulk properties prevail in the other two dimensions. Finally, two-dimensional confinement of charge carriers (i.e. providing the exciton with only one-dimensional mobility) results in "quantum well wires". The importance of size and dimensionality quantizations is that they result in altered mechanical, chemical, electrical, optical, magnetic, electro-optical, and magneto-optical properties.[2-5] For example, quantum dots can be tuned by changing their diameters to absorb and emit light at desired wavelengths. This property renders the construction of finely tunable and efficient semiconductor lasers to be feasible. Semiconductor quantum dots can, in principle, be designed to capture a single electron at a time. Realization of this concept will not only provide answers to fundamental questions in quantum physics, but it will also open the door to the construction of ultrahigh density integrated circuits and information storage devices based on the presence or absence of individual electrons.[6]

The electronic structure of nanofabricated, 200-Å-diameter quantum dots was measured for the first time in 1987.[7] The method involved electron beam lithography of a chip containing a buried layer of quantum-well material, metal deposition on the resultant surface, and the removal of the resist from and the etching away of the chip, except where it was protected by the metal layer. This and other alternative solid-state methods do not lend themselves to large scale production or to the construction of quantum dots smaller than 100 Å.

Development of a new generation of nanostructured devices requires innovative approaches which are based on a fundamental understanding of the chemistry and physics involved at the molecular level. Chemists have become increasingly involved in the various aspects of materials science. They have designed new synthetic methodologies for established materials and have created new ones with unique physical and chemical properties. Colloid chemistry is particularly well suited for advancing materials science since

- it has matured[8] and has become quantitative and predictive thanks to both the vast number of new techniques being utilized and theories being developed; and,

- significantly, many biominerals, which constitute mother nature's response to advanced materials, can be considered to be colloidal systems.

Construction of nanostructured advanced materials, based on modern "wet" colloid chemistry, has been the long-term research objective of our laboratories. Our approach has been based on membrane mimetic chemistry[9] and inspired by biomineralization.[10-15] Advantage has been taken, in this approach, of membrane mimetic systems to provide chemical and spacial control for the *in situ* generation and stabilization of ultrasmall catalytic, semiconducting and magnetic particles and particulate films. Reversed micelles,[16-22] surfactant vesicles,[23-26] bilayer lipid membranes,[27-29] Langmuir-Blodgett (LB) films,[30,31] and monolayers[32-42] have been used as membrane-mimetic systems.

SEMICONDUCTORS UNDER MONOLAYERS

We have found monolayers to be particularly useful as templates for the *in situ* generation of nanocrystalline particulate films since they mimic two-dimensional crystal growth at biological surfaces[14,15] and since precise crystallographic information has become available on the orientation and packing of surfactant headgroups from X-ray diffraction studies using synchrotron sources.[43-48] There are several advantages to using monolayer matrices for nanoparticulate film generation. First, stable, well-characterized and long-lasting monolayers can be formed from a large variety of surfactants. Second, monolayer surface areas and charges are two-dimensionally controllable and the composition of the aqueous subphase is readily variable. Third, monolayers, along with the particulate films grown under them, can be conveniently transferred to solid supports (ie. to substrates). Differences between inorganic particles generated between the headgroups of LB films[31] and particulate films formed under monolayers floating on an aqueous subphase[35-42] should be recognized. Available space between the headgroups of LB films limits the growth of the particles to 40-60 Å. Conversely, particulate films up to several thousand Å can be grown under monolayers; and, subsequent to their transfer to substrates, the surfactant monolayer can, if desired, be removed.

Experimental set-ups used for the generation and *in situ* monitoring of nanocrystalline particulate films are illustrated in Figure 1. The arrangement shown in the top allows the injection of a precursor gas (H_2S, H_2Se, NH_3, for example, while that in the bottom permits the generation of the desired gaseous precursor (Na_2S + dilute acid, for example, for generating H_2S). Facilities were available for determining surface pressure *vs.* surface area and surface potential *vs.* surface area isotherms in the film balance placed under the glass cover. Arrangements were also made for the continuous monitoring of reflectivities; angle-dependent reflectivities; Brewster-angle and fluorescence microscopies; and non-linear optical parameters.

Figure 1. Schematics of the experimental arrangements used for the generation of semiconductor particles at the negatively charged, surfactant headgroup-aqueous subphase interface and that used for the in situ monitoring of reflectivities. P = polarizer and D = detector.

Evolution of a nanocrystalline particulate film, illustrated by the formation of sulfide semiconductor particulate films (Figure 2), has been discussed in terms of the following steps:[35]

a) formation of metal-sulfide bonds at a large number of sites at the monolayer-aqueous interface;

b) downward growth of well-separated nanocrystalline metal sulfide particles;

c) coalescence of clusters into interconnected arrays of semiconductor particles;

d) formation of the "first layer" of a porous sulfide semiconductor particulate film composed of 20- to 40-Å-thick, 30- to 80-Å-diameter particles;

e) diffusion of fresh metal ions to the monolayer head group area;

f) formation of a "second layer" of the porous sulfide semiconductor particulate film (by using steps a, b, and c); and

g) build-up of "subsequent layers" of the sulfide semiconductor particulate film (by using steps a, b, and c) up to a plateau thickness (ca. 300 Å for CdS and ca. 3500 Å for ZnS) beyond which the film cannot grow.

The presence of a monolayer with an appropriate surface charge is an essential prerequisite for sulfide semiconductor particulate film formation. Infusion of H_2S over an aqueous metal-ion solution, in the absence of a monolayer, resulted in the formation of large, irregular, and polydispersed metal-sulfide particles which precipitated in the bulk solution before settling to the bottom of the trough. Furthermore, no sulfide semiconductor particulate film formation could be observed upon the infusion of H_2S to a positively charged monolayer (dioctadecyldimethyl-ammonium bromide, for example) floating on an aqueous metal-ion ($CdCl_2$, for example) subphase.

Figure 2. Proposed schematics for the initial and subsequent growth of a monolayer-supported, porous, SQSPF. The d_x and d_y dimensions are in the plane and the d_z dimension is normal to the plane; they refer to the earliest observable particles. d'_x, d'_y, and d'_z are dimensions in the plane and are normal to the plane; they refer to particles observed at later stages of their growth.

To-date, cadmium sulfide, zinc sulfide, lead sulfide, cadmium selenide, and lead selenide semiconductor particulate films have been grown, in situ, under mono-layers.[33,35-42,49] Absorbances (A) increased linearly with increasing thicknesses of the CdS and ZnS particulate films. Absorption coefficients, σ-values, were calculated from

$$\sigma = A/d'_s \tag{1}$$

and determined to be 2.4×10^5 cm^{-1} at 239 nm and 5.8×10^4 cm^{-1} at 475 nm for the CdS particulate film. These values agreed well with that determined for electro-deposited CdS films (σ (435 nm) $= 2.06 \times 10^4$ cm^{-1}). Similarly, an absorption coefficient of 5.8×10^4 cm^{-1} at 315 nm was determined for the ZnS particulate film. Knowledge of absorption coefficients facilitated the assessment of direct band-gap energies, E_g, from

$$(\sigma h\omega)^2 = (h\omega - E_g)C \qquad\qquad (2)$$

where $h\omega$ is the photon energy. Typical plots of the data, as determined by Eq. 2, are shown in Figure 3.[36] Values of E_g for CdS particulate films of d'_s = 63 Å, 125 Å, 163 Å, 204 Å, and 298 Å were assessed to be 2.54 eV, 2.48 eV, 2.46 eV, 2.44 eV, and 2.43 eV (Figure 3). Henglein's published E_g vs. particle-size curve[2] was used to estimate the average diameter of the 63-Å-thick CdS particles to be ca. 50 Å.

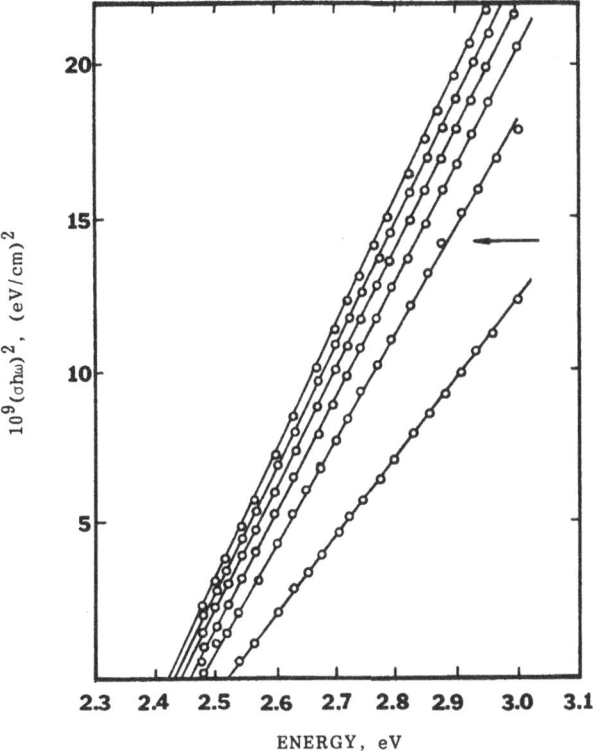

Figure 3. Plots of $(\sigma h\omega)^2$ against energy for 63-Å-, 125-Å-, 163-Å-, 204-Å-, 263-Å-, and 298-Å-thick (in the order shown by the arrow) CdS particulate films.

Increasing the thickness of the CdS particulate film resulted in progressively decreased direct band-gaps and, hence, in progressively larger CdS particles. The thickest CdS particulate film studied exhibited a direct band-gap equal to that reported for bulk CdS semiconductors (2.4 eV).[2] A direct band-gap of 3.75 eV was assessed for the 359-Å-thick ZnS particulate film.

Prolonged heating of the semiconductor particulate films at high temperatures resulted in pronounced changes in their absorption spectra. The absorbance of a 192-Å-thick CdS particulate film (vacuum dried at 10^{-3} torr for three days) decreased upon heating at 490° C for 5, 15, and 25 minutes.[36,37,49] Five minutes of heating shifted the direct band-gap from 2.47 eV to 2.40 eV (or to an absorption edge of 515 nm) equal to that of bulk CdS. A similar behavior was noted for 359-Å-thick ZnS particulate films; heating at 300° C for 15 minutes shifted the direct band-gap from 3.75 eV to 3.64 eV (or to an absorption edge of 340 nm). Prolonged heating of semiconductor particulate films have, therefore, two important consequences. First, their properties become similar to those found for bulk semiconductors. Second, they are annealed to the substrate. Annealed semiconductor particulate films could not be washed or wiped away from their substrates. In contrast, vertical dipping of untreated semiconductor particulate films into water resulted in a partial loss of material from the subphase. Interestingly, annealing of semiconductor particulate films, prepared under thiol surfactants, resulted in band-gap shifts to higher energy.[50] Transmission electron micrographs of 30- to 50-Å-thick CdS particulate films indicated the presence of 20- to 80-Å-diameter particles possessing a relatively narrow size distribution and average diameters of 47 Å.[34,36] HOPG was established, using scanning tunneling microscopy (STM), to provide an atomically flat surface with periodic roughnesses in the order of 1 Å. In two-dimensional STM images of HOPG-supported ZnS and CdS particulate films, the presence of 10- to 20-Å-thick, 30- to 40-Å-diameter ZnS and 20- to 30-Å-thick, 40- to 50-Å-diameter CdS particles is clearly discernable. The widths of the semiconductor particles observed by STM agree well with the corresponding diameters determined by transmission electron microscopy.[34,36]

Electrical and photoelectrical measurements were carried out on CdS particulate films deposited on glass substrates or teflon sheets.[36] The resistivity (ρ) of a semiconductor particulate film, measured between two parallel copper electrodes, is given by

$$\rho = R \, \frac{Ld'_s}{a} \tag{3}$$

where R is the measured resistivity in Ω, L is the length of the copper electrodes, a is the distance between them, and d'_s is the thickness of the semiconductor particulate film. Resistivities of 200- to 300-Å-thick CdS particulate films were determined to fall in the range of $(3-6)10^7 \, \Omega$ cm. These values represent the measurement of 10 samples of different thicknesses and may be attributed, in part, to the presence of different amounts of water in the films. The ρ values determined for CdS particulate films are some six orders of magnitude higher than those observed for materials having intrinsic conductivity.

The dark resistance of CdS particulate films was found to decrease exponentially with increasing temperature.[36] Illumination decreased the resistivity (i.e. increased the conductivity) of CdS particulate films by some two orders of magnitude and matched the absorption spectrum of the corresponding CdS particulate film nicely. Photoconductivity originates, therefore, in the production of conduction band electrons, e^-_{CB}, and valence band holes, h^+_{VB}, during band-gap irradiation of CdS:

$$CdS \xrightarrow{h\nu} e^-_{CB} + h^+_{VB} \tag{4}$$

Steady-state irradiation of CdS particulate films also resulted in the development of photovoltage. Irradiation by a 10-nsec, 343-nm laser pulse gave rise to a transient photovoltage. The magnitude of the photovoltage (1-8 mV) was found to increase linearly with the energy of the laser pulse (0.1-1.0 mJ). The rise time of the transient signal, corresponding to Eq. 4, was faster than the response time of the instrument used (10 nsec). The decay time of the signal was on the order of 3×10^{-4} seconds. This decay corresponds to charge recombination.

ORIENTED PARTICULATE GROWTH UNDER MONOLAYERS: MOLECULAR RECOGNITION

Molecular recognition between the monolayer headgroups and incipient semiconductor nanocrystallites can, in many ways, be regarded as mimicking biomineralization[10-13] and represents an important milestone in the realization of the potential of a colloid-chemical approach to band-gap engineering. Lead sulfide (PbS) particulate films composed of highly oriented, equilateral-triangular crystals have been in situ generated by the exposure of AA-monolayer-coated, aqueous, lead-nitrate $[Pb(NO_3)_2]$ solutions to H_2S (Figure 4).[41] AA monolayers, in their solid states, consist of $CH_3(CH_2)_{18}COOH$ molecules two-dimensionally arrayed at the air-water interface. Spread over the aqueous subphase, the carboxyl or the carboxylate groups of AA are aligned perpendicularly to the water surface. The alkyl chains of AA, fully extended in the air in a planar zig-zag conformation, are oriented approximately normal to the surface in a triangular lattice of hexagonal close packing with a lattice constant of a = 4.85 Å.[44,51] Combined synchrotron X-ray reflection and diffraction data established a structural model for AA monolayers at air-water interfaces. The model required the hydrocarbon chains to be well packed in a pseudohexagonal lattice and tilted toward their nearest neighbor.[52]

Rationalization of the packing of the AA headgroups at the water-air interface is, unfortunately, less than straightforward.[45] The absence of information concerning the extent of headgroup ionization (at a bulk pH of 5.5), counterion binding, and the degree of hydration hinders the interpretation of experimental results and the development of a reliable theoretical approach for predicting headgroup organization at the monolayer-subphase interface. Using the experimentally determined value for the surface area of one AA molecule (20.0 Å^2/molecule) permitted assessment of the lattice constant to be 4.81 Å (a) and the $d_{(100)}$ spacing to be 4.16 Å ($d_{(100)}$ = a·sin60°). These values are in good agreement with those determined for AA monolayers by synchrotron X-ray scattering (a = 4.85 Å and d_{100} = 4.13 Å).[44] They are also similar to those determined for cadmium stearate (a = 4.89 Å and $d_{(100)}$ = 4.20 ± 0.10 Å) and other fatty acid monolayers.[44,51]

Reliable assessment of the arrangement and crystallinity of AA-monolayers supported by a Pb^{2+} subphase is equally elusive. Our data is best accommodated in terms of an AA:Pb^{2+} = 3:4 ratio (Figure 5). Grazing incidence X-ray diffraction measurements of lead arachidate monolayers demonstrated the existence of long-range ordering (250 Å) of Pb^{2+}.[51]

PbS is known to crystallize in a cubic crystalline lattice with a lattice constant of a = 5.9458 Å. Atomic coordinates are (0,0,0) and (½,½,0) for Pb and (½,½,½) and (0,0,½) for S. The nearest-neighbor separation of Pb-Pb and S-S atoms of 4.20 Å matches the $d_{(100)}$ network spacing of the AA monolayer. This fit implies the alignment of PbS along its (111) plane to the (100) plane of the AA monolayer (Figure

5). A comparison of the interatomic Pb to Pb distance of the (111) plane of the PbS crystal (4.20 Å) with that of the $d_{(100)}$ spacing of the AA monolayers (4.16 Å) revealed a mismatch of only 1% between these two crystals.

Epitaxial growth of PbS under well-compressed AA monolayers is explicable in terms of the geometrical complementarity between PbS and the AA headgroups. The strong intrinsic electrostatic interaction results in a very high Pb^{2+} concentration at the monolayer interface. The extremely low solubility of PbS in water (K_{SP} = 8.81 x 10^{-29} at 25° C) favors its rapid and random nucleation. However, the presence of the

Figure 4. Transmission electron micrograph (TEM) at limiting aperture coverage of PbS crystals formed by the slow (30 min.) infusion of H_2S to an AA monolayer in the Lauda film balance (kept at π = 26 mNm^{-1} surface pressure) floating on an aqueous 5.0 x 10^{-4} M $Pb(NO_3)_2$ solution. The PbS particulate film was deposited on a formvar-coated, 200-mesh copper grid.

monolayer acts to drastically diminish the reaction rate, limiting the encounter of the PbS precursors. The rate and amount of H_2S can also be controlled. These measures have ensured the formation of a critically sized nucleus and the subsequent ion-by-ion heteroepitaxial growth of the PbS crystals.

Significantly, equilateral-triangular PbS crystals have been grown under compressed monolayers in the same orientation (see Figure 4).

Epitaxial growth of nanocrystalline lead selenide particles under AA monolayers has also been demonstrated.[42] Exposure of an AA-monolayer-coated $Pb(NO_3)_2$ solution to H_2Se resulted in the formation of reasonably uniform crystals which constituted the particulate film. A typical three-dimensional, 470 nm-by-470 nm image of PbSe crystallites is shown in Figure 6.[42] The thickness of these crystals was

Figure 5. top) Three-dimensional representation of the PbS crystal lattice and (111) plane. bottom) Schematic two-dimensional representation of the proposed overlap between Pb^{2+} ions and AA headgroups; $\bigcirc$ = AA headgroup, $\bullet$ = Pb^{2+}, and $\circledcirc$ = Pb^{2+} and AA headgroups. A unit cell is highlighted by the dotted area which is enclosed by heavy lines.

determined to be in the order of 65 Å by sectioning techniques. The equilateral triangular PbSe crystals were not entirely uniform; along their corners they were somewhat thicker than in their middles (Figure 6). Interestingly, almost all crystals were aligned in their edge directions. Electron diffraction of a 2-μm-diameter area (Figure 6) showed diffraction spots corresponding to the {220}, {422}, and {440} planes, indicating that all crystals line up and have their (111) planes perpendicular to the electron beam.

Figure 6. Three-dimensional side view of a PbSe particulate film imaged on a 470-nm x 470-nm section by atomic force microscopy (AFM).

In some cases, particularly at high monolayer surface pressures, formation of rod-like PbSe particles was observed (Figure 7).[42] The thickness of the rods was typically 10 nm or less and the length was on the order of 100 nm. In addition to the rods, there were also many dot-like particles in the 10-nm size range (Figure 7). Generation of rod-like semiconductor particles under monolayers provides a potential approach to nanofabricated quantum wires.

GROWTH OF MIXED SEMICONDUCTOR PARTICULATE FILMS UNDER MONOLAYERS: AN APPROACH TO SUPERLATTICE FORMATION

Methods have also been developed for the formation of composite and sandwich semiconductor particulate films.[53] Composite semiconductor particulate films have been prepared by the infusion of hydrogen sulfide across well-compressed monolayers, prepared from a thiol surfactant, which coated a mixture of aqueous lead nitrate and zinc nitrate solutions. The composition of the semiconductor was determined by the concentration ratios of the lead and nitrate ions in solutions and by their binding constants to the thiol surfactants.

Figure 7. TEM of a PbSe particulate film. The film was formed by the 25-minute infusion of H_2Se (50 μL) over an AA monolayer, kept at 40 mN/m, which was floating on an aqueous 5.0×10^{-4} M $Pb(NO_3)_2$ solution in the Lauda trough. The PbSe particulate film was transferred to an amorphous-carbon-coated 200-mesh copper grid. The bar = 200 nm.

A sandwich semiconductor is like a cherry and its stone; it contains one type of material as its core and a second material as its shell. Sandwich semiconductor particulate films have been prepared by the sequential formation of lead sulfide and zinc sulfide (or zinc sulfide and lead sulfide) under monolayers prepared from $(C_{18}H_{37})_2N^+(CH_2CH_2SH)_2Br^-$.[53] Absorption spectra of ZnS-, PbS-, and PbS-coated ZnS are illustrated in Figure 8.

Excitation by visible light into composite or sandwich semiconductor particulate films results in an initial charge separation in the smaller band-gap semiconductor. If the conduction band of the second semiconductor lies at a lower energy than that in the first material, the electrons are promptly transferred into it. Thus, efficient charge separation can be brought about by the utilization of composite and/or sandwich semiconductor particulate films. Selective excitation into lead sulfide in a zinc-sulfide-coated lead sulfide semiconductor particulate film resulted in electron transfer into ZnS, which was detected by the development of a blue color due to reduced methylviologen (Figure 9).[53] Importantly, irradiation into ZnS semiconductor particles did not result in methylviologen reduction and sandwich semiconductor particulate films

11

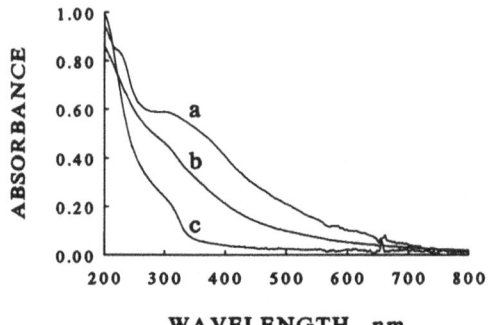

WAVELENGTH, nm

Figure 8. Absorption spectra of ZnS (a), ZnS(PbS) (b), and PbS (c) generated *in situ* under monolayers prepared from $C_{18}H_{37}N(CH_2CH_2SH)_2$. The subphase contained 2.5×10^{-4} M $Zn(NO_3)_2$ (a) or $Pb(NO_3)_2$ (c). The ZnS(PbS) composite semiconductor particulate film was prepared by exchanging Zn^{2+} with Pb^{2+} subsequent to the formation of the ZnS particulate film. Semiconductor particulate films were formed by H_2S (250 μL) infusion over the monolayers for 60 (a and b) and 30 (c) minutes.

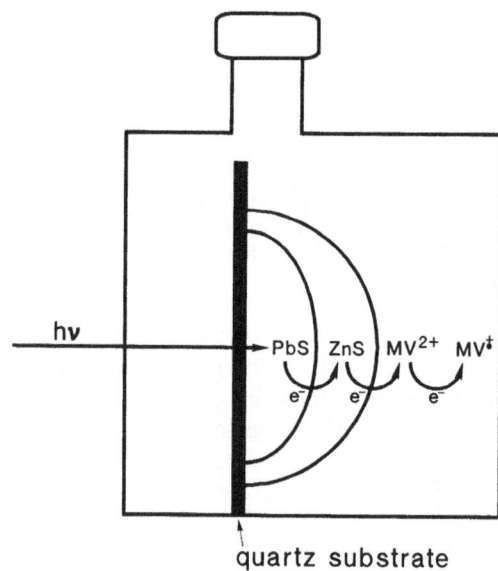

quartz substrate

Figure 9. Schematics of sequential photoelectron transfer from PbS to ZnS to MV^{2+}.

were twice as efficient in mediating photoelectron transfer than PbS particulate films which were prepared under identical conditions (Figure 10). These results are encouraging since they open the door to superlattice construction.

CONCLUSION

Semiconductor particulate film preparation under monolayers permits convenient thickness, morphology, and dimensionality controls. Manipulation of the surfactant composition and organization, as well as the chemistries at the monolayer interphase and subphase, permits the generation of periodically and spatially modulated superlattices. The relatively simple experimental procedure is versatile and allows the convenient monitoring of the physical and chemical changes which accompany the *in*

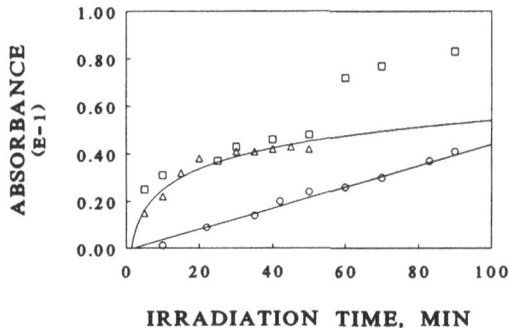

IRRADIATION TIME, MIN

Figure 10. Photoelectron transfer to methyl viologen. Absorption spectra of MV$^{+\bullet}$, generated in the steady-state irradiation of PbS (c) and ZnS(PbS) (b) semiconductor particulate films (deposited onto quartz substrates) in 4.0 x 10^{-4} M aqueous methylviologen solution which contained 0.01% (v/v) benzyl alcohol at different irradiation times, using a 200 W Hg/Xe lamp with a 450 nm cut-off filter. Prior to irradiation, the solution was deaerated by Ar bubbling for 1 hour.

situ generation of nanocrystalline particulate films. Importantly, monolayer-supported nanocrystalline particulate films can be readily transferred onto solid substrates. The method developed here enables the characterization of the same system at an aqueous-solid interface (*ie.*, at an electrolyte-solid state junction) and in the solid state. It provides, therefore, a molecular-level insight into currently utilized epitaxies. Most significantly, the colloid chemical approach to bandgap engineering will open the door to construction of highly innovative nanofabricated electronic devices.

ACKNOWLEDGMENTS

Support of this work by a grant from the National Science Foundation is greatly appreciated. The true credit is due, of course, to my co-workers (whose names appear in the cited joint publications) for their creative, skillful, and dedicated work.

REFERENCES

1. G.A. Ozin, Nanochemistry: synthesis in diminishing dimensions, *Adv. Mater.* 4:612 (1992).
2. A. Henglein, Mechanism of reactions on colloidal microelectrodes and size quantization effects, *Top. Curr. Chem.* 143:115 (1988).
3. A. Henglein, Physicochemical properties of extremely small colloidal metal and semiconductor particles, *Chem. Rev.* 89:1861 (1989).
4. M.G. Bawendi, M.L. Steigerwald, and L.E. Brus, The quantum mechanics of larger semiconductor clusters ("quantum dots"), *Ann. Rev. Phys. Chem.* 41:477 (1990).
5. Y. Wang, Nonlinear optical properties of nanometer-sized semiconductor clusters, *Acc. Chem. Res.* 24:133 (1991).
6. A. Chiabrera, E. Di Zitti, F. Costa, and G.M. Bisio,, Physical limits of integration and information processing in molecular systems, *J. Phys. D: Appl. Phys.* 22:1571 (1989).
7. M.A. Reed, Quantum dots, *Scientific American* January:118 (1993).
8. A. Adamson. "Physical Chemistry of Surfaces," John Wiley & Sons, New York (1990).
9. J.H. Fendler. "Membrane Mimetic Chemistry," John Wiley, New York (1982).
10. P. Westbroek and E.W. de Jong. "Biomineralization and Biological Metal Accumulation," Reidel, Dordrecht, Holland (1983).
11. K. Simkiss and K.M. Wilbur. "Biomineralization," Academic Press, San Diego (1989).
12. H.A. Lowenstam and S. Weiner. "On Biomineralization," Oxford University Press, New York (1989).
13. L. Addadi and S. Weiner, Control and design principles in biological mineralization, *Angew. Chem. Int. Ed. Eng.* 31:153 (1992).
14. S. Mann, D.D. Archibald, J.M. Didymus, B.R. Heywood, F.C. Meldrum, and V.J. Wade, Biomineralization: biomimetic potential at the inorganic-organic interface, *MRS Bull.* 17(10):32 (1992).
15. S. Mann, J. Webb, and R.J.P. Williams. "Biomineralization, Chemical and Biological Perspectives," VCH, Weinheim, Germany (1989).
16. M. Meyer, C. Wallberg, K. Kurihara, and J.H. Fendler, Photosensitized charge separation and hydrogen generation in reversed micelle entrapped platinized colloidal cadmium sulfide, *J. Chem. Soc. Chem. Commun.* 90 (1984).
17. P. Lianos and J.K. Thomas, Cadmium sulfide of small dimensions produced in inverted micelles, *Chem. Phys. Lett.* 125:299 (1986). P. Lianos and J.K. Thomas, Small CdS particles in inverted micelles, *J. Colloid Interface Sci.* 117:505 (1987).
18. T.F. Towey, A. Khan-Lodhi, and B.H. Robinson, Kinetics and mechanism of formation of quantum-sized cadmium sulphide particles in water-aerosol-OT-oil microemulsions, *J. Chem. Soc. Faraday Trans.* 86:3757 (1990).
19. L. Motte, C. Petit, L. Boulanger, P. Lixon, and M.P. Pileni, Synthesis of cadmium sulfide in situ in cadmium bis(ethyl-2-hexyl) sulfosuccinate reverse micelle, *Langmuir* 8:1049 (1992).
20. T. Dannhauser, M. O'Neil, K. Johansson, D. Whitten, and G. McLendon, Photophysics of quantized colloidal semiconductors dramatic luminescence enhancement by binding of simple amines, *J. Phys. Chem.* 90:6074 (1986).
21. M.L. Steigerwald, A.P. Alivisatos, J.M. Gibson, T.D. Harris, R. Kortan, A.J. Muller, A.M. Thayer, T.M. Duncan, D.C. Douglas, and L.E. Brus, Surface derivatization and isolation of semiconductor cluster molecules, *J. Am. Chem. Soc.* 110:3046 (1988).
22. M.A. Marcus, W. Flood, and M.L. Steigerwald, Structure of capped CdSe clusters by EXAFS, *J. Phys. Chem.* 95:1572 (1991).
23. Y.M. Tricot and J.H. Fendler, *In situ* generated colloidal semiconductor CdS particles in dihexadecylphosphate vesicles: quantum size and asymmetry effects, *J. Phys. Chem.* 90:3369 (1986).
24. H.-C. Youn, Y.-M. Tricot, and J.H. Fendler, Photochemistry of colloidal cadmium sulfide at dihexadecylphosphate interfaces: electron transfer to methylviologen and colloidal rhodium, *J. Phys. Chem.* 91:581 (1987).
25. H.J. Watzke and J.H. Fendler, Quantum size effects of *in situ* generated colloidal CdS particles in dioctadecyldimethylammonium chloride surfactant vesicles, *J. Phys. Chem.* 91:854 (1987).
26. H.-C. Youn, S. Baral, and J.H. Fendler, Dihexadecyl phosphate, vesicle-stabilized and *in-situ*-generated mixed CdS and ZnS semiconductor particles. Preparation and utilization for photosensitized charge separation and hydrogen generation, *J. Phys. Chem.* 92:6320 (1988).
27. S. Baral, X.K. Zhao, R. Rolandi, and J.H. Fendler, Formation and characterization of microcrystalline semiconductor particles on bilayer lipid membranes, *J. Phys. Chem.* 91:2701 (1987).

28. X.K. Zhao, S. Baral, R. Rolandi, and J.H. Fendler, Semiconductor particles in bilayer lipid membranes (BLMs). Formation, characterization, and photoelectrochemistry, *J. Am. Chem. Soc.* 110:1012 (1988).

29. X.K. Zhao, P.J. Herve, and J.H. Fendler, Magnetic particulate thin films on bilayer lipid membranes (BLMs), *J. Phys. Chem.* 93:908 (1989).

30. X.K. Zhao, S. Xu, and J.H. Fendler, Ultrasmall magnetic particles in Langmuir-Blodgett films, *J. Phys. Chem.* 94:2573 (1990).

31. S. Xu, S.K. Zhao, and J.H. Fendler, Ultrasmall semiconductor particles sandwiched between surfactant headgroups in Langmuir-Blodgett films, *Adv. Mater.* 2:183 (1990).

32. Y. Yuan, I. Cabasso, and J.H. Fendler, Size-quantized, semiconductor-particle-mediated photo-electron transfer in ultrathin, phosphonate-functionalized, polymer-blend membranes, *Macromol.* 23:3198 (1990).

33. K.C. Yi and J.H. Fendler, Template-directed semiconductor size quantization at monolayer-water interfaces and between the headgroups of Langmuir-Blodgett films, *Langmuir* 6:1519 (1990).

34. X.K. Zhao, Y. Yuan, and J.H. Fendler, Size-quantized semiconductor particles formed at monolayer surfaces, *J. Chem. Soc. Chem. Commun.* 1248 (1990).

35. X.K. Zhao, S. Xu, and J.H. Fendler, Semiconductor particles formed at monolayer surfaces, *Langmuir* 7:520 (1991).

36. X.K. Zhao and J.H. Fendler, Semiconductor particulate films on solid supports, *Chem. Mater.* 3:168 (1991).

37. X.K. Zhao and J.H. Fendler, Size quantization in semiconductor particulate films, *J. Phys. Chem.* 95:3716 (1991).

38. X.K. Zhao, L.D. McCormick, and J.H. Fendler, Scanning tunneling microscopic, optical, and scanning tunneling spectroscopic characterization of size-quantized cadmium selenide particulate films *in situ* generated at monolayer interfaces, *Langmuir* 7:1255 (1991).

39. X.K. Zhao, L.D. McCormick, and J.H. Fendler, Electrical and photoelectrochemical characterization of CdS particulate films by scanning electrochemical microscopy, scanning tunneling microscopy and scanning tunneling spectroscopy, *Chem. Mater.* 3:922 (1991).

40. X.K. Zhao, L.D. McCormick, and J.H. Fendler, Preparation-dependent rectification behavior of lead sulfide particulate films, *Adv. Mater.* 4:93 (1992).

41. X.K. Zhao, J. Yang, L.D. McCormick, and J.H. Fendler, Epitaxial formation of PbS crystals under arachidic acid monolayers, *J. Phys. Chem.* 96:9933 (1992).

42. J. Yang, J.H. Fendler, T.-C. Jao, and T. Laurion, Electron and atomic force microscopic investigations of lead selenide crystals grown under monolayers, *J. Electron Microsc. Tech.* in press (1993).

43. K. Kjaer, J. Als-Nielsen, C.A. Helm, P. Tippmann-Krayer, and H. Möhwald, Synchrotron x-ray diffraction and reflection studies of arachidic acid monolayers at the air-water interface, *J. Phys. Chem.* 93:3200 (1989).

44. J. Als-Nielsen and H. Möhwald, Synchrotron x-ray scattering studies of Langmuir films, *in:* "Handbook on Synchrotron Radiation," S. Ebashi, M. Koch, and E. Rubinstein, eds., Elsevier Science Publishers, B.V., The Netherlands (1991).

45. D. Jacquemain, S.G. Wolf, F. Leveiller, M. Deutsch, K. Kjaer, J. Als-Nielsen, M. Lahav, and L. Leiserowitz, Two-dimensional crystallography of amphiphilic molecules at the air-water interface, *Angew Chem. Int. Ed. Engl.* 31:130 (1992).

46. D. Möbius and H. Möhwald, Structural characterization of monolayers at the air-water interface, *Adv. Mater.* 3:19 (1991).

47. D. Jacquemain, F. Leveiller, S.P. Weinbach, L. Leiserowitz, K. Kjaer, and J. Als-Nielsen, Crystal structures of self-aggregates of insoluble aliphatic amphiphilic molecules at the air-water interface, *J. Am. Chem. Soc.* 113:7684 (1991).

48. P. Tippmann-Krayer and H. Möhwald, Precise determination of tilt angles by x-ray diffraction and reflection with arachidic acid monolayers, *Langmuir* 7:2303 (1991).

49. Y. Yuan, I. Cabasso, and J.H. Fendler, Preparation of ultrathin, size-quantized semiconductor particulate films at oriented mono- and poly(vinylbenzyl)phosphonate interfaces and their characterization on solids, *Chem. Mater.* 2:226 (1990).

50. K. Yi and J.H. Fendler, Unpublished results (1992).

51. H. Möhwald, Phospholipid and phospholipid-protein monolayers at the air/water interface, *Annu. Rev. Phys. Chem.* 41:441 (1990).

52. M. Sze, "Physics of Semiconductor Devices," John Wiley & Sons, New York (1981).

53. K. Yi and J.H. Fendler, Unpublished results (1993).

AMPHIPATHIC CHITOSAN SALTS

Thomas Rathke and Samuel M. Hudson

Fiber and Polymer Science Program
Box 8301   College of Textiles
North Carolina State University
Raleigh, NC   27695-8301

INTRODUCTION

The interaction of polyelectrolytes with amphipathic molecules, such as lipids, is widely considered to play a key role in the formation of biological membranes. Mesophase formation by polyelectrolytes may also play a role in the self assembly of these membranes. There have been some theoretical treatments of polyelectrolyte mesophases and there is some experimental evidence for DNA cholesteric phases[1-4]. In this article we consider the behavior of the biopolymer chitosan, which is the partially N-deacetylated form of the polysaccharide chitin.

Chitosan is a copolymer of 2-acetamido-2-deoxy-β-D-glucopyranose (GlcNAc) and 2-amino-2-deoxy-β-D-glucopyranose (GlcN) residues. At acidic pH values, the GlcN unit forms a cationic site, and chitosan may then be soluble in aqueous solution, depending on the degree of N-acetylation (d.a.). The cationic polymer can then be considered as a heterogeneously structural polyelectrolyte. Chitosan is also a plentiful, highly versatile biopolymer, under consideration for a number of applications[5].

The objective of this work was to screen the formation of a number of amphopathic salts of chitosan with various surfactants. An amphoteric surfactant was identified which formed a soluble complex with chitosan that was easily converted into a film structure. The formation of these films and a preliminary report on their physical properties and morphology is reported.

EXPERIMENTAL

MATERIALS

Chitin was purchased from Sigma Chemical Company, USA. Chitosan was obtained by treating it twice with 50% w/v NaOH in deionized water for 2 hrs at 100°C under nitrogen. After each treatment the samples were washed until neutral.

Hydrobromic acid titration was used to determine the degree
of N-deacetylation, by the method of Domszy and Roberts[6].
The surfactants are shown in Table 1. These reagents
were used as received.

FORMATION OF AMPHIPATHIC SALTS AND THEIR FILMS

To convert chitosan into films, the polymer was
dissolved in 5% v/v aq. acetic acid at 5% w/v concentration.
To the chitosan solution, a stoichiometric amount of
surfactant was added. In most cases a precipitate was
observed and collected. Those salts which stayed in aqueous
solution were cast into films with a doctor knife onto glass
plates and allowed to air dry. Chitosan/surfactant
solutions and films were observed under crossed polars using
an Olympus microscope, Model BHSP.

Table 1. Surfactants screened for film casting with
chitosan.

Anionic Surfactants

Sodium Lauryl Sulfate	Sodium Octyl Sulfate
Xylene Sulfonate	Sodium Octadecyl Sulfate
Sodium Cetylsulfuric acid	Tamol 731
Emersol 153NF Stearic Acid	1,2-Ethanesulfonic Acid (disodium salt)
Sodium Cumenesulfonate	Igepon TK-32
Tryfac 5560	Aerosol OT
Igepon T-77	Igepon TC-42
Sodium Sulfanilic Acid	Sodium Benzene Sulfonate

Cationic Surfactants

Dodecyltrimethylammonium Bromide

Nonionic Surfactants

Triton X-100
Polyoxethylene 4 Lauryl Ether
Polyoxethylene 10 Lauryl Ether
Polyoxethylene 23 Lauryl Ether

Amphoteric Surfactant

Dihydrogenated Tallow Dimethyl Ammonium Methyl Sulfate
Taurine
N-Octyl-N,N Dimethyl-3-Ammonio-1-Propane-Sulfonate
N-Dodecyl-N,N Dimethyl-3-Ammonio-1-Propane-Sulfonate (DAP)
N-Octyldecyl-N,N Dimethyl-3-Ammonio-1-Propane-Sulfonate

RESULTS AND DISCUSSION

Table 1 shows the list of surfactants that were
screened to check for binding with chitosan. Of particular
interest was whether chitosan/surfactant complexes could

form meso-phase systems. Whereas chitosan itself can form gel-like mesophase solutions as shown in Figure 1, and as reported previously by Ogura et.al.[7], no birefringent solutions were directly observed for any of these chitosan/surfactant systems reported here. In fact, as expected, anionic surfactants when added to the dissolved chitosan gave an inhomogeneous solution, forming precipitates. This was due to the cationic-anionic binding in the acidic solution. These salts are slightly soluble in some organic solvents, particularly dimethyl sulfoxide. The cationic surfactants gave homogeneous solutions, but the films obtained were too weak to effectively remove from the glass plate. Birefringent solutions were not observed at high concentrations. The non-ionic surfactants gave homogeneous solutions which had a turbid appearance, but no birefringence. Again, the resultant films were waxy and had no strength.

The most interesting behavior is that given by the amphoterics. The amphoterics at and below a hydrocarbon tail of 12 carbons, gave homogeneous solutions while those with a larger tail gave a non-homogeneous solution or did not dissolve. Films cast from N-dodecyl-N,N-dimethyl-3-aminopropane sulfonate (DAP) yielded the best tensile properties. The chitosan/DAP solutions were not birefringent at solid levels of 15% w/v. In fact, it was observed that DAP clarified solutions of chitosans, at low concentrations, that were still N-acetylated greater than 20%, which can contain undissolved gels. The amphoterics apparently bind tightly with chitosan, bringing into solution, marginally soluble material. The tensile properties of the films obtained from these solutions are described by Rathke and Hudson[8].

Photomicrographs of as-cast chitosan films, and those containing DAP, are shown in Figures 2-4. Figure 2 shows the appearance of a chitosan film cast from 5% aq. acetic acid. These undrawn films are generally not birefringent, with a minimum of morphological detail. Mechanically, they are somewhat brittle, with a strain to failure of 3%. In contrast is the appearance of the chitosan/DAP films, shown in Figures 3 and 4. Figure 3, shows a film that was cast with a molar ratio of 1 to 8 for DAP to Chitosan glucoside units. These films are plastic and will draw at room temperature upto 30% in elongation. The film in Figure 3 shows excess DAP on the surface and is highly birefringent. Figure 4, shows the underlying morphology of these systems, after the excess DAP is washed off with methanol. This film has also been drawn 30%. A morphology suggestive of focal conic type mesophases is observed. During the slow drying of these films from aqueous solution, there may be the development of a mesophase as the solution becomes concentrated, and has time for development of the orientation.

The cooperative binding of chitosan with another surfactant, dodecyl sulfate has been described by Wei and Hudson[9]. We assume that a similar phenomena occurs with DAP, and leads to the formation of oleophilic regions within these films. By exposing these films to oil soluble dyes,

Figure 1. A photomicrograph of a cholesteric chitosan
solution, obtained at a concentration of 10% solids in 90%
formic acid. The scale bar is 100 microns.

Figure 2. A photomicrograph of an as-cast film of chitosan
from aqueous acetic acid. The scale bar is 100 microns.

Figure 3. A photomicrograph of a drawn, unwashed film of chitosan/DAP showing the presence of DAP surfactant which has bloomed to the surface. The scale bar is 100 microns.

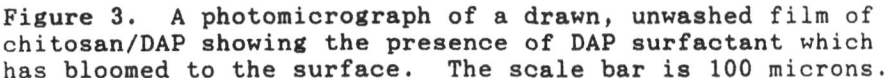

Figure 4. A photomicrograph of a drawn and washed film of chitosan/DAP showing the bi-phasic morphology of these films. The scale bar is 100 microns.

the presence of these regions was shown. For example, the film described for Figures 3 and 4 above, was exposed to a suspension of C.I. Disperse Blue 56 Dye (Color Index 63285). This dye is an organic soluble azo compound, typically used to dye hydophobic fibers such as polyethylene terephthalate (PET). In this experiment, the chitosan/DAP film along with a chitosan control were placed in the Disperse Blue 56 dye bath and allowed to equilibrate for 72 hours at room temperature. Upon removal from the bath and after washing and drying the films, the absorbance of the films at 627nm was determined on a UV-VIS spectrophotometer. The chitosan/DAP film had an absorbance of 0.142 and was noticeably stained blue by the dye. The chitosan controls had an absorbance of 0.03, with no noticeable discoloration by the dye. This is taken to indicate the presence of hydrophobic regions within these films.

In conclusion, water soluble chitosan/amphoteric salt complexes have been found. The salt involving N-dodecyl-N,N-dimethyl-3-aminopropane sulfonate (DAP), is found to yield strong, clear flexible films. These films exhibit a different morphology than for films of pure chitosan. Additionally, the presence of hydrophobic regions within the films is revealed by the affinity of chitosan/DAP films to oleophilic dyes.

ACKNOWLEDGEMENTS: The support of the North Carolina Textile Foundation for this work is gratefully acknowledged.

REFERENCES

1. A. Stroobants, et.al., Effect of electrostatic interaction on the liquid crystalline phase transition in solutions of rod-like polyelectrolytes, Macromol. 19:2232 (1986).

2. G.J. Vroege, The isotropic nematic phase transition and other properties of a solution of semiflexible polyelectrolytes, J. Chem. Phys. 90:4560 (1989).

3. I.A. Nyrkova and A.R. Khokhlov, Liquid crystal ordering in polyelectrolyte solutions, Biophysika 31:771 (1986).

4. A.R. Khohklov, Theories on liquid crystal ordering of polymers based on the Onsager approach, in: "Fundamentals of Liquid Crystalline Polymers," A. Ciferri, ed., VCH Publishers, New York, NY (1991). (1991).

5. P. Sanford, Chitosan, a natural cationic biopolymer: Commercial applications, Prog. Biotechnol. 3:363 (1987).

6. J. Domszy and G. Roberts, Evaluation of infrared spectroscopic techniques for analyzing chitosan, Makromol. Chem. 186:1671 (1985).

7. Ogura et.al., Liquid crystalline phases based on
    chitosan and its derivatives, in: "Proceedings of the
    Second International Chitin and Chitosan Conference",
    S. Hirano and S. Tokura, eds., Japan Chitin Chitosan
    Society, Tottori, Japan (1982).

8. T.D. Rathke and S.M. Hudson, A novel way to control
    tensile properties of chitosan films, in: "Industrial
    Polysaccharides," M. Yalpani, ed. ACS Symposium Series,
    American Chemical Society, Washington, DC (in press).

9. Y.C. Wei and S.M. Hudson, The binding of sodium dodecyl
    sulfate with cross-linked chitosan fibers, in:
    "Proceedings of the fifth international conference
    on chitin and chitosan.", C. Brine, ed., Elsevier
    Publishers, New York, NY (1992).

CHEMICAL AUTOPOIESIS: SELF-REPLICATION OF MICELLES AND VESICLES

Peter Walde, Pascale Angelica Bachmann, Peter Kurt Schmidli, and
Pier Luigi Luisi

Institut für Polymere
Eidgenössische Technische Hochschule
Universitätstrasse 6
CH-8092 Zürich, Switzerland

INTRODUCTION

Micelles and vesicles display a few characteristic features which are present in the structures of the living. First of all, they are self-assembling, i.e. they are examples of self-organization which is under thermodynamic control. Furthermore, micelles and vesicles are bounded structures, i.e. they have a closed interface (boundary) which discriminates an inside from an outside. This boundary acts as a semipermeable membrane, in the sense that it permits the input and the output of low molecular weight compounds with a certain degree of specificity. Also, the inside defines a geometrically well distinct microcompartment, and reactions occurring inside or at the interface of the aggregates are somewhat different from those occurring in the bulk solvent.

All these analogies with living structures have attracted the attention of chemists and biologists over the years. Thus, vesicles have been considered as simple models for the lipidic shell matrix of contemporary biological cells[1]; and they have been also regarded as the most probable precursors for protocells[2,3] - in turn, the formation of protocells is considered an important step in prebiotic chemistry[3-8].

One interesting question is whether micelles and vesicles can also be used to model some dynamic properties of the living cells, such as self-replication. Self-replication of linear chemical structures has attracted much attention over the last couple of years thanks to the work by von Kiedrowski[9-11] and Rebek Jr.[12-14]. In our work - which will be reviewed here - the emphasis is rather on self-replication of bounded structures. We define the term self-replication as an increase of the number of the aggregates owing to reactions which take place within the boundary of the aggregates itself.

The emphasis on boundary, and on the above definition of self-replication as a process occurring within the boundary, sets our work within the frame of autopoiesis. This term (from the greek 'auto-', self; and '-poìesis', formation) was coined in the seventies by Maturana and Varela in an attempt to define the minimal life[15,16]. An autopoietic unity is such a minimal life structure, and is an unity which is self-generating and self-perpetuating as a consequence of its own activities within a boundary of its own making[15-17]. Self-replication was not foreseen in the original theory of autopoiesis, and work in our group[18] has recently shown that when self-replication takes place within the boundary of a self-replicating structure, it can be taken as a criterium for autopoiesis.

A theoretical paper on self-replicating reverse micelles was presented a couple of years ago[19], and later on this concept was experimentally investigated[20-23]. Later on, the self-replication of bounded structures was extended to vesicles (liposomes)[23,24].

The concept of autopoiesis in connection with bounded systems has been considered also by other groups[25,26].

As already mentioned, in this article we will review our work on self-replicating micelles and vesicles, emphasizing mostly the experimental aspects. The final section contains our concluding remarks and an outlook. For a more theoretical discussion on the implications of our work, the reader is referred elsewhere[18].

SELF-REPLICATING REVERSE MICELLES

First of all, let us remind that reverse micelles are dynamic nanometer-sized surfactant aggregates in an apolar organic solvent, containing in the interior of the micelles a polar microdomain[1]. The surfactant molecules themselves build the shell of the micelles - they are the components of the boundary between the polar pool and the bulk solvent. If these micelles host a chemical reaction which takes place at the boundary of the micelles and which produces new surfactant molecules - and in turn more micelles - we consider this micellar system to be self-replicating. (The term self-replication is defined as an increase of the number of the particles which is caused by a reaction taking place within the boundary of the parent particles, Fig. 1.)

In order to follow experimentally a change in the concentration of reverse micelles as the reaction proceeds; and in order to follow possible changes in the size of the micelles, two methodologies have been applied: time-resolved fluorescence quenching[27-29] and quasielastic light scattering[30,31].

In the case of the fluorescence method a fluorescent probe molecule and a quencher compound are first solubilized in the aqueous interior of the reverse micelles. After a pulsed excitation of the fluorescent probe for about 3 nanoseconds, the time-dependency of the decrease in fluorescence intensity is measured, usually within the first microsecond after excitation. By analyzing this decrease, the intramicellar quenching extent can be obtained which is directly proportional to the quencher concentration (which is known) and inversely proportional to the concentration of micelles. Since this fluorescence method allows directly the determination of the number of micelles, it is a rather important technique for studying the self-replication of micelles.

With quasielastic light scattering measurements, information on the size and the polydispersity of micelles (and other surfactant aggregates, such as vesicles) can be obtained. This technique is based on the measurement of the diffusion coefficient of the micelles by analyzing the time dependence of the intensity fluctuations in scattered light due to the Brownian motion of the micellar particles.

In the following, we will report on the self-replication of reverse micelles as investigated by using three different systems. In all three cases surfactant molecules are the components of the micellar boundary which separates aqueous microcompartments from an apolar solvent. Due to a simple chemical or enzymatic reaction that takes place within the boundary of the micelles, new surfactant molecules are formed which in turn leads to an increase in the concentration of micelles.

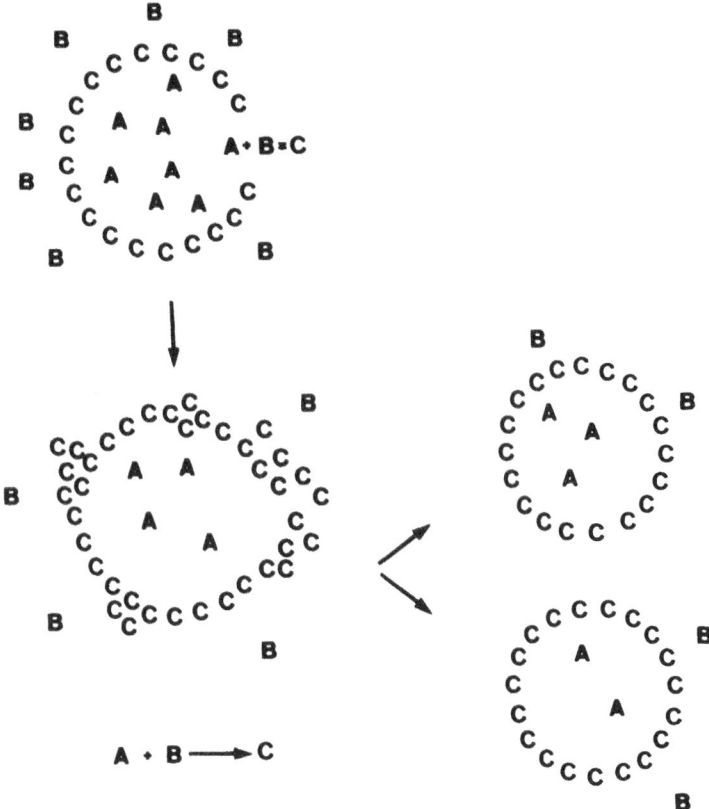

Fig. 1. Schematic general representation of self-replicating reverse micelles. The micelles host a reaction which takes place at the boundary of the micelles leading to the formation of C, the component of the boundary. A and B are present in excess and due to a fast micellar exchange, A,B and C rapidly redistribute over all the micelles present in solution[19].

Non-Enzymatic Ester Hydrolysis

The first system consists of reverse micelles formed in isooctane/1-octanol (9:1, v/v) by sodium octanoate and aqueous LiOH at w_o=9.2 (w_o=[H_2O]/[octanoate]). The octanoate concentration was initially 50 mM; and LiOH was 23 mM with respect to the total volume of the solution (overall), corresponding to a concentration in the water pool of 2.86 M. In this system 1-octanol is cosolvent as well as cosurfactant with approximately 14 % (or 90 mM) of the total 1-octanol being cosurfactant[20]. The reverse micelles have a spherical shape and they are rather monodisperse with an average hydrodynamic radius of 2.27 nm, as determined by quasielastic light scattering.

This reverse micellar system has been made self-replicating for a period of a couple of days by adding 25 mM octanoic acid octyl ester (overall). This ester is solubilized preferentially in the bulk organic solvent and to some extent at the micellar interface where the ester will be hydrolyzed by the hydroxyl ions present in the polar interior of the micelles. This hydrolysis yields octanoate - the surfactant - and octanol - the cosurfactant. As the reaction time progresses, the total number of surfactant/cosurfactant molecules increases which in turn leads to an increase in the number of micelles. The reaction scheme is illustrated in Fig. 2. Fig. 3 shows the progress of the reaction as conveniently followed by Fourier transform infrared spectroscopy: as the ester is hydrolyzed, the intensity of its C=O (st) band around 1744 cm^{-1} decreases, while simultaneously the intensity of the C=O (st) band of the octanoate at 1570 cm^{-1} increases. The general principle of the infrared methodology has been described in an other context[32].

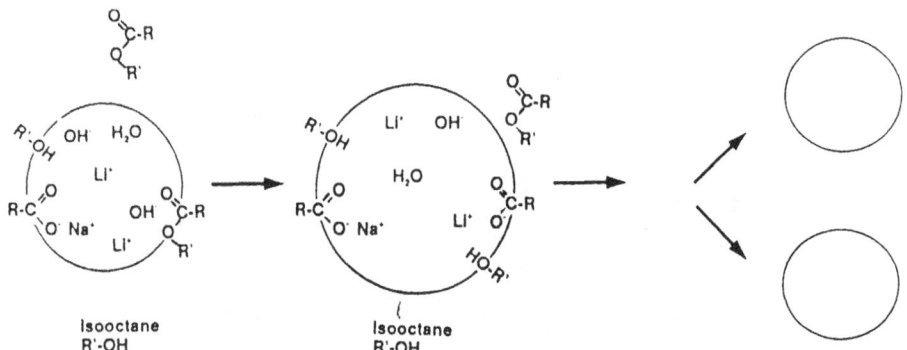

Fig. 2. Schematic representation of a self-replicating reverse micellar system which is based on a non-enzymatic ester hydrolysis[20,21]. The chemical reaction is: CH$_3$-(CH$_2$)$_6$-COO-(CH$_2$)$_7$-CH$_3$ + LiOH → CH$_3$-(CH$_2$)$_6$-COO$^-$Li$^+$ + HO-(CH$_2$)$_7$-CH$_3$. R-: CH$_3$-(CH$_2$)$_6$-; R'-: CH$_3$-(CH$_2$)$_7$-.

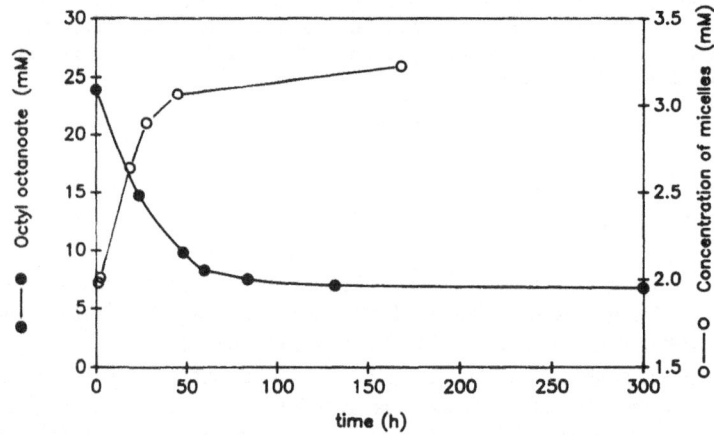

Fig. 3: Self-replicating reverse micelles. Decrease in octyl octanoate concentration and increase in the concentration of reverse micelles as a function of reaction time at 25 °C. Starting conditions: 50 mM sodium octanoate, 24 mM octyl octanoate and 23 mM LiOH in isooctane/1-octanol (9:1, v/v) at w$_o$=9.2[20,21].

Using the technique of time-resolved fluorescence quenching[27] with 1-pyrenesulfonic acid as fluorescence probe and sodium iodide as quencher, it is possible to monitor changes in the concentration of reverse micelles as the concentration of ester and octanoate change. Starting with 50 mM octanoate and 25 mM ester, the concentration of micelles will increase from 1.91 mM to 3.03 mM after 300 hours at room temperature with an overall yield of 72 %[20]. Since the amount of water remains constant, the size of the reverse micelles decreases during the reaction, in our case from a hydrodynamic radius of initially 2.27 nm to 2.05 nm[20]; at the same time, the mean number of octanoate molecules per micelle decreases from 26.1 to 22.4[20].

Enzyme-Catalyzed Ester Hydrolysis

The hydrolysis of fatty acid esters, such as those used in the system just described, is catalyzed by one particular class of hydrolytic enzymes, namely lipases[33]. It was therefore interesting to test whether a modification of our first system would bring us to a replicating reverse micellar system, in which the actual replication would be initiated and driven by an enzyme. For this, sodium octanoate was again used as surfactant and two different lipases, one from *Chromobacterium viscosum* and the other from *Humicola lanuginosa*. Since both lipases not only catalyze the hydrolysis of ester substrates but also their synthesis from the corresponding acid and alcohol - particularly in low water systems - octanol has been replaced as cosolvent and cosurfactant by octylamine. The solvent system used was then isooctane/octylamine (85:15, v/v).

In the presence of 50 mM sodium octanoate and 1 M water (w_o=20) 8 % (or 70 mM) of the total octylamine is acting as cosurfactant[21]. Trioctanoyl glycerol was used as substrate to be hydrolyzed at a concentration of 100 mM. Due to its low solubility in water, the substrate will be localized mainly in the bulk solvent. The hydrodynamic radius of the micelles in this system - as determined by quasielastic light scattering - was 4.54 nm[21]. In the presence of lipase (20 µg/ml overall), the substrate molecules are converted into products as time progresses. And, since both lipases are not soluble in the bulk organic solvent mixture, all lipase-catalyzed substrate transformations will take place at the boundary of the reverse micellar aggregates. These substrate transformations will yield different products due to hydrolysis and aminolysis reactions. In a complete hydrolysis reaction, all ester bonds of trioctanoyl glycerol will be broken, yielding finally for each substrate molecule one molecule of glycerol and three molecules of octanoate. In a side reaction, octylamine takes the place of water, acting as a second nucleophile which leads to the formation of N-octyloctanamide. The whole reaction is illustrated in Fig. 4.

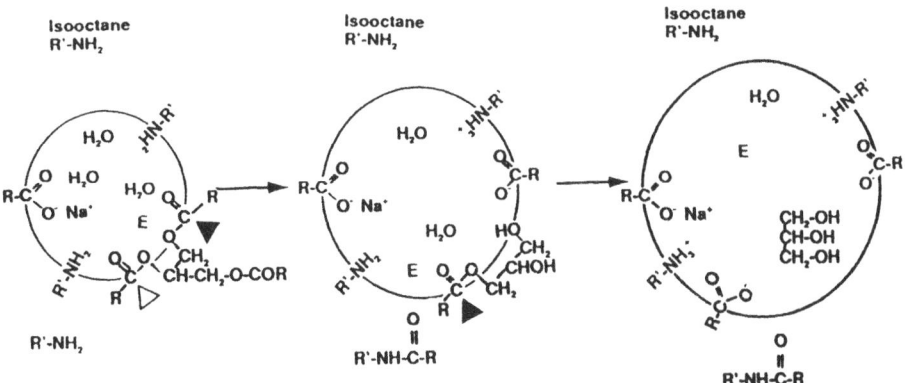

Fig. 4. Schematic representation of a self-replicating reverse micellar system which is based on a lipase catalyzed ester-hydrolysis[21]. Hydrolysis (marked with a filled triangle): trioctanoyl glycerol + 3 H_2O → glycerol + 3 octanoic acid. Aminolysis (marked with an open triangle): trioctanoyl glycerol + 3 octylamine → glycerol + 3 N-octyloctanamide. R- and R'- as in Fig. 1.

The decrease in ester concentration and the formation of octanoate and N-octyloctanamide can be followed and quantified with Fourier transform infrared spectroscopy, most conveniently between 1800 and 1300 cm^{-1}, in the region of the C=O(st) bands. Fig. 5 shows the progress of the reaction in the case of *H. lanuginosa*

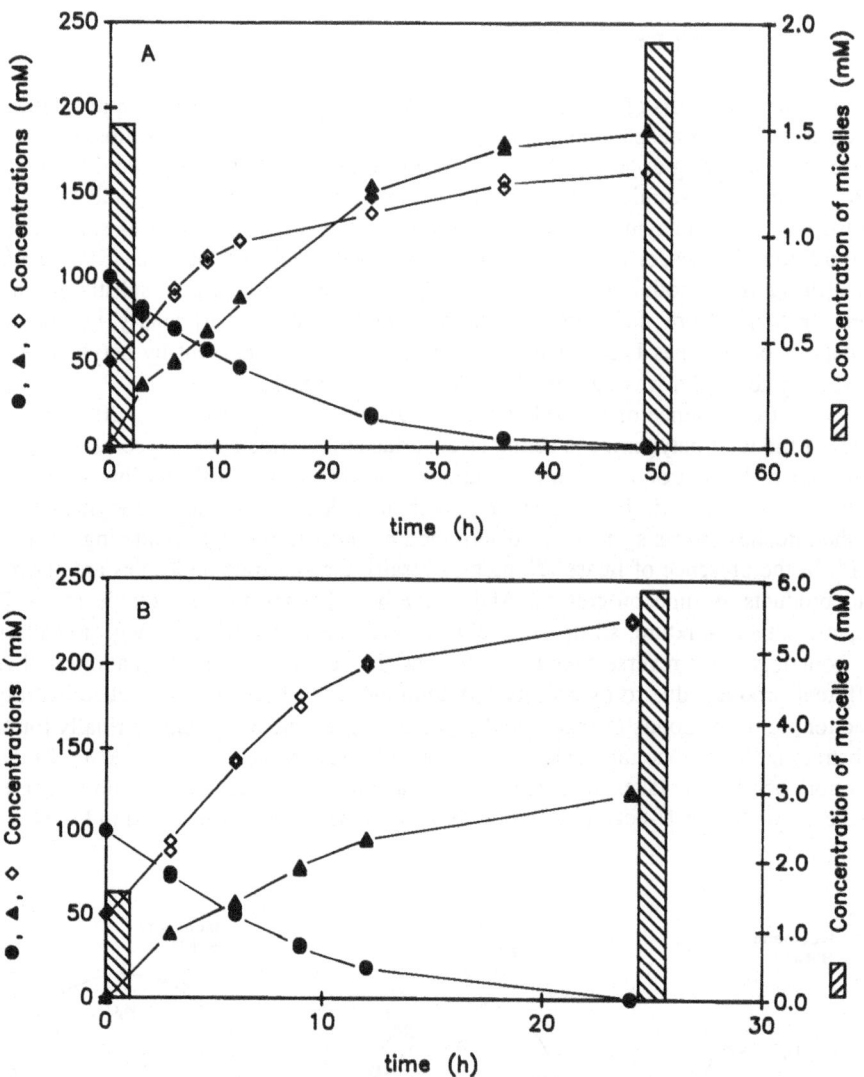

Fig. 5. Time progress of enzymatically driven self-replicating reverse micelles. Decrease in trioctanoyl glycerol (●), formation of octanoate (◇) and N-octyloctanamide (▲) and increase in reverse micelle concentration as a function of time at 25 °C. Starting conditions: 50 mM sodium octanoate, 100 mM trioctanoyl glycerol and 20 µg lipase/ml in isooctane/octylamine (85:15, v/v) at w_o=20. A: with lipase from *H. lanuginosa*; B: with lipase from *C. viscosum*.

lipase (Fig. 5A) and of *C. viscosum* lipase (Fig. 5B). At reaction equilibrium, the concentration of reverse micelles has increased by a factor of 1.3 and 3.8, respectively: from initially 1.52 mM to 1.91 mM (for *H. lanuginosa* lipase) and to 5.86 mM (for *C. viscosum* lipase).

Octanol Oxidation

In both self-replicating reverse micellar systems described so far, surfactant (and cosurfactant) molecules are formed at the micellar interface by converting ester substrates that are mainly localized in the bulk solvent. The conversion of the solvent itself - more precisely one component of the solvent - into surfactant molecules is the basic chemical principle of the system which we summarize in this sub-chapter.

Once more, sodium octanoate is the surfactant and isooctane/1-octanol (85:15, v/v) the solvent[21]. The alcohol again is both cosolvent as well as cosurfactant. Upon solubilizing sodium permanganate inside the water pools of the octanoate reverse micelles, octanol will be oxidized to octanoate. Since sodium permanganate is insoluble in the bulk solvent, the reaction again will take place at the boundary of the reverse

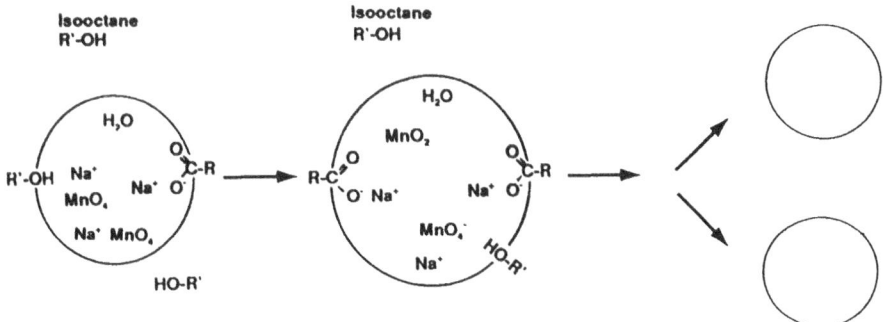

Fig. 6. Schematic representation of a self-replicating reverse micellar system which is based on the oxidation of octanol[21]. The chemical reaction is $3\ CH_3\text{-}(CH_2)_7\text{-}OH + 4\ NaMnO_4 \rightarrow 3\ CH_3\text{-}(CH_2)_6\text{-}COO^-Na^+ + 4\ MnO_2 + NaOH + 4\ H_2O$. R- and R'- as in Fig. 1.

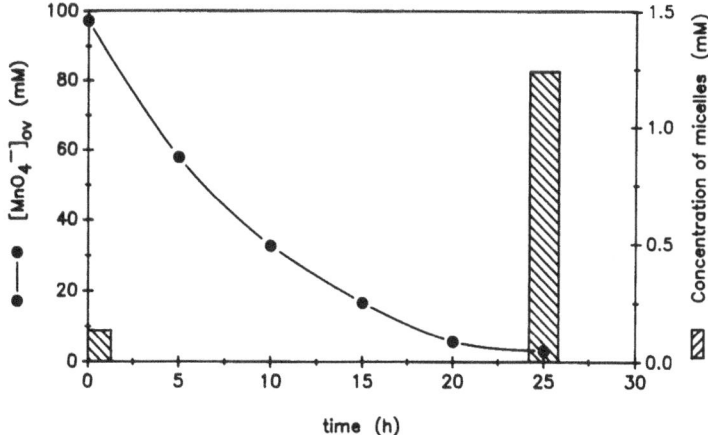

Fig. 7. Changes of permanganate concentration and micelle concentration in self-replicating reverse micelles which are based on the oxidation of octanol. Starting conditions: 50 mM sodium octanoate, 97.2 mM sodium permanganate in isooctane/1-octanol at $w_o=30$ and 25 °C[21].

micelles. As the alcohol is oxidized, permanganate is reduced as can be followed by analyzing spectrophotometrically the decrease in the absorption at 546 nm. Again, changes in size and concentration of the reverse micelles have been followed by the methods of quasielastic light scattering and time-resolved fluorescence quenching. Fig. 6 illustrates the system schematically and Fig. 7 shows the decrease in the permanganate concentration with reaction time at 25 °C for a system with the following initial conditions: 50 mM sodium octanoate in isooctane/1-octanol (85:15, v/v), 97.2 mM sodium permanganate (overall) and w_o=30. Under these conditions, all permanganate initially present in the reaction system was reduced after about 30 minutes and the insoluble inorganic products were eliminated. During this time, the concentration of octanoate molecules doubled, from 50 mM to almost 100 mM and the concentration of reverse micelles increased by a factor of 9.4 (from 0.132 mM at the beginning to 1.24 mM after completion of the reaction)[21]. This enormous and on a first glance surprising increase in the number or reverse micelles can be explained by the observed drastic changes in the size of the micelles which are mainly a consequence of the constance of the water content during the reaction[21].

SELF-REPLICATING AQUEOUS MICELLES

The same physical methods applied for reverse micelles have also been used to detect changes during the replication of aqueous micelles: time-resolved fluorescence spectroscopy and quasielastic light scattering (see above).

Furthermore, the same type of chemical reactions used in the case of self-replicating reverse micelles have also been used for aqueous micelles. In the following, we describe one such system which is based on the oxidation of an alcohol. In addition, a novel concept will be described where - starting from an aqueous/organic two phase system - self-replicating micelles are first created by a simple chemical reaction taking place at the water/oil interface.

Octanol Oxidation

Sodium octanoate in aqueous solution forms micelles above its cmc of about 340 mM. These micelles can host 1-octanol[34] which itself is water insoluble. After addition of sodium permanganate, the solubilized alcohol will slowly be oxidized to octanoate at the micellar interface, octanoate being the surfactant of which the micelles are composed (Fig. 8). As time progresses, the number of octanoate molecules

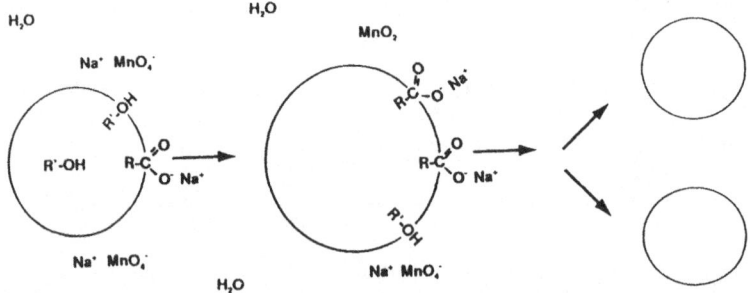

Fig. 8. Schematic representation of a self-replicating aqueous micellar system which is based on the oxidation of octanol[21]. The chemical reaction is the same as that of Fig. 5. R- and R'- as in Fig. 1.

increases until all octanol is oxidized, leading to an increase in the number of octanoate micelles[21].

Starting with 1.2 M sodium octanoate, 0.48 M 1-octanol and 0.64 M sodium permanganate, the concentration of micelles increases after 20 hours at room temperature from 14 mM to 20 mM, as again determined by time-resolved fluorescence quenching, after removal of insoluble MnO_2[21].

Formation of Micelles and Their Self-Replication based on Ester Hydrolysis

In all the experiments described so far, the replicating units - aqueous or reverse micelles - were initially already present. In other words, the micro-compartmentalization for the self-replication was given by the initial reaction conditions. A novel and also more interesting situation would be one in which the micro-compartment is not initially present, but is instead itself first created chemically. In the following, one such system will be described[23]:

This system consists initially of two clear phases, a lower aqueous phase (20 ml 3 N NaOH) and a supernatant phase of 6 ml octanoic acid ethyl ester in a 100 ml round bottom flask. While the aqueous phase is stirred, the whole system is kept under reflux slightly below 100 °C. As time progresses, octanoic acid ethyl ester is hydrolyzed to yield sodium octanoate and ethanol. The velocity of this hydrolysis reaction shows a dependency on the reaction time as illustrated in Fig. 9. This octanoate-time profile can be rationalized in the following way:

Initially, the ester substrates are hydrolyzed at the interface between the two phases, and since the interfacial area is relatively small (20 cm^2 ca.), the reaction proceeds rather slowly. The bimolecular rate constant is of the order of 2×10^{-4} M^{-1}s^{-1}. As more and more octanoate molecules are formed, the critical concentration for micelle formation will be reached (0.1 M under the high salt conditions used). And, as soon as the first octanoate micelles are formed, octanoic acid ethyl ester molecules will be

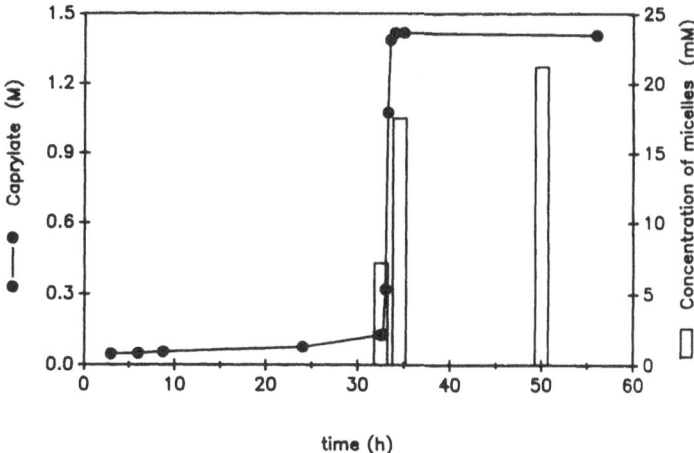

Fig. 9. Time progress of octanoate concentration and micelle concentration following the hydrolysis of octanoic acid ethyl ester by NaOH in a biphasic system at 100 °C ca. Starting conditions: 20 ml 3 N NaOH (stirred at 150 rpm) and 6 ml octanoic acid ethyl ester in a 100 ml round bottom flask[23].

solubilized by the micelles which leads to a considerable increase in the interfacial reaction area and consequently also to an increase in the rate of hydrolysis. The bimolecular rate constant increased by a factor of about 900 to 0.19 $M^{-1}s^{-1}$. Each octanoate micelle can host up to 5 ester molecules as determined by turbidimetric titration[23]. After completion of the reaction, all ester molecules are hydrolyzed and the initial two phase system has converted into a single aqueous phase, containing 21 mM micelles. The onset of the exponential phase in Fig. 9 varies with the stirring speed used. An increase in the stirring speed by a factor of 3 decreases the onset time from 34 hours to 7-8 hours[23]. A diminution of the onset time is also observed if sodium octanoate is already present to some extent from the beginning[23].

In any case, all the system has remarkable properties despite of its rather "primitive" composition. It contains at maximum only five compounds: water, sodium hydroxide, octanoic acid ethyl ester, octanoate and ethanol; composed of only four different elements: H, O, C and Na. The heaviest molecule present has a relative molecular mass of 172 g/mol.

Another interesting property of the system is the fact that the self-replicating microcompartments created by the simple hydrolysis reaction are aqueous micelles which can be transformed into vesicles by lowering the pH of the solution to ca. 6.5[5,35,36] by either titrating with HCl or CO_2[23]. This property relates the system to questions of the possible origin of protocells within chemical and biological evolution.

SELF-REPLICATING VESICLES

As mentioned above another approach to a self-replicating chemical system is the use of vesicles as microcompartments. By introducing enzymes, which are able to form new surfactant molecules, into the closed bilayer structure of the vesicles one obtains a system where the enzymatic reaction that takes place within the boundary of the vesicles leads to an increase in the concentration of the vesicles.

In the investigated system the four enzymes from pig liver microsomes involved in the so-called salvage pathway synthesis of phosphatidylcholine (lecithin), Fig. 10, have been inserted into phosphatidylcholine vesicles, and the extent of phosphatidylcholine production of these 'proteovesicles' have been studied. Details about the preparation and characterization of the proteovesicles are given elsewhere[24]. An important observations are that the proteovesicles have a spherical shape and that they are rather monodisperse with an average hydrodynamic radius of 26±0.2 nm, as determined by quasielastic light scattering. Concerning the self-replication process the four-step synthesis of phosphatidylcholine according to Fig. 10 is, however, more important. The time progress of this enzymatic synthesis of PC is shown in Fig. 11. The starting observation are that the substrates of the enzymatic reactions are water soluble, and that the product is an amphiphilic molecule which - due to its low solubility in water - will be located in the bilayer of the proteovesicles. The incorporation of newly synthesized phosphatidylcholine has then to be considered. If the fatty acid residues in the alkyl moieties of the produced phosphatidylcholine are short (ten carbon atoms or less), one expects an increase of the spontaneous curvature of the vesicles, based on geometrical considerations[24]. Due to this effect and due to the increase of the total interface during the reaction, the synthesis of short-chain phosphatidylcholine will thus lead to an increase in the number of the proteovesicles.

This theoretical prediction has been tested experimentally by the synthesis of dihexanoyl phosphatidylcholine in proteovesicles formed by soybean phosphatidylcholine, in which the average fatty acid chain contains 16 carbon atoms. A 10% reaction yield in

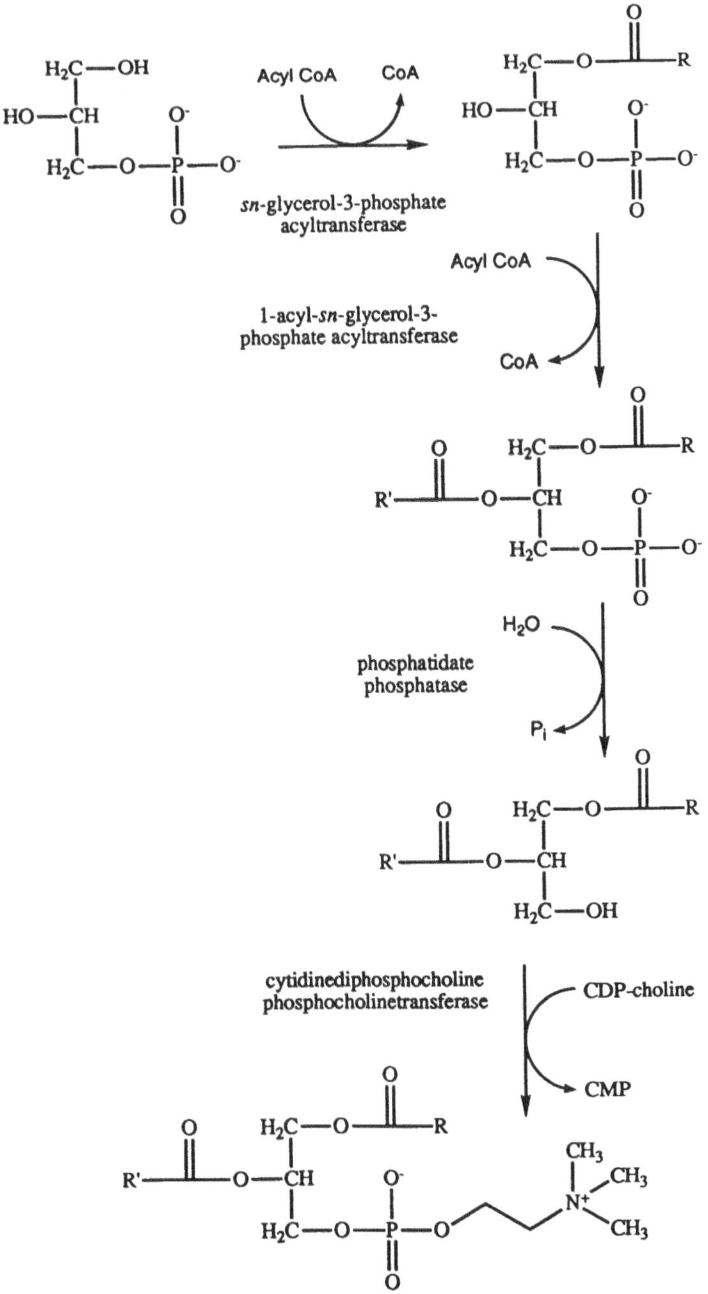

Fig. 10. Salvage pathway for the synthesis of phosphatidylcholine.

short chain phosphatidylcholine (with respect to the initially present soybean phosphatidylcholine) leads to a decrease of the mean hydrodynamic radius of the vesicles from 26±0.2 nm to 23±0.3 nm[24]. Unfortunately, a direct determination of the concentration of vesicles in this complex system is at the moment not possible by using the method of time-resolved fluorescence quenching[28].

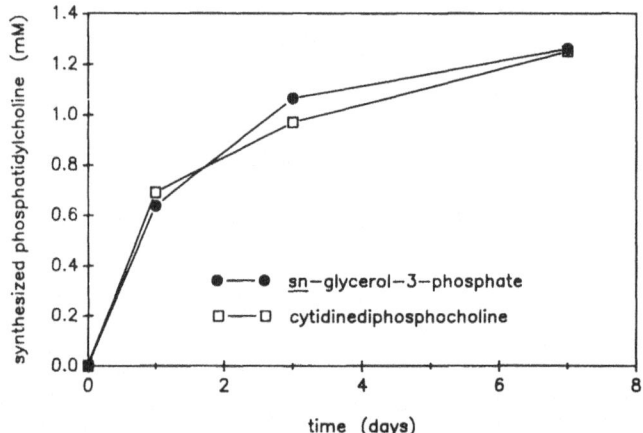

Fig. 11. Proteovesicle-mediated synthesis of phosphatidylcholine (mainly 1-palmitoyl-2-oleyl-*sn*-glycero-3-phosphocholine)[24]. Starting conditions: 13 mM soybean phosphatidylcholine in 70 mM NaCl, 20 mM Tris/HCl, pH 7.4 containing per ml 2 mg pig microsomal protein, 10 mg *sn*-glycerol-3-phosphate, 10 mg palmitoyl coenzyme A, 10 mg oleoyl coenzyme A, and 10 mg cytidinediphosphocholine.

CONCLUDING REMARKS AND OUTLOOK

Reactions which lead to self-replication until now are restricted to the interfacial area of the aggregate, i.e. to the boundary itself. The boundary is of course integer part of the aggregate, and therefore the definition of self-replication is fulfilled. It would be aesthetically nice to develop self-replication based on reagents which are both soluble only in the water pool of reverse micelles; or both in the hydrophobic interior of an aqueous micelle. Due to the particular chemical nature of a surfactant molecule, this maybe to prove very difficult if not impossible.

One should also notice that the fidelity of self-replicating reverse micelles is not completely guaranteed: in fact, during self-replication, reverse micelles become smaller and smaller. This is due to the fact that we are working under condition of constant overall water concentration, and freshly formed surfactant brings then to a decrease of w_o, i.e. of the radius of the micelles. In principle, this problem could be circumvented by continuously supplying water from the outside: in our first theoretical paper[19] it was proposed to do so with a Windsor II system, i.e. a biphasic system with the reverse micellar phase supernatant to an aqueous phase. However, we were till now unable to realize one such system (there are mostly difficulties with the stability of a biphasic system; and/or with the fact that in such a case the w_o value is fixed, and assumes only the equilibrium w_o value).

Another fidelity problem is apparent in the enzymatically driven self-replication schemes: enzymes are obviously not being replicated, and the freshly made aggregates which do not contain enzyme are not able to self-replicate. The characteristics of an autopoietic system in which self-generation is caused by a catalyst which is not able to replicate has been theoretically considered by Varela[37].

Despite these limitations, our work clearly shows that it is possible, and actually not so difficult, to devise surfactant molecular aggregates hosting reactions which lead to more copies of the same aggregate. Among those systems, reverse micelles appear to give the highest self-replication rates, probably partly due to their higher monodispersity. Aqueous micelles, on the other hand, are biologically more interesting. The biological relevance is increased by the observation that aqueous micelles can be transformed into vesicles by rather simple chemical operations, such as pH adjustment in the case of fatty acids[35,36,38] or by dialysis of certain micellar systems[39-42]. This can actually bring one to suggest that micelles could have played an important role during the prebiotic formation of protocells[6,19].

Vesicles (liposomes) will be the focus of attention of our future work on self-replication. Two different lines of research on this area have originated from our previous work. One line of research is exemplified by the lecithin liposomes. In this case, the emphasis is on complex enzymatic liposomal machineries, by which more and more complex cellular systems can be mimicked. Due to this initial complexity, such systems cannot be of much use for prebiotic chemistry.

The second line of research with vesicles has instead prebiotic chemistry as the central issue. This is the line of research which emphasizes simplicity - as exemplified by the self-replicating octanoate micelles. In this case, we have no enzyme, but simple reactions which might have initiated self-assembly and self-replication under prebiotic conditions.

A final comment should bring us back to the notion of autopoiesis. As already mentioned in the introduction, autopoiesis, being based on bounded structures and on self-replication of a structure within its own boundary, sets the definition of the minimal life. The question then which can be asked in connection with self-replicating micelles and liposomes is whether, and to what extent, the bounded structures described in this article can be considered as expressions of the minimal living. We tend to answer affirmatively to such a question, but this is within the limits of the self-imposed definition of autopoiesis. This definition of the minimal life is of course arguable and may leave one unsatisfied; however one should give then an alternative scientific definition of the living.

REFERENCES

1. J.H. Fendler. "Membrane Mimetic Chemistry," John Wiley & Sons, New York (1982).
2. H.J. Morowitz, B. Heinz, and D.W. Deamer, The chemical logic of a minimum protocell, *Origins Life Evol. Biosphere 18*: 281-287 (1988).
3. J. Oro and A. Lazcano, A holistic precellular organization model, *in*: "Prebiological Self Organization of Matter," C. Ponnamperuma and F.R. Eirich, eds., A. DEEPAK Publishing, Hampton, Virginia, pp. 11-34 (1990).
4. D.W. Deamer and J. Oro, Role of lipids in prebiotic structures, *BioSystems 12*: 167-175 (1980).
5. W.R. Hargreaves and D.W. Deamer, Origin and early evolution of bilayer membranes, *in*: "Light Transducing Membranes: Structure, Function, and Evolution," D.W. Deamer, ed., Academic Press, New York, pp. 23-59 (1978).

6. J. Nagyvary and J.H. Fendler, Origin of the genetic code: a physical-chemical model of primitive codon assignments, *Origins of Life 5*: 357-362 (1974).
7. D.W. Deamer, Role of amphiphilic compounds in the evolution of membrane structure on the early earth, *Origins of Life 17*: 3-25 (1986).
8. H. Yanagawa, Y. Ogawa, K. Kojima, and M. Ito, Construction of protocellular structures under simulated primitive earth conditions, *Origins Life Evol. Biosphere 18*: 179-207 (1988).
9. G. von Kiedrowski, Ein selbstreplizierendes Hexadesoxynucleotid, *Angew. Chem. 98*: 932-934 (1986).
10. G. von Kiedrowski, B. Wlotzka, and J. Helbling, Sequenzabhängigkeit matrizengesteuerter Synthesen von Hexadesoxynucleotid-Derivaten mit 3'-5'-Pyrophosphatverknüpfung, *Angew. Chem. 101*: 1259-1261 (1989).
11. G. von Kiedrowski, B. Wlotzka, J. Helbling, M. Matzen, and S. Jordan, Parabolisches Wachstum eines selbstreplizierenden Hexadesoxynucleotids mit einer 3'-5'-Phosphoamidat-Bindung, *Angew. Chem. 103*: 456-459 (1991).
12. J. Jr. Rebek, Molecular recognition and the development of self-replicating systems, *Experientia 47*: 1096-1104 (1991).
13. J.S. Nowick, Q. Feng, T. Tjivikua, P. Ballester, J. Jr. Rebek, Kinetic studies and modelling of a self-replicating system, *J. Am. Chem. Soc. 113*: 8831-8839 (1991).
14. V. Rotello, J.-I. Hong, and J. Jr. Rebek, Sigmoidal growth in a self-replicating system, *J. Am. Chem. Soc. 113*: 9422-9423 (1991).
15. F.J. Varela, H.R. Maturana, and R. Uribe, Autopoiesis: the organization of living systems, its characterization and a model, *BioSystems 5*: 187-196 (1974).
16. G.R. Fleischaker, Autopoiesis: the status of its system logic, *BioSystems 22*: 37-49 (1988).
17. G.R. Fleischaker, Three models of a minimal cell, *in*: "Prebiological Self Organization of Matter," C. Ponnamperuma and F.R. Eirich, eds., A. DEEPAK Publishing, Hampton, Virginia, pp. 235-247 (1990).
18. P.L. Luisi, Defining the transition to life: self-replicating bounded structures and chemical autopoiesis, in preparation.
19. P.L. Luisi and F.J. Varela, Self-replicating micelles - a chemical version of a minimal autopoietic system, *Origins Life Evol. Biosphere 19*: 633-643 (1989).
20. P.A. Bachmann, P. Walde, P.L. Luisi, and J. Lang, Self-replicating reverse micelles and chemical autopoiesis, *J. Am. Chem. Soc. 112*: 8200-8201 (1990).
21. P.A. Bachmann, P. Walde, P.L. Luisi, and J. Lang, Self-replicating micelles: aqueous micelles and enzymatically driven reactions in reverse micelles, *J. Am. Chem. Soc. 113*: 8204-8209 (1991).
22. P.A. Bachmann, P.L. Luisi, and J. Lang, Self-replicating reverse micelles, *Chimia 45*: 266-268 (1991).
23. P.A. Bachmann, P.L. Luisi, and J. Lang, Autocatalytic self-replicating micelles as models for prebiotic structures, *Nature*, in press (1992).
24. P.K. Schmidli, P. Schurtenberger, and P.L. Luisi, Liposome-mediated enzymatic synthesis of phosphatidylcholine as an approach to self-replicating liposomes, *J. Am. Chem. Soc. 113*: 8127-8130 (1991).
25. H. Schwegler and K. Tarumi, The "protocell": a mathematical model of self-maintenance, *BioSystems 19*: 307-315 (1986).
26. L. Margulis and R. Guerrero, From origins of life to evolution of microbial communities: a minimalist approach, *in*: "Prebiological Self Organization of Matter," C. Ponnamperuma and F.R. Eirich, eds., A. DEEPAK Publishing, Hampton, Virginia, pp. 261-278 (1990).

27. J. Lang, A. Jada, and A. Malliaris, Structure and dynamics of water-in-oil droplets stabilized by sodium bis(2-ethylhexyl) sulfosuccinate, *J. Phys. Chem. 92*: 1946-1953 (1988).

28. J. Lang, The time-resolved fluorescence quenching method for the study of micellar systems and microemulsions: principle and limitations of the method, *in*: "The structure, Dynamics and Equilibrium Properties of Colloidal Systems," D.M. Bloor and E. Wyn-Jones, eds., Kluwer, Dordrecht (1990).

29. A. Verbeeck and F.C. DeSchryver, Fluorescence quenching in inverse micellar systems: possibilities and limitations, *Langmuir 3*: 494-500 (1987).

30. G.D.J. Phillies, Quasielastic light scattering, *Anal. Chem. 62*: 1049A-1057A (1990).

31. M. Zulauf and H.-F. Eicke, Inverted micelles and microemulsions in the ternary system H$_2$O/aerosol-OT/isooctane as studied by photon correlation spectroscopy, *J. Phys. Chem. 83*: 480-486 (1979).

32. P. Walde and P.L. Luisi, A continuous assay for lipases in reverse micelles based on Fourier transform infrared spectroscopy, *Biochemistry 28*: 3353-3360 (1989).

33. B. Borgström and H.L. Brockman, eds., "Lipases", Elsevier, Amsterdam (1984).

34. P. Eckwall, Composition, properties and structures of liquid crystalline phases in systems of amphiphilic compounds, *in*: "Advances in Liquid Crystals, Vol. 1", G.H. Brown, ed., Academic Press, New York, pp. 1-142 (1975).

35. W.R. Hargreaves and D.W. Deamer, Liposomes from ionic, single-chain amphiphiles, *Biochemistry 17*: 3759-3768 (1978).

36. D.P. Cistola, J.A. Hamilton, D. Jackson, and D.M. Small, Ionization and phase behavior of fatty acids in water: application of the Gibbs phase rule, *Biochemistry 27*: 1881-1888 (1988).

37. F.J. Varela, "Principles of Biological Autonomy," North Holland, New York (1979).

38. W. Li and T.H. Haines, Uniform preparations of large unilamellar vesicles containing anionic lipids, *Biochemistry 25*: 7477-7483 (1986).

39. J. Brunner, P. Skrabal, and H. Hauser, Single bilayer vesicles prepared without sonication. Physico-chemical properties, *Biochim. Biophys. Acta 455*: 322-331 (1976).

40. M. Ueno, C. Tanford, and J.A. Reynolds, Phospholipid vesicle formation using nonionic detergents with low monomer solubility. Kinetic factors determine vesicle size and permeability, *Biochemistry 23*: 3070-3076 (1984).

41. P. Schurtenberger, N. Mazer, and W. Känzig, Micelle to vesicle transition in aqueous solutions of bile salt and lecithin, *J. Phys. Chem. 89*: 1042-1049 (1985).

42. S. Almog, T. Kushnir, S. Nir, and D. Lichtenberg, Kinetic and structural aspects of reconstitution of phosphatidylcholine vesicles by dilution of phosphatidylcholine-sodium cholate mixed micelles, *Biochemistry 25*: 2597-2605 (1986).

SIMPLE MODELS FOR THE STRATUM CORNEUM LIPIDS

Stig E. Friberg and Zhuning Ma

Center for Advanced Materials Processing
and Department of Chemistry
Clarkson University
Potsdam, New York 13699–5814

ABSTRACT

Models for stratum corneum lipids using lipid combinations were prepared in which ceramides were replaced by lecithin. All the combinations including the one with a composition identical to that of intact stratum corneum were lamellar liquid crystals.

Low angle x–ray diffraction was used to determine interlayer spacing and the composition was adjusted to give identical lipid dimensions and water penetration.

Replacing the ceramides by lecithin resulted in a structure with crystallization at higher water content than the original stratum corneum lipids. This deficiency was improved by partial replacement of the nonpolar lipids with soy–bean oil.

INTRODUCTION

The stratum corneum layer serves important functions being the mammalian body's defence against dehydration, excessive water uptake and against chemical and biological attack (1–4). The lipid part of the stratum corneum is responsible for the different barriers (5–8) and the research interest in the individual lipid structures as well as their molecular organization has been intense. Analysis methods are still developed (9) although by now, in spite of arguments about specifics (10,11) the structure and amount of the main constituents appear well established (12–14). Hence, at present, part of the research is focussed on special structures (15–22) while the main effort remains on structure/function relationships (25) especially the use of keratinocyte cultures (26–29).

The essential feature of the stratum corneum lipid barrier to water evaporation is not the presence of specific lipids, but the structural organization (30–34) and, hence, a large number of models have been suggested. All these models are, in principle, used to measure water transport across the bilayer.

However, the recent contribution by Potts (35) indicates that transversal transport in the bilayer structure may be more relevant to the function of the stratum corneum as barrier to water transport. This means that the relative dimensions of the hydrophilic and hydrophobic layers and the amount of water penetration into the latter are the decisive factors for the water transport.

With this in mind, we found a need for a stratum corneum model with as exact as possible a reproduction of these factors. Low angle x-ray diffraction is a useful tool for such a determination (36). X-ray data is available for intact stratum (37–39). The first investigation of this kind (37) gave an interlayer spacing of 65Å. A later publication gave a different value, 131Å, (38), but recent contributions by Bowstra (39, 40), have demonstrated that one of the structures indeed has an interlayer spacing of 65Å. The higher value of 120Å (39) became dominant first after denaturation of human stratum corneum. There is no doubt that the lipid structure is complex (40) and that a final discussion of its water interaction should await more complete models.

With this in mind, the aim of the present contribution is limited to an evaluation of the potential to exchange the ceramides with less complex and, hence, less expensive compounds, while retaining the interaction with water. We have chosen to replace the ceramides with lecithin and to compensate for the changed polar properties by modifying the characteristics of the triglycerides.

EXPERIMENTAL

MATERIALS

The materials for the models are given in Table I. The composition of the three models are shown in Table II. The free fatty acids were partially neutralized by ammonia (NH_3/FA molar ratio = 0.65) because this is the amount of NH_3 in the stratum corneum (41).

Table I. The materials for the lipids models.

Free Fatty Acids	Source	Purity
Oleic Acid	Sigma	99%
Linoleic Acid	Sigma	99%
Palmitoleic Acid	Sigma	99%
Palmitic Acid	Fisher Scientific	RG
Stearic Acid	Sigma	99%
Myristic Acid	Sigma	99.5%
Arachidic Acid	Sigma	99%
Phosphatidglethanolamine	Avanit Polar Lipids	>99%
Cholesterol	Aldrich	95%
Cholesteryl Sulfate	Research Plus	89+%
Cholesteryl Oleate	Aldrich	97%
Triolein	Sigma	99%
Squalene	Aldrich	99%
Pristane	Aldrich	98%
Ceramides (III)	Sigma	99%
Lecithin	Soy Extracts	Commercial
Soybean Oil		Commercial

Table II

Lipid Models, Compositions			
compounds	ceramides model %(W)	lecithin model %(W)	soy bean oil model %(W)
*FFA	19.7	19.7	19.7
stearic acid	9.9	9.9	9.9
myristic	3.8	3.8	3.8
palmitic	36.8	36.8	36.8
arachidic	0.3	0.3	0.3
oleic	33.1	33.1	33.1
linoleic	12.5	12.5	12.5
palmitoleic	3.6	3.6	3.6
cholesterol	17.3	17.3	17.3
cholesteryl oleate	6.0	6.0	
cholesteryl sulfate			
sodium salt	5.0	5.0	5.0
PE.	4.0	4.0	4.0
triolein	13.5	13.5	
squalene	6.0	6.0	6.0
pristane	2.0	2.0	0
soy bean oil	0	0	6.0
ceramides	26.5	0	0
lecithin	0	26.5	26.5
TOTAL	100.0	100.0	84.5
W% for one phase			
L.L.C.	15---40+	33---49	28---43
W% for 65	35%⁻	35%⁻	38%
water penetration	−9%	3%	−16%
d₀	42	43.5	41

*Free Fatty Acids were neutralized partially by ammonia solution (0.65 mol.), which was called host with water penetration 34% about, d_o39.

PREPARATION OF MODELS

The lipids were added with water containing the appropriate amount of ammonia into capped vials with a constriction and thoroughly mixed by repeated centrifugation of the content through the restriction. Homogenity was monitored by observation in optical microscope between crossed polarizers.

LOW ANGLE X–RAY DIFFRACTION

A small amount was added to a thin wall capillary and introduced into the x–ray equipment (Kiessig low angle cornea Richard Seiffert) both ends. Ni–filtered radiation was used and the reflections recorded by a Teunelec position sensitive gas ionization detector system (PSD–1100). The optical microscope was Olympus BH, equipped with a optical camera (Olympus C–35A).

RESULTS

Fig. 1 gives the essential information about the different models. The ceramide model showed a stable lamellar liquid crystal between 16.2 and 39.2% water. Replacing the ceramides by lecithin changed the range to 33.0 – 49.0% while the addition of soybean oil gave a range of 27.9 – 43.0%.

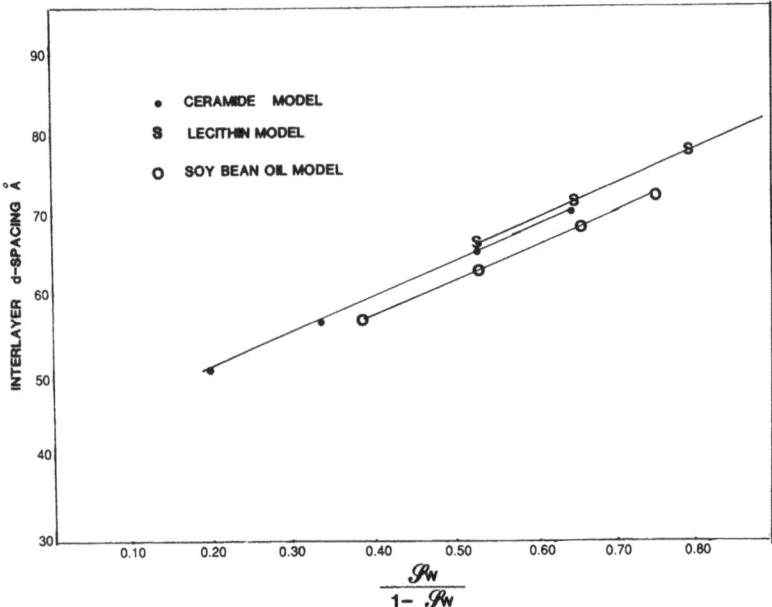

Figure 1. Interlayer spacing versus water/lipid volume ratio, $\phi_w/(1-\phi_w)$, for the three models.

All the models were lamellar liquid crystals as shown by the optimal microscopy pattern with the sample between crossed polarizers, Fig. 2.

All three models gave a similar increase of the interlayer spacing from the one of the carboxylic acid/carboxylate combination, the "host". The soybean model gave a value 2.4A lower than the ceramide model while the lecithin model showed numbers 1.3A in excess of it. The variation with water content was identical for the three models.

DISCUSSION

The results reveal the potential for models exactly to mimic the behavior relative to water by models with the exact composition of stratum corneum as given by Lampe et. al. (15). They also demonstrate the problems faced during the attempts and illustrate the superior quality of the original stratum corneum lipids. A few words about the importance of the stability range may be justified.

The most critical factor is the minimum water content with retained stability of the liquid crystalline structure, because the subsequent lipid crystallization causes a strong

CERAMIDE MODEL

LECITHIN MODEL

SOY BEAN OIL MODEL

Figure 2. Optical microscopy photographs of model samples viewed between crossedpolarizers.

enhancement of the water evaporation rate (34). In this aspect the lecithin model, that crystallizes already when the water content is 35%, is obviously inferior. This value should be compared to the water content in stratum corneum in vivo, ~10%, when the layer becomes brittle. The model with soybean oil is an improvement bringing the water content down to 27% before crystallization is initiated. However, even this value is considerably in excess of 10%.

The second important feature to be examined is the degree to which the two modified models mimic the structural organization of the complete model. The two structure features of interest are the location of the lipids and the degree of penetration by the water, Fig. 3. Both these factors are obtained from the low angle x–ray results, Fig. 1.

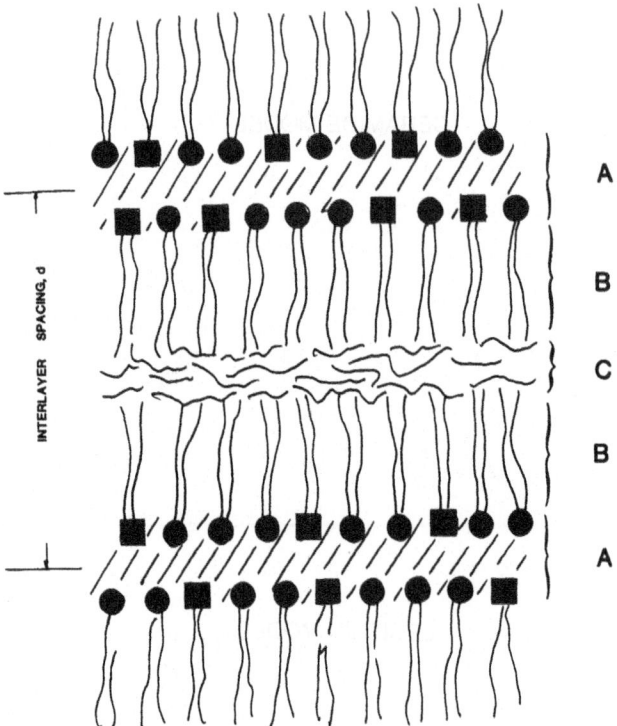

Figure 3. A lamellar liquid crystalline structure consists of three zones
A = Aqueous solution between surfactant polar group layers
B = Surfactant layers
C = Space between mirroring methyl groups of the surfactant

The variation of interlayer spacing, d, with the water/amphiphilic ratio $\phi_w/1-\phi_w$, R is a straight line

$$d = d_o\{1+(1-\alpha)R\}$$

in which d_o represents the dimension of the lipid parts, Fig. 3, and α is the fraction of water penetrating into the hydrophobic part, Fig.3.

The d_o values, Table II, do not vary to a significant degree and are at a level to be expected from lipids with chain lengths found in biological systems. With 18 carbon atoms the length of a fully extended chain is at the level of 22Å and an interlayer spacing of 41–44Å indicates only a small fraction of the lipids in the zone C, Fig. 3.

The penetration of water is small varying from -0.16 to +0.03, Table II. The negative values may at first appear surprising; they show that the increase of interlayer spacing is greater than corresponds to the volume of the added water. The explanation is that the presence of enhanced amounts of water leads to a transfer of compounds from zone B, Fig. 3 to either zone A or zone C, Fig. 3. The low angle x–ray diffraction patterns cannot distinguish between these two mechanisms. However, the essential information, the fact that the penetration of water is extremely small for all three models, is obvious from the x–ray results.

It should be noted that the stratum corneum lipid structure in situ does not show swelling with added water (39,40). The models presented in this article are at obvious variance with that fact. The reason for this difference is to be found in the influence of the amino acids in stratum corneum (41). Their partition between the corneocytes and the lipids as well as their influence on the water/lipid interaction is treated in a later publication.

This information may initially appear to limit the value of the present contribution or even make its use highly doubtful. However, a conclusion to that effect must be considered premature. The interaction between the purely liquid part and water is, of course, an essential property and in this article used as a gague to evaluate the influence of exchanging the ceramides with lecithin.

It should be noted that the low angle x–ray diffraction patterns of these models offers a unique opportunity to determine not only the water penetration, but more importantly, the location of any chemical compound entering the stratum corneum structure. The value of such information for pharmaceutical and cosmetical applications is immediately evident.

REFERENCES

1. A.M. Kligman, Stratum Corneum lipids, in "Epidermis", W. Montagna, ed., Academic Press, New York–London, Chap.1.
2. P.M. Elias, Structure and function of the stratum corneum permeability barrier, Drug Development Research 13, 97 (1988).
3. A.G. Matoltsy, Keratinization, J. Invest. Dermatol. 67, 20 (1976).
4. R.J. Schuplein, R.L.Bronaugh, Transdermal transport, in "Biochemistry and Physiology of the Skin" (L.A. Goldsmith, Ed), Oxford Physiology Univ. Press, N.Y. 1983, p. 1255.
5. P.M. Elias, Lipids and the epidermal permability barrier, Arch Dermatol Res 270, 95 (1981).
6. G.L.Flynn, Mechanism of percutaneous absorption from physicochemical evidence, in "Percutaneous Absorption" (R.L.Bronaugh, H.I. Maibach, Eds.) Marcel Dekker, N.Y. (1985) p.17.
7. L. Landmann, The epidermal permeability barrier, Anat. Embryol 178, 1 (1988).
8. B.D. Anderson, P.V. Raykav, Solute structure–permeability relationships in human stratum corneum, J. Invest. Dermatol. 93, 280 (1989).
9. R.H.Guy, R.Fleming, Membrane moldels for skin penetration studies, Chem. Rev. 88, 455 (1988).
10. B.C. Melnik, J. Hollman, E.Erler, B.Verhoeven, G. Plewig, Microanalytical screening of all major stratum corneum lipids by sequential high–performance thin–layer chromatography, J. Invest. Dermatol 92 231 (1989).
11. J.T.Bartz, P.W. Wertz, D.T.Downing, The origin of alkanes found in human skin surface lipids, J. Invest. Dermatol. 93, 723 (1989).

12. P.M. Elias, M.L. Williams, S.J. Rehfeld, N-alkanes in the skin: function or fancy?, Arch Dermatol. 126, 868 (1990).

13. P.W. Wertz, D.T. Downing, Glycolipids in mammalian epidermis: structure and function in the water barrier, Science 217, 1261 (1982).

14. M.A. Lampe, A.L. Burlingame, J. Whitney, M.L. Williams, B.E. Brown, E.Roitman and P.M. Elias, Human stratum corneum lipids: characterization and regional variations, J. Lipid Res. 24, 120 (1983).

15. M.A. Lampe, M.L. Williams, P.M. Elias, Human epidermal lipids: characterization and modulation during differentiation, J. Lipid Res. 24, 131 (1983).

16. M.E. Stewart, D.T.Downing, Unusual cholesterol esters in the sebum of young children, J. Invest. Dermatol. 95, 603 (1990).

17. P.W. Wertz, D.T.Downing, Free sphingosine in human epidermis, J. Invest. Dermatol. 94, 159 (1990).

18. P.W. Wertz, K.C. Madison, D.T. Downing, Covalently bound lipids of human stratum corneum, J. Invest. Dermatol. 92, 109 (1989).

19. K.C.Madison, D.C. Swartzendruber, P.W. Wertz, D.T. Downing, Lamellar granule extursion and stratum corneum, J. Invest. Dermatol 90, 110 (1988).

20. D.J. Donaldson, J.T. Mahan, Keratinocyte migration and the extracellular matrix, J. Invest. Dermatol. 90, 623 (1988).

21. P.W.Wertz, D.T.Downing, Metabolism of linoleic acid in porcine epidermis, J. Lipid Res. 31, 1839 (1990).

22. P.M. Elias, Retinoid effects on the epidermis, in "Pharmacology" (M.J.Rand, C.Raper, Eds.) Elsevier, N.Y. (1987) p. 697.

23. S.J. Orlow, A.K. Chakrabarty, J.M. Pawelek, Retinoic acid is a potent inhibitor of inducible pigmentation in marine and hamster melanoma cell lines, J. Invest. Dermatol 94, 461 (1990).

24. P.W. Wertz, D.C. Swartzendruber, D.J. Kitko, K.C. Madison, D.T. Downing, The role of the coneocyte lipid envelopes in cohesion of the stratum corneum, J. Invest. Dermatol. 93, 169 (1989).

25. B.D. Anderson, P.V. Raykar, Solute structure-permeability relationship in human stratum corneum, J. Invest Dermatol. 93, 280 (1989).

26. K.A.Holbrook, H.Hennings, Extracellular calcium regulates and terminal differentiation of cultured mouse epidermal cells, J. Invest. Dermatol. 81,11 (1983).

27. M.Regnier, J.Schweizer, S.Michel, C.Bailly, M.Prunierase, Expression of high molecular weight (67k) keratin in humas keratinocytes cultured on dead de-epidermized dermis., Exp. Cell Res. 165, 63 (1986).

28. V.H.W.Mak, M.B.Cumpstone, A.H.Kennedy, C.S.Harmon, R.H.Guy, and R.O.Potts, Barrier function of human keratinocyte cultures grown at the air-liquid interface, J.Invest. Dermatol. 96, 323 (1991).

29. K.C.Madison, D.C.Swartzendruber, P.W.Wartz and D.T.Downing, Marine keratinocyte cultures grown at the air/medium interface synthesize stratum corneum lipids and "Recycle" linoleate during differentiation, J. Invest. Dermatol. 93, 10 (1989).

30. G.Grubauer, K.R.Feingold, R.M.Harris, P.M.Elias, Lipid content and lipid type as determinants of the epidermal permeability barrier, J. Lipid Res. 30, 89-96, (1989).

31. S.E.Friberg, I. Kayali, T.Suhery, L.D.Rhein and F.A.Simion, Keratinization, J. Pharm. Sci. (In Press).

32. R.H. Guy, R. Fleming, Transport acress a phospholipid barrier, J. Colloid I. Sci. 83, 130 (1981).

33. W. Abraham, D.T. Downing, Preparation of model membrane for skin permeability studies using stratum corneum lipids, J. Invest. Dermatol. 93, 809 (1989).

34. S.E. Friberg, I. Kayali, Water evaporation rate from a model of the stratum corneum, J. Pharm. Sci. 78(8), 639 (1989).
35. R.O.Potts, personal information
36. K.Fontell, X–ray diffraction by liquid crystals amphiphilic systems, in "Liquid Crystals and Plastic Crystals", Vol. 1 (G.W.Gray and P.A.Winsor, Ed:s) Ellis Harwood Chichester (1974) p.80.
37. S.E.Friberg and D.W.Osborne, Small angle x–ray diffraction pattern of stratum corneum and a model structure for its lipid., J. Disp. Sci. Tech. 6, 486 (1985).
38. S.H.White, D.Mirejovsky and G.I.King, Structure of lamellar lipid domains and corneocyte envelopes of marine stratum corneum, Biochemistry 27, 3785 (1988).
39. J.A.Bouwstra, M.A.deVries, G.S. Gooris, W.Bras, J.Brussee and M.Ponec, Thermodynamic and structural aspects of the skin barrier, J.Contr. Rel,16, 209–220 (1991).
40. J.A.Bouwstra, G.S.Gooris, J.A.van der Spek and W.Bras, Structural investigations of human stratum corneum by small angle x–ray scattering, J. Invest. Derm. (in press).
41. V.E.Schwarz, Biochemische stigmata menschlicher hautoberglache im alter, Z.Kin. Chem. Klin. Biochem. 12.Jg. (1974) s.93–97.

PHOTOTHERMAL EFFECT IN ORGANIZED MEDIA: PRINCIPLES AND APPLICATIONS

Chieu D. Tran

Department of Chemistry
Marquette University
Milwaukee, Wisconsin 53233

I. INTRODUCTION

Photothermal phenomena, including the thermal lens, photothermal deflection and photothermal refraction [1-19], are based on the temperature rise that is produced in an illuminated sample by nonradiative relaxation of the energy absorbed from a laser [1-19]. In the thermal lens effect, the radially symmetric, Gaussian intensity distribution (TEM$_{oo}$) of the laser beam generated heat which is strongest at the center of the beam because that is where the beam intensity is strongest. Consequently, a lens-like optical element is formed in the sample owing to the temperature gradient between the center of the beam and the bulk material. Under continuous illumination, the thermal lens reaches a steady-state strength which is given as the reciprocal of the focal length:

$$\frac{1}{f(\infty)} = \frac{1.205 \ PA \ (dn/dT)}{\pi \ k \ \omega^2} \tag{1}$$

where P and ω are the power and the beam spot size of the excitation laser at the sample; A is the absorbance of the sample; dn/dT and k are temperature coefficient of the refractive index and thermal conductivity of the solvent, respectively. For weak absorbing species, the thermal lens signal which is measured as the relative change in the beam center intensity in the far field, $\Delta I_{bc}/I_{bc}$, is given by:

$$\frac{\Delta I_{bc}}{I_{bc}} = \frac{1.206P \ (dn/dT)A}{\lambda k} \tag{2}$$

where λ is the wavelength of the probe beam.

The relative change in the beam intensity, when determined by conventional absorption techniques for a weakly absorbing species having absorbance A, is

$$(I_o-I)/I_o = 1 - 10^{-A} = 2.303 \ A \tag{3}$$

It is therefore clear from eq 2 and 3 that the sensitivity of the thermal lens technique is relatively higher than that of conventional absorption methods. The sensitivity enhancement factor E is calculated from eq 2 and 3 to be:

$$E = \{P \ (dn/dT)\} \ /1.91 \ \lambda k \tag{4}$$

It is evidently clear from eq. 4 that the sensitivity of the thermal lens technique is not only directly proportional to the excitation laser power P but also depends on the thermal physical properties of the solvent used. This property enables the technique to have relatively higher sensitivity than conventional absorption methods. According to this equation,when a 632.8 nm laser beam of 50 mW is used for the measurement, the sensitivity of thermal lens in CCl_4 is estimated to be 237 times higher than that by conventional absorption techniques [1-19]. In fact, samples (in gaseous, liquid or solid form) whose absorbances are as low as 10^{-7} have been detected using this ultrasensitive technique [1-19].

As mentioned earlier, the thermal lens technique also depends on the thermal physical properties of the solvent employed. Higher sensitivity per unit excitation laser power, i.e., high E/P value, is predicted to be achieved if the measurements are performed in solvent having high dn/dT and low thermal conductivity k value. Generally, nonpolar solvents are good media for thermal lens measurements owing to their high dn/dT and low k values[20-23]. Water, which is the universal solvent for biochemical and biological samples is the worst medium for thermal lens technique because it has a very low dn/dT and high k values. In fact, it has been calculated and experimentally verified by us that at the same laser intensity, thermal lens measurements in CCl_4 and n-pentane are 38 and 40 times higher than those in water [2,4,24]. The dependency of the thermal lens signal on the thermal physical properties of solvent can, in principle, be exploited to enhance the sensitivity and selectivity of the technique. One such possibility is based on the used of organized assemblies such as micelles, reversed micelles and crown ethers to modify the thermal physical properties, namely dn/dT and k values, of the environmental around the analyte in order to enhance its thermal lens signal intensity.

This chapter is not an extensive review on the current status of the thermal lens technique but rather an overview on recent advances which involve the use of organized media to enhance the sensitivity and selectivity of the technique. Initially, the principles of the technique will be described. Subsequently, specific utilization of organized media to enhance the sensitivity and selectivity of the technique will be reported after a brief description of the instrumentation for the thermal lens measurements. Finally, novel applications which are based on the unique features of the thermal lens technique, namely the use of the thermal lens technique for structural study of liquids and solutions will be delineated.

II. THEORY

The following theory is described for the thermal lens technique. It can be easily modified to explain other related techniques such as the photothermal deflection and photothermal refraction.

A molecule is excited into either vibrational or electronic excited state by absorption of light. Subsequently the excited molecule releases the excitation energy in the form of heat via nonradiative relaxation processes. If the radiative processes of the molecule is negligible compared to nonradiative, the heat generated equals the excitation energy.

The heat Q generated per unit length and unit time within the irradiated sample is expresses as [25-27]:

$$Q(r) = \frac{2(\ln 10) PC\varepsilon e^{-\frac{2r^2}{\omega^2}}}{\pi\omega^2} \tag{5}$$

where r is the distance from the beam center, ω is the laser beam radius, ε is the absorption coefficient of the absorbing species with concentration C, and P is the laser power.

The increase in temperature ΔT can be calculated for any time during and after excitation, by use of appropriate Green's function for the heat equation describing the situation. The time dependence of temperature change within the sample can be calculated as:

$$\Delta T(r, t) = \frac{2(\ln 10) PC\varepsilon}{\pi\omega^2 \rho c_p} \int_o^\infty \frac{e^{-\frac{2r^2/\omega^2}{1+2(t-\tau)/t_c}}}{1+2(t-\tau)/t_c} d\tau \tag{6}$$

where c_p is heat capacity of the sample, ρ is the density, t is the time after the laser beam onset, τ is the excitation period, t_c is the characteristic time constant which can be calculated from the sample's density, heat capacity, thermal conductivity k and beam radius, i.e.,

$$t_c = \frac{\omega^2 \rho c_p}{4k} \tag{7}$$

The temperature changes during excitation can be obtained by integrating eq. 6 from $\tau=0$ to t, where $t \leq \tau$. After the excitation is off, Q(r, t) equals to zero and the thermal lens strength starts to decrease. This is due to the dissipation of heat into the environment. In such situation τ is taken as the upper limit for the integration in eq. 6. The following discussion will be focused mainly on the build-up of the thermal lens and its maximum value.

The change in the refractive index induced by the temperature increase can be written as:

$$\Delta n = \Delta T \frac{dn}{dT} \tag{8}$$

where n is the refractive index.

The heat generated in eq. 5 and the temperature change in eq. 6 are generalized for any light source. For the case of the thermal lens measurement, an expression relating the effect of refractive index change on the laser beam intensity in the far field is needed. To obtain such an expression, approximations are required since presently, it is not possible to analytically solve the integral in eq. 6 [25-27]. Currently, there are two different approximations, namely the parabolic and the aberrant approximations [25-27]. In the parabolic model the thermal lens is assumed to have a parabolic refractive index distribution [27]. The aberrant model takes into account the aberrant nature of thermal lens [25,26]. This is accomplished by use of the diffraction theory of aberrations [25,26]. While the mathematical treatment for the parabolic model is relatively simple, it is based on the relatively unrealistic (i.e., no aberrant in the thermal lens). Furthermore, experimental results have indicated that the aberrant model is more appropriate than the parabolic model. As a consequence, only the aberrant approximation is described here [26].

In the aberrant model, the treatment of the thermal lens effect on the beam propagation is based not on the focal length of thermal lens but rather on the phase shift at the input plane of the laser beam or the optical path length variation $\phi(r,t)$ which is induced by the thermal lens [25,26]. The induced phase shift is proportional to the sample length ℓ and the difference in refractive index in the beam center ($r = 0$) and at radius r. It can, therefore be derived in terms of temperature increase as expressed in eq. 8, i.e.,

$$\phi(r,t) = \frac{dn}{dT}[T(0,t) - T(r,t)]\ell \tag{9}$$

If the induced phase shift is small (i.e., $2\pi\phi/\lambda \ll 1$) and if approximations such as $Z_2 \gg \omega$, $Z_2 \gg Z_1$ are made (Z_1 and Z_2 are the distance from the sample to the beam waist and to the detector, respectively), the diffraction integral can be described as:

$$U_{bc}(t) = \frac{(2\pi)^{1/2}U_0\omega e^{-\frac{2\pi Z_2 i}{\lambda}}i}{\lambda Z_2}\int_0^\infty (1 - \frac{2\pi\phi i}{\lambda}) e^{-(1+\xi t)u}du$$

$$\tag{10}$$

where U_{bc} is the complex phase and amplitude of the waves on the axis at the output plane at the position of detector; U_0 is the initial value of U_{bc}, and the ratio r^2/ω^2 was substituted by u. By using eq. 6 for temperature change, the diffraction integral will have the final form:

$$\tag{11}$$

$$U_{bc}(t) = K_0\int_0^\infty \{1 - \frac{\theta i}{t_c}\int_0^t [1 - e^{-\frac{2u}{1+2(t-\tau)/t_c}}]d\tau\}e^{-(1+\xi i)u}du$$

All pre-integral constants are included in K_0.

Eq. 11 can be solved by performing the integration over u and then over τ. There is no need to assume the parabolic function for the temperature distribution, as in the case of parabolic model. The solution for the beam center intensity variation, I(t), is found as the square of the absolute value of complex phase and amplitude of waves, $|U_{bc}|^2$. For $t \leq \tau$ it has the form:

$$I(t) = I(0)[1 - \theta\arctan[\frac{2\xi}{3+\xi^2+(9+\xi^2)t_c/2t}]$$

$$+\{\frac{\theta}{2}\arctan[\frac{2\xi}{3+\xi^2+(9+\xi^2)t_c/2t}]\}^2 \tag{12}$$

$$+(\frac{\theta}{4}\ln\{\frac{[(2+t_c/t)(3+\xi^2)+6t_c/t]^2+16\xi^2}{(9+\xi^2)^2(2+t_c/t)^2}\})^2]$$

where I(0) is the initial beam center intensity, $\xi = Z_1/Z_c$ and

$$\theta = -\frac{2.303 PA (dn/dT)}{\lambda k}$$

(13)

In thermal lens experiments, θ is usually less than 0.1. Therefore terms equal to θ^2 can be omitted. Under these conditions, the steady state thermal lens, expressed in terms of the relative beam center intensity is:

$$\frac{I(0) - I(\infty)}{I(\infty)} = -1 + [1 - \theta \arctan(\frac{2\xi}{3 + \xi^2})]^{-1}$$

(14)

Differentiating eq. 13 with respect to ξ reveals that the thermal lens strength is maximum when the sample is positioned $Z_c(3)^{1/2}$ from the beam waist. Therefore, at $\xi = 3^{1/2}$ and for small θ values, relative change in beam center intensity $\Delta I/I$ is linearly proportional to θ:

$$\frac{\Delta I}{I} = \theta \arctan(3^{-1/2}) = \frac{\theta}{1.91}$$

(15)

Substitute eq. 13 into eq. 14 to obtain:

$$\frac{\Delta I}{I} = \frac{1.206 PA (dn/dT)}{\lambda k}$$

(16)

III. INSTRUMENTATION

In general thermal lens can be categorized into two types: single beam and double beam or pump/probe.

In the single beam system, a laser beam serves as the excitation (pump) as well as the monitor (probe). The systematic diagram of the system is shown in Figure 1. As illustrated, the sample is illuminated by a cw laser whose amplitude is modulated by either a mechanical chopper or an electronic shutter. A PIN photodiode, placed behind a pinhole, is used to measure the change in the laser beam center intensity as it passes through the sample and is being absorbed. A lens placed between the laser and the sample is used to facilitate the alignment and to optimize the thermal lens signal. It was found that change in the distance between the lens and the sample or strictly speaking, the distance between the laser beam waist and the sample can increase, decrease or even diminish the thermal lens signal. For example, no signal could be observed when the sample is placed at the beam waist. It has been theoretically predicted and subsequently verified by experimental results that thermal lens signal intensity is maximum when the distance between the beam waist and the sample (i.e., Z_1) equals to $3^{1/2}$ time the confocal distance Z_c [1,28]. From the definition $Z_c = \pi \omega_o^2/\lambda$, it is clear that Z_c can be calculated from the laser wavelength λ and the laser beam spot size at the beam waist ω_o. Since it is relatively time consuming to determine the ω_o value accurately, in practice, the alignment is performed by placing the lens on a translation stage and varying the Z_1 distance until maximum thermal lens signal is obtained. The thermal lens signal intensity was also found to be dependent on the distance between the sample and the detector (i.e., Z_2) and the aperture of the pinhole. However, for each Z_2 value, there is a certain size of the pinhole which gives a maximum

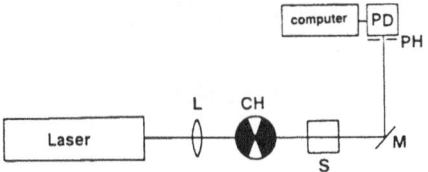

Figure 1.   Schematic diagram of the single beam thermal lens spectrometer: L, lens; CH, chopper; S, sample; M, mirror; PH, pinhole; PD, PIN photodiode.

and constant thermal lens signal for a given sample. Therefore, the selection of Z_2 and the aperture size is relatively simple, i.e., it can be easily performed by initially deciding the Z_2 value (which normally is about 1.5 to 3.0 m) and subsequently adjusting the size of the pinhole until maximum signal is obtained.

Since a single laser beam is used as excitation and monitoring, the data acquisition and analysis for the single beam instrument is relatively time consuming. Generally, the beam center intensity of the laser beam is recorded as a function of time.   Subsequent curve fitting into eq. 12   will provide θ and t_c [1,24,29]. Alternatively, θ value can also be calculated from [intercept]$^{-1}$ of the plots of $I_{bc}(t)/I_{bc}(0) - I_{bc}(t)$ versus $1/t$ where $I_{bc}(0)$ and $I_{bc}(t)$ are the beam center intensity at time $t = 0$ and $t = t$ [29]. It is thus evidently clear that due to this time consuming and cumbersome, the single beam system has not been extensively used.

The schematic diagram for a pump/probe thermal lens instrument is shown in Fig. 2. Two laser beams are used in this apparatus: one which has a relatively stronger power and is absorbed by the sample, is used as the pump beam while the weaker one (and is normally not absorbed by the sample) is used as the probe beam. These two beams can be from two different lasers or from the same laser. Two lenses are used to focus the pump and the probe beam. The distance from the sample to the waist of the pump beam is different from that to the probe beam. It was found that the sample should be at the waist of the pump beam in order to receive maximum optical power (power of the beam is inversely proportional to the square of the beam spot size). Similar to the case of the single beam instrument, the distance from the sample to the probe beam waist should be equal to $3^{1/2} Z_c$ [28].

Data acquisition for the double beam instrument is relatively simple than the single beam apparatus.   In the double beam, only the pump beam is modulated.   As a consequence, phase locking devise such as the lock in amplifier can be used for data acquisition.

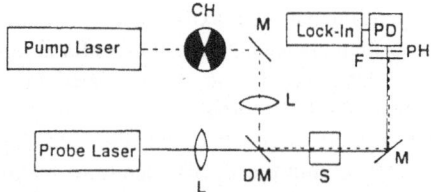

Figure 2.   Schematic digram of the pump/probe thermal lens spectrometer: L, lens; DM, dichroic mirror; CH, chopper; S, sample; M, mirror; F, filter; PH, pinhole;

Table I. Relative Thermal Lens Signal Intensity of Tb^{3+} in Different Media

Surfactant[a]	Solvent	Relative Signal Intensity
	Water	1
	dioxane	22
AOT	dioxane	25
AOT	n-nonane	15
AOT	n-heptane	21
AOT	n-pentane	23
AOT	cyclohexane	20
AOT	carbon tetrachloride	38

[a] Concentrations of AOT used were 0.20 M.

IV. APPLICATIONS

1. Reversed Micelles

Amphiphilic surfactant molecules form micelles in aqueous solution and reversed micelles in nonpolar solvent. In the reversed micelle, the surfactant's polar headgroups outline a water soluble polar core of the micelle shielded by the nonpolar hydrocarbon tails [30-34]. Consequently, the reversed micelle contains a small water pool in which a water soluble analyte can be dissolved. Because the size of the water pool containing the dissolved analyte is relatively small, the thermo-optical properties of the micellar solution are expected to be similar to those of the bulk nonpolar solvent. As a consequence, the solubilization of such analytes in reversed micelle is expected to provide an enhancement in the thermal lens compared to that possible in bulk water alone.

Table I shows the effects of solvents and reversed micelles on the thermal lens signal intensity of terbium chloride [4]. The lanthanide ion was chosen because of its high hydrophilicity: it is very soluble in water, sparingly soluble in polar solvents and insoluble in nonpolar solvents [4]. As expected, the thermal lens signal amplitude of Tb^{3+} in methanol is 10 times higher than that in water. The enhancement is further increased to 15 times when Tb^{3+} is in 1-butanol. Substantial enhancement was found, however, when Tb^{3+} is solubilized in AOT reversed micelles. Sodium bis(2-ethylhexyl) sulfosuccinate or Aerosol-OT (AOT) was used to investigate the effect of reversed micelles because this anionic surfactant can form reversed micelles in a variety of nonpolar solvents and can solubilized large amounts of water [30,34].

It is interesting to note that in spite of the fact that the Tb^{3+} ions are soluble in the water pool of the reversed micelles, the thermal lens signal intensities of the Tb^{3+} are much higher than that in pure water. The enhancements were found to be dependent on the particular nonpolar solvent in which the reversed micelles are formed. Under identical conditions, Tb^{3+} exhibits 38-fold enhancement when it is solubilized in AOT/CCl_4 reversed micelles as compared to that in pure water. The enhancement factor decreases to 23 in AOT/n-pentane, 20 in AOT/cyclohexane, 21 in AOT/n-heptane and 15 in AOT/nonane.

Taking advantage of the fact that AOT forms reversed micelles in dioxane [30,34] and terbium chloride is soluble in this solvent, thermal lens measurements were also performed for Tb^{3+} in pure dioxane and for Tb^{3+} solubilized in the water pool of AOT/dioxane reversed micelles. The thermal lens of Tb^{3+} in pure dioxane is 22 times more than that in pure water and is the same, within experimental error, with that of Tb^{3+} in AOT/dioxane (25X; Table I). These results clearly demonstrate that the surfactant molecules do not produce observable effect on the dn/dT and k values of the dioxane solution. Additional information can be obtained from the relative thermal lens signal intensity of Tb^{3+} in AOT/CCl_4, AOT/hydrocarbons and AOT/cyclohexane reversed micellar systems compared to that in pure water. It is known that AOT forms its largest reversed micelle in cyclohexane as the aggregation number, N, in this solvent is between 45 and 65 [30,34]. Medium size micelles can be found in hydrocarbons (N = 25 - 31) with the CCl_4 provides the smallest (N = 17) [30,34]. Any micellar effect, if existent, is expected to provide a larger enhancement for the medium leading to large size micelles whereas smaller enhancements are expected for the smaller size micelles. However, the observed enhancement values are in the reversed order: AOT/CCl_4 > AOT/hydrocarbons > AOT/cyclohexane.

Information of the micellar effect on the thermal conductivity, k, is relatively scarce and inconclusive. It seems that in the absence of aggregation such as the case of AOT in CCl_4 without any solubilized water, the surfactant has no observable effect on k because the k values were found to be the same with and without the surfactant [4]. Adding water to the solution leads to the formation of AOT reversed micelles in CCl_4 [30-34]. The thermal conductivity of the mixture was found to dependent on its chemical composition [4]. The information is still inconclusive at the moment but it seems to suggest that at low solubilized water concentration, the k value for the mixture is higher than that for pure CCl_4 solution [4]. This, according to eq. 3, would result in lowering the thermal lens signal intensity. Because the observed thermal lens signals were in fact, enhanced by AOT/CCl_4 reversed micelles, the reversed micellar effect on the k value of the solution is very small at low solubilized water concentration.

Collectively, these results seem to suggest that the effect of the reversed micelles on the dn/dT and k values of the bulk nonpolar solvents is rather small. The thermal lens signal magnitude of Tb^{3+} is AOT reversed micellar systems is governed mainly by the thermo-optical properties of the nonpolar solvent even though the metal ions are solubilized in the water pool. This deduction can be further explained in terms of the relative size of the water pool and the laser beam spot size. The concentrations of the solubilized water in the reversed micellar systems used in this work were kept very low (ω = [H_2O]/[AOT] < 1) so that the diameter of those water pools is estimated to be less than 20Å [30-34]. This value is smaller than the spot size of the laser beam at the sample which is estimated to be 0.10 mm. Therefore, the measured thermal lens is strongly effected by the thermo-optical properties of the bulk nonpolar solvent of the reverse micelle.

2. Crown Ethers

Crown ethers are group of macrocyclic ligands whose chemistry is of particular interest because of their ability to form complexes with a variety of metal ions [30-32,35-43]. This complex formation is of extreme importance in analytical chemistry, particularly separation science because it enables the selective extraction of metal ions from aqueous solution into organic phase [35-43]. It is thus possible to enhance not only the sensitivity but also the selectivity of the thermal lens detection of such metal ions as lanthanide ions. This anticipation is based on the fact that crown ether (host) can selectively form an inclusion complex with a metal ion (guest) and this complex formation will introduce the hydrophobicity into the analyte (i.e., metal ion). As a consequence, the complex will be extracted from the thermo-optically poor water into a nonpolar solvent that has relatively better thermo-optical properties. The thermal lens signal intensity of the extracted inclusion complexes is expectedly higher than that in pure water. Selectivity enhancement in this case would depend on the size selectivity of the macrocyclic host, i.e., the formation of inclusion complexes between the crown ethers such as 18-crown-6, 15-crown-5 and 12-crown-4 and the lanthanide ion can only be achieved when the size of the metal ion is comparable to the cavity size of the crown ether.

Recently, we were able to exploit this possibility to develop a novel method to enhance the sensitivity and selectivity of the thermal lens detection of lanthanide ions [44,45]. In this method, the rare earth ions were selectively extracted from water to organic solvent with the use of crown ether, e.g., 18-crown-6, 15-crown-5 as synergistic extractant and lithium benzoate as the counter ion. For instance, by use of 15-crown-5, neodymium ion (Nd^{3+}) was extracted from aqueous solution into organic phase such as ethyl acetate, heptanol, octanol, 1,2-dichroroethane, methylene chloride and chloroform [44]. Compared to that in water, the thermal lens signal intensity of Nd^{3+} was substantially enhanced when it is extracted into the organic phase: up to 24-fold enhancement was observed in dichloroethane, 20X in $CHCl_3$, 18X in CH_2Cl_2 , 14X, 13X and 11X in octanol, heptanol and ethyl acetate, respectively [44].

The induced selectivity is demonstrated by experiments in which a series of lanthanide ions ranging from Pr^{3+} to Er^{3+} were extracted from the aqueous phase into the organic phase using different ethers, namely 12-crown-4 (12-C-4), 15-crown-5 (15-C-5) and 18-crown-6 (18-C-6). As listed in the Table II, only small amounts of lanthanide ions (between 8% and 12%) were extracted into the chloroform phase when 12-C-4 was used as the extracting agent. The low extraction yields can be explained in terms of the mismatching in sizes of the crown ether and the lanthanide ions; i.e., the cavity of 12-C-4, which is between 1.2 and 1.5 Å, is too small to accommodate the lanthanide ions whose ionic diameters are between 1.76 and 2.02 Å [46]. Substantial improvement in the extraction yield can be accomplished when 15-C-5 or 18-C-6 was used as the extracting agent. For instance, up to 39% and 41% of Er^{3+} ions can be extracted from the aqueous phase into the chloroform phase with the use of 15-C-5 and 18-C-6, respectively. The enhancement in the extraction yield is probably due to the fact that the cavities of these crown ethers are sufficiently large (the cavities of 15-C-5 and 18-C-6 are 1.7-2-2 and 2.6-32 Å, respectively [47]) to accommodate the lanthanide ions and thus enable them to be extracted into the organic phase. As expected, these two crown ethers also offer noticeable selectivity in the extraction processes: at the same extraction conditions as those for the Er^{3+} ions, only 10% and 28% of the Pr^{3+} ions were extracted into the organic phase by 15-C-5 and 18-C-6, respectively. The size selection provided by the 15-C-5 is relatively better than that by the 18-C-6 even though, for the same ion, the extraction yields obtained by using the latter are higher than the former. This is probably because the cavity of the 18-C-6 is larger and hence more flexible than that of 15-C-5. It can, therefore, adopt a variety of conformations to accommodate the lanthanide ions having different sizes. Consequently, a rather low selectivity is obtained. The larger size of 18-C-6 also makes it more lipophilic than 15-C-5 and hence enables it to extract more metal ion from the aqueous phase into the organic phase.

Table II. Size Effect on the Extractions of Lanthanide Ions by Crown Ether

Metal Ion	Extraction Yield, %		
	12-Crown-4	15-Crown-5	18-Crown-6
Pr^{3+}	6	10	28
Nd^{3+}	6	14	29
Sm^{3+}	8	20	30
Eu^{3+}	a	25	31
Tb^{3+}	a	30	35
Dy^{3+}	a	33	37
Ho^{3+}	11	31	38
Er^{3+}	12	39	41

[a]Signals were too small to be measured accurately.

V. UNIQUE CHARACTERISTICS OF THERMAL LENS TECHNIQUE

Up to now, the dependency of the thermal lens signal on the thermal physical properties of solvent has only been exploited to enhance the sensitivity and selectivity of the technique. Conceivably, one can also take advantage of this dependency in the reverse way, i.e., to use the thermal lens technique as a precise and sensitive method for the determination of the thermal physical properties of solvents, namely their dn/dT and k values. These two thermal physical properties can be individually determined by measuring the transient of the thermal lens. This is because, as mentioned earlier, the thermal lens, when recorded as the time dependent change in the far field beam center intensity ($I_{bc}(t)$) after the onset of excitation laser illumination ($I_{bc}(0)$), is given by eq 12. The thermal time constant t_c which depends on the spot size, ω, of the excitation beam in the sample, the density, ρ, specific heat capacity C_p and thermal conductivity, k, of the solvent, is given by:

$$t_c = (\omega^2 \rho C_p)/4k \qquad (17)$$

Thus, θ and t_c values can be obtained by nonlinear curve fitting of the thermal lens transient to eq. 12. Substitution of the t_c value into eq. 19 will yield either the thermal conductivity or the heat capacity of the micellar system. The dn/dT value can then be calculated from eq. 13 using the θ and k values. The accuracy of the method can be evaluated by performing the thermal lens measurement on a solvent whose thermal conductivity, density and specific heat capacity values are known and comparing the t_c values obtained by thermal lens method with the literature values.

The dn/dT and k values of various solutions including aqueous and organic solvents such as methanol, carbon tetrachloride, chloroform, DMF, DMSO and heptane were accurately determined by use of this method [48-50]. In addition to their scientific and

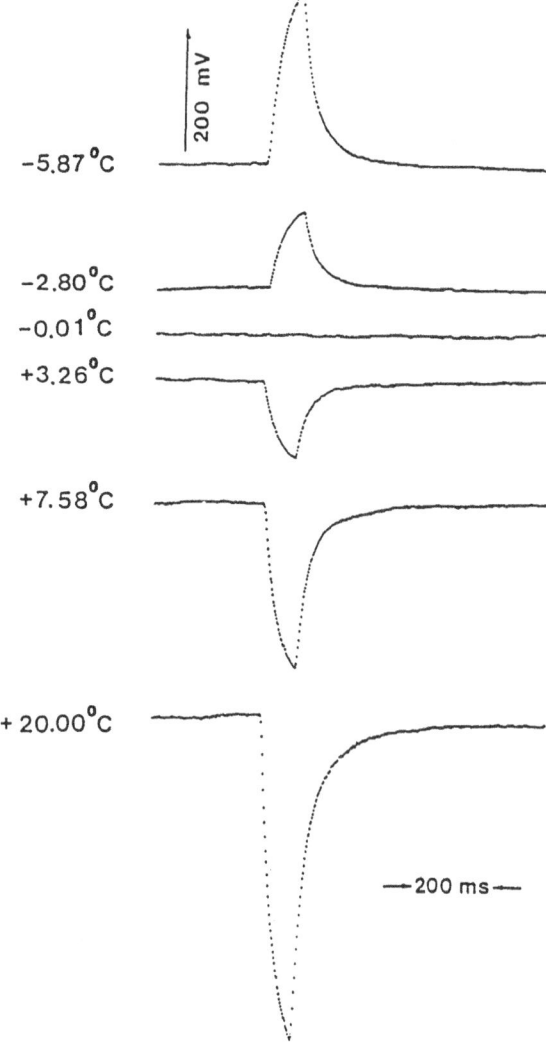

PD, PIN photodiode; Lock-in, lock-in amplifier.

Figure 3.   Thermal lens signals of aqueous solution at different temperatures.

technological importance, the obtained dn/dT and k values can provide insight into the structure of solutions.   Of particular interest is the information on the effect of the temperature, electrolytes and surfactants on the structure of water.

From the study of the temperature effect on thermal lens measurements in water, it was found that the magnitude and sign of the thermal lens signal intensity are strongly dependent on the temperature of the aqueous solution [48,49].   As shown in Figure 3, dependent on whether the measurements are performed at temperatures lower or higher than -0.01°C, the photoinduced thermal lens can have either a positive (converging) or negative (diverging) focal length.   At precisely -0.01 ± 0.04°C, no thermal lens signal could be observed.   This is because the dn/dT values of water are positive at T < -0.01°C, negative at T > -0.01°C, and equal to zero (i.e., maximum refractive index) at T = -0.01°C [48,49].

Using the method described above, the k and dn/dT values of water at different temperatures were calculated from the corresponding thermal lens transient signals. The results are shown in Table III [50]. As listed, within experimental error, the k value remains constant at different temperatures. The dn/dT value, however, undergoes significant change as the temperature varies. Increasing temperature leads to an increase in the dn/dT value. For instance, increase the temperature by 16 degrees (from 1.26°C to 17.98°C) leads to an 14 fold increase in the dn/dT value [48-50]. This observation can be explained in terms of the structure of water.

The structure of water is known to be greatly affected by the temperature of the solution. Increasing the temperature of the aqueous solution will have the "hydrogen bond breaking effect" [51-53]. As a consequence, the water has less hydrogen bonds, becomes less structure (or less order) and hence can have large changes in the density and in the refractive index with temperature (i.e., higher dρ/dT and dn/dT values).

Effects of electrolytes and surfactants on the thermal lens signal have also been investigated [50]. The study was undertaken because it has been reported, based on the IR, Raman and NMR studies, that adding certain electrolytes or surfactants into water has the same effect as heating up the solution [51-60]. Therefore, it may be possible to enhance the sensitivity of the thermal lens technique by adding electrolytes to the aqueous solution. Since there are many types of electrolytes and surfactants, namely, the so-called "water structure former" and "structure breaker", it is important to systematically investigate this possibility and whether it is possible to elucidate the enhancement mechanism [51-60]. It was found that electrolytes did, indeed provide up to a 2-fold enhancement in the thermal lens signal [50]. Higher enhancement (up to 8-fold) can also achieve when surfactants were added to the solution [50]. In both cased, the enhancement was found to be due to the effect of the electrolyte not on the thermal conductivity but rather on the dn/dT of the solution. Interestingly, the effect of electrolytes and surfactants seems to be different from the temperature effect [48,49]. This conclusion was derived when the change in dn/dT upon the addition of electrolytes or surfactants was deconvoluted into two parts. The first, which results from the difference in the specific refractivity and dn/dT of the electrolyte or surfactant solution as compared to that of pure water, has a stronger effect on the enhancement than the second, which is caused by the alteration of the hydrogen-bonding network of water [50].

CONCLUSION

Taken together, it is clear and unequivocally evident that the thermal lens is very sensitive to the thermal physical properties and structure of the environment. This unique feature can be exploited either to enhance the sensitivity of the technique or to investigate the structure of the medium. Up to 38X enhancement in the signal intensity can be achieved when reversed micelles are used to improve the thermal physical properties of the environment around the solute (i.e., from pure water which is the poor thermal physical medium to organic solvent which has relatively better thermal physical properties). Normal micelles can also enhance, at a relatively lesser extend, the thermal lens signal intensity. Different from the reversed micelles case, the enhancement in this case is, however, due to the disrupting in the structure of water by surfactant molecules. Substantial improvement in the sensitivity and selectivity can be accomplished when crown ethers are used to form complexes with metal ion and to facilitate the extraction of the ion pair complex formed from the thermal physically poor water to the thermal physically good organic solvent. The size effect of the crown ether is the origin of the enhanced selectivity.

The thermal lens technique can also be used to gain insight into the structure of medium. This is possible because the thermal conductivity k and specially dn/dT values are known to be very sensitive to the structure of the environment and these two values can be accurately determined by use of the thermal lens technique. A variety of potential applications can be realized using this unique property of the thermal lens technique. They include the use of the technique to study the structure of the water pool of the reversed micelles, and the kinetic determination of enzymatic reactions which are initiated by enzymes entrapped in the water pool. In fact it has been theoretically demonstrated and experimentally confirmed recently that the thermal lens technique can be use for the kinetic determination of reactions as fast as a few microseconds [61,62]. The thermal lens technique alleviates limitation which is currently imposed on the conventional fast kinetic techniques such as stopped flow, temperature jump. The present restriction stems from the fact that conventional techniques which are based on the measurement of absorbances by tranmission method, have low sensitivity. As a consequence, higher concentration of the reagent is needed in order to provide a measurable signal. The use of high concentration produces an unwanted effect, namely, it accelarates the rate of the reaction. Therefore, stopped flow and temperature jump are not suitable for the kinetic determinations chemical reactions of species which have low absorbances and fast reaction rates (order of few microseconds). The thermal lens technique, due to its inherent high sensitivity, does not suffer from this limitation.

ACKNOWLEDGMENT: The author is grateful to his former and present harworking coworkers, specifically Dr. Mladen Franko and Weifeng (Kerry) Zhang, for their dedications. Financial support for this work, provided by the National Institutes of Health, National Center for Research Resources, Biomedical Research Technology Program, is gratefully acknowledged.

REFERENCES

1. J. M. Harris and N. J. Dovichi, Thermal Lens Calorimetry, Anal. Chem. 52:695A (1980).
2. C. D. Tran, Simultaneous Enhancement of Fluorescence and Thermal Lensing by Reversed Micelles, Anal. Chem. 60:182 (1988).
3. M. Franko and C. D. Tran, Development of a Double Beam, Dual Wavelength Thermal Lens Spectrometer for Simultaneous Measurement of Absorption at Two Different Wavelengths, Anal. Chem. 60:1925 (1988).
4. C. D. Tran and T. van Fleet, Micellar Induced Simultaneous Enhancement of Fluorescence and Thermal Lensing, Anal. Chem. 60:2478 (1988).
5. M. Franko and C. D. Tran, Simultaneous Determination of Two Component Mixtures and pHs by Dual Wawelength Thermal Lens Spectrometry, Appl. Spectrosc. 43:661 (1989).
6. C. D. Tran and M. Franko, Dual Wavelength Thermal Lens Spectrometry as a Sensitive and Selective Method for Trace Gas Analysis, J. Phys. E: Sci. Instrum. 21:586 (1989).
7. W. B. Jackson, N. M. Amer A. C. Boccara and D. Fournier, Photothermal Deflection Spectroscopy and Detection, Appl. Opt. 20:1333 (1981).
8. M. A. Olmstead, N. M. Amer, S. Kohn, D. Fournier and A. C. Boccara, Photothermal Displacement Spectroscopy: An Optical Probe for Solid and Surfaces, Appl. Phys. A 32:141 (1983).
9. C. D. Tran, He-Ne Laser Intracavity Photothermal Beam Deflection Spectrometry, Anal. Chem. 58:1714 (1986).

10. C. D. Tran, Intracavity He-Ne Laser Photothermal Deflection as a Sensitive Method for Trace Gas Analysis, Appl. Spectrosc. 40:1108 (1986).

11. C. D. Tran, Helium-Neon Laser Intracavity Photothermal Beam Deflection Densitometer, Appl. Spectrosc. 41:512 (1987).

12. C. D. Tran, Development of a Double Beam, Dual Wavelength Thermal Lens Using a Helium-Neon Laser, Analyst 112:1417 (1987).

13. N. Teramae and J. D. Winefordner, Double-Beam Lens Spectrocopy with Dual-Beam Configuration Based on Photo-Differential Detection, Appl. Spectrosc. 41:164 (1987).

14. N. J. Dovichi, T. G. Nolan and W.A. Wiemer, Theory for Laser Induced Photothermal Refraction, Anal. Chem. 56:1700 (1984).

15. C. D. Tran and M. Xu, Dual Wavelength Photothermal Refraction Spectrometry for Small Volume Samples, Appl. Spectrosc. 43:1056 (1989).

16. M. Xu and C. D. Tran, Multiwavelength Thermal Lens Spectrophotometer, Anal. Chim. Acta 235:445 (1990).

17. C. D. Tran and M. Xu, Ultrasensitive Thermal Lens-Circular Dichroism Spectropolarimeter for Small Volume Samples, Rev. Sci. Instrum. 60:3207 (1989).

18. M. Xu and C. D. Tran, Thermal Lens-Circular Dichroism Spectropolarimeter, Appl. Spectrosc. 44:962 (1990).

19. M. Xu and C. D. Tran, Thermal Lens-Circular Dichroism Detector for High Performance Liquid Chromatography, Anal. Chem. 62:2467 (1990).

20. Y. S. Toulukian, P. E. Liley and S. K. Saxena, "Thermophysical Properties of Matter, T.P.R.C. Data Series", IFI-Plenum Press, New York (1970).

21. R. C. Weast, "CRC Handbook of Chemistry and Physics" CRC Press Inc., Boca Raton, Florida (1985).

22. N. V. Tederberg, "Thermal Conductivity of Gases and Liquids", MIT Press, Cambridge (1965).

23. J. Timmermans, "Physics-Chemical Constants of Pure Organic Compounds", Vol. 1 and 2, Elsevier, Amsterdam (1965).

24. M. Franko and C. D. Tran, Thermal Lens Effect in Electrolyte and Surfactant Media, J. Phys. Chem. 95:6688 (1991).

25. R. Vyas and R. Gupta, Photothermal Lensing Spectroscopy in a Flowing Medium: Theory, Appl. Opt. 27:4701 (1988).

26. S. J. Sheldon, L. V. Knight and J. M. Thorne, Laser Induced Thermal Lens Effect: A New Theoretical Model, Appl. Opt. 21:1663 (1982).

27. C. A. Carter and J. M. Harris, Comparison of Models Describing the Thermal Lens Effect, Appl. Opt. 23:476 (1984).

28. T. Berthoud, N. Delorme and P. Mauchien, Beam Geometry Optimization in Dual Beam Thermal Lensing Spectrometry, Anal. Chem. 57:1216 (1985).

29. M. Franko and C. D. Tran, Water as a Unique Medium for Thermal Lens Measurements, Anal. Chem. 61:1660 (1989).

30. J. H. Fendler, "Membrane Mimetic Chemistry", Wiley, New York (1982).

31. D. W. Armstrong, Micelles in Separations: a Practical and Theoretical Review, Sep. Purif. Methods, 14:212 (1985).

32. W. L. Hinze and D. W. Armstrong "Ordered Media in Chemical Separations" American Chemical Society, Washington, D.C. (1987).

33. D. W. Armstrong, Optical Isomer Separation by Liquid Chromatography, Anal. Chem. 59:84A (1987).

34. J. H. Fendler and E. J. Fendler, "Catalysis in Micellar and Macromolecular Systems", Academic Press, New York (1975).

35. B. S. Mohite and S. M. Khopkar, Solvent Extraction Separation of Strontium as 18-Crown-6 Complex with Picrate Ion, Anal. Chem. 59:1200 (1987).

36. L. M. Tsay, J. S. Shih and S. C. Wu, Solvent Extraction of Rare Earth Metals with Crown Ethers, Analyst, 108:1108 (1983).

37. W. Wenji, C. Bozhong, J. Zhong-Kao and W. J. Ailing, Radioanal. Chem. 76:49 (1983).

38. V. K. Manchanda and C. A. Chang, Solvent Extraction Studies of Europium(III), Ytterbium(III), and Lutetium(III) with Ionizable Macrocyclic Ligands and Thenoyltrifluoroacetone, Anal. Chem. 59:813 (1987).

39. Y. Hasegawa and S. Haruna, Ion-pair Extraction of Lanthnoid(III) with Crown Ethers and Picrate Ion, S. Solv. Ext. Ion Exch. 5:255 (1987).

40. Y. Hasegawa, M. Masuda, K. Hirose and Y. Fukuhara, Solvent Extraction of Europium(III) with Crown Ethers and Picrate Ions, Solv. Ext. Ion Exch. 2:451 (1984).

41. H. F. Aly, S. M. Khalifa, J. D. Navratil and M. T. Saba, Extraction of Some Lanthanides and Actinides by 15-Crown-5 Thenoyltrifluoroacetone Mixture in Chlorotorm, Solv. Ext. Ion Exch. 3:623 (1985).

42. D. D. Ensor, G. R. McDonald and C. G. Pippin, Extraction of Trivalent Lanthanides by a Mixture of Didodecylnaphthalenesulfonic Acid and a Crown Ether, Anal. Chem. 58:1814 (1986).

43. J. Tang and and C. M. Wai, Solvent Extraction of Lanthanides with a Crown Ether Carboxylic Acid, Anal. Chem. 58:3235 (1986).

44. C. D. Tran and W. Zhang, Thermal Lens Detection of Lanthanide Ions by Solvent Extraction Using Crown Ethers, Anal. Chem. 62:830 (1990).

45. C. D. Tran and W. Zhang, Luminescence Detection of Rare Earth Ions by Energy Transfer from Counter Anion to Crown Ether-Lanthanide Ions Complexes, Anal. Chem. 62:835 (1990).

46. J. -C. Bunzli in "Handbook on the Physics and Chemistry of Rare Earths", Vol. 9, Chapt. 60, K. A. Gschneidner and L. Eyring eds., North-Holland, Amsterdam (1987).

47. K. L. Cheng, K. Ueno and T. Imamura, "Hankbook of Organic Analytical Reagents", CRC Press, Boca Raton, Florida (1982).

48. M. Franko and C. D. Tran, Temperature Effect on Photothermal Lens Phenomena in Water: Photothermal Focusing and Defocusing, Chem. Phys. Lett. 158:31 (1989).

49. M. Franko and C. D. Tran, Water as a Unique Medium for Thermal Lens Measurements, Anal. Chem. 61:1660 (1989).

50. M. Franko and C. D. Tran, Thermal Lens Effect in Electrolyte and Surfactant Media, J. Phys. Chem. 95:6688 (1991).

51. K. A. Hartman, Jr. The Structure of Water and the Stability of the Secondary Structure in Biological Molecules. An Infrared and Proton Magnetic Resonance Study, J. Phys. Chem. 70:270 (1966).

52. J. Stangret and J. Kostrowicki, IR Study of Aqueous Metal Perchlorate Solution, J. Sol. Chem. 17:165 (1988).

53. G. E. Walrafen, M. S. Hokmabadi and W. H. Yang, Raman Isosbestic Points from Liquid Water, J. Chem. Phys. 85:6964 (1986).

54. F. Rull and J. A. de Saja, Effect of Electrolyte Concentration on the Raman Spectra of Water in Aqueous Solutions, J. Raman Spectrosc. 17:167 (1986).

55. M. M. Marciacq-Rauselot and M. Lucas, Nuclear Magnetic Resonance Chemical Shift of the Water Proton in Aqueous Solutions at Various Temperatures. Some Thermodymic Properties of These Solutions, J. Phys. Chem. 7:1056 (1973).

56. R. Bhanumathi and S. K. Vijayalakshamma, ^{1}H NRM Chemical Shifts of Solvent Water in Aqueous Solutions of Monosubstituted Ammonium Compounds, J. Phys. Chem. 90:4666 (1986).

57. R. E. Verrall in "Water - a Comprehensive Treatise", Vol. 3, F. Franks ed., Plenum Pres, New York (1974).

58. F. Frank and D. S. Reid in "Water a Comprehensive Treatise", Vol. 2, Chapter 5, F. Franks ed., Plenum Press, New York (1973).

59. M. Klose and J. U. Naberuchin, "Water - Struktur und Dynamik", Akademie Verlag, Berlin (1986).
60. D. Eisenberg and W. Kauzmann, "The Structure and Properties of Water", Oxford University Press, Oxford, U.K. (1969).
61. M. Franko and C. D. Tran, Thermal Lens Technique for Sensitive Kinetic Determinations of Fast Chemical Reactions. Part I. Theory, Rev. Sci. Instrum. 62:2430 (1992).
62. M. Franko and C. D. Tran, Thermal Lens Techniques for Sensitive Kinetic Determinations of Fast Chemical Reactions. Part II. Experimental, Rev. Sci. Instrum. 62:2438 (1992).

ROLE OF POLYPYRROLE IN IMPROVING THE "COMMUNICATION"

ABILITY OF METALLIC ELECTRODES WITH ORGANIC MOLECULES

Long Jiang and Qingxiu Chen

Institute of Photographic Chemistry
Academia Sinica, Beijing, 100101
China

INTRODUCTION

In recent years, the development of the nanometer scale sensing device based on a bio- or biomimetic membrane coated on a metallic substrate for use in biosensor, molecular electronics and biomolecular electronics has been a research area of great interest[1]. A crucial problem tackled in these works is the electrical communication between biological species with metallic electrodes.

Heller et al[1] reported the direct electrical communication between chemically modified enzymes and metal electrodes. Take glucose oxidase electrode as an example. They suggested that the enzyme's redox centers were located deep in the enzyme even if the enzyme is adsorbed on the electrode and the distances between redox centers and the surface of the electrode exceed the distance across which electrons are transferred at measurable rates. To enable electron transfer from redox centers of enzymes to the metal electrode, the authors used a modified enzyme and attached an array of electron-transfer relays to it. (Figure 1)

It shows that the polypyrrole here has the same function as the relays in Heller's model. In our previous work[5] on LB film enzyme electrode, it has been found that the polypyrrole could markedly improve the electrical communication ability between the metallic electrode and the enzyme even though the enzyme was not immobilized in polypyrrole (Figure 3).

The difference between polypyrrole and a metallic electrode is that the former is uneven, flexible, and perhaps possesses some side chain which can penetrate into the enzyme shell to contact with the redox center of enzyme.

It was known that organized assemblies of monolayers such as the multilayer LB film are what is used to investigate the energy and electron transfer in planned molecular architecture.[6-8]

It was also reported[9] that linear molecules with an extended system of conjugated double bonds, such as polyacetylene, polythiophene, and poly(phenylene-vinylene), could be incorporated into the LB film, through which electron transfer will be possible. As mentioned by many authors, undoped conjugated molecules are insulators; they are semiconductors with a fairly large band gap in an undoped state and conduct only upon doping.[10-11] This means excited electrons can pass

Advances in the Applications of Membrane-Mimetic Chemistry
Edited by T.F. Yen *et al.*, Plenum Press, New York, 1994

through such molecules. In order to gain a deeper insight into this phenomenon, the LB film of arachidic acid and those containing three species of electroactive organic molecules was used as an enzyme model to see the electrical communication behavior between metal and organic molecules. This property gives us a hint that we can use such molecules to mimic the redox centers and LB film matrix as the shell of an enzyme. In this paper, cyclic voltammetry has been adopted to investigate the mixed LB films of cadmium arachidate with an "electroactive organic molecule" deposited on a platinum electrode. In order to improve the "communication" ability between metal electrode and organic molecules, an electroconductive polymer -- polypyrrole, has been used to modify the Pt electrode.

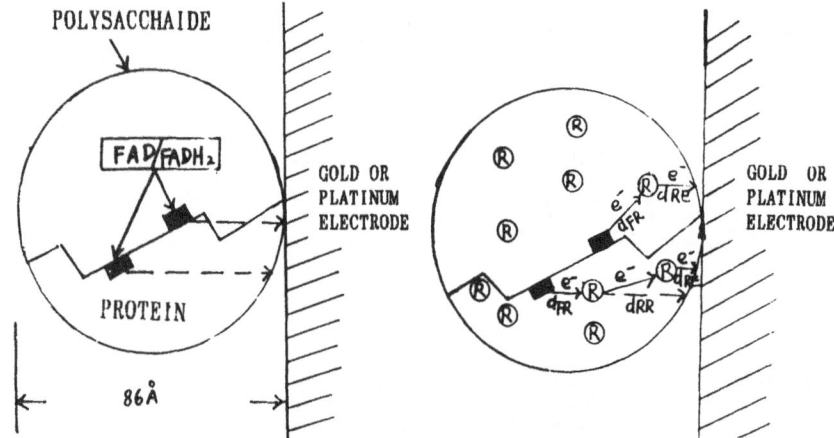

Figure 1. Schematic drawing of the glucose oxidase molecule, showing the electron-transfer distances involved in the various steps of moving an electron from its two FAD/FADH centers to a metal electrode. Left: the enzyme before modification. Right: the modified enzyme, after chemical attachment of an array of electron-transfer relays (Y. Degani and A. Heller[2]).

Almost at the same time Aizawa et al. [3] and Umana et al. [4] reported a kind of immobilized enzyme electrode --- glucose oxidase/polypyrrole system, made by electropolymerization of pyrrole in the presence of the enzyme. In these systems, the neutral form of PP(I) is an insulator, but it is easily electroxidized to its conducting form(II) [4]

the electron transfer process as follows [3].

Figure 2. Polypyrrole-Glucose oxidase membrane (M. Aizawa et al.[3]).

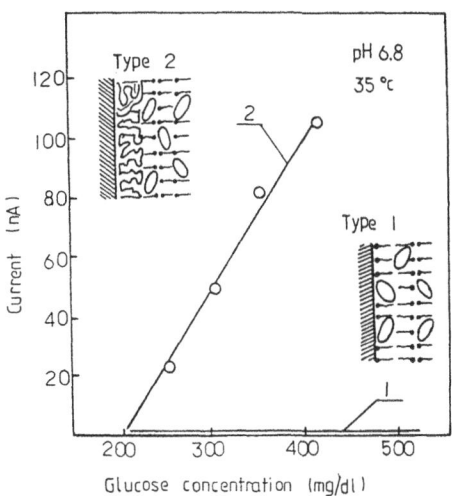

Figure 3. Current response against glucose concentration of electrodes of types (1) and (2): type (1), electrode with GOD-immobilized lipid LB films; type (2), GOD-immobilized LB film coated on polypyrrole membrane.

EXPERIMENTAL

Terthiophene(I), 2-formaldehyde-terthiophene and 4-(N, N-dimethylamine)- 4'-nitroazobenzene(III) were kindly given by Prof. Z.Q. Yao and Y.Q. Shen (Fig. 4). Water used in the subphase was obtained by deionization followed by ultrafiltration. LB film deposition was conducted with a HBM-SS LB film deposition apparatus and a moveable shiplike plate floating on the subphase according to Kuhn's principle[12] . Arachidic acid and electroactive compounds were dissolved in chloroform and were deposited at a surface pressure of 30 mN/m on a platinum electrode. Since it was reported[13] that terthiophene could be incorporated with arachidic acid in a molar ratio of 1:5 and were orientated nearly vertical to the film plane, we prepared our sample according to this ratio without further identification of its orientation.

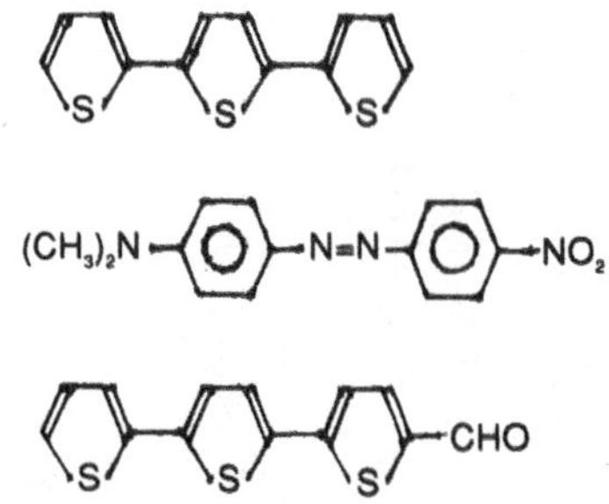

Figure 4. The structure of compounds obtained by deionization is followed by ultrafiltration.

Polypyrrole was made by electric polymerization carried out at 0.65 V with respect to the saturated calomel electrode(SCE) on a platinum plate having a surface area of 2 x (0.5 x 0.7) cm which was dipped in an aqueous solution of 1M KCl and 0.1M pyrrole. According to the report in reference 14, there are different states on the Pt electrode surface with potential scans. In order to get repeatable Pt surface, all the Pt plates were treated at +1.5V vs. SCE for 2 minutes before use. Cyclic voltammograms were conducted using a voltammeter made in China, type DHX-II.

A trielectrode cell was used to carry out the experiments where a Pt plate covered by LB films acted as a working electrode. SCE with a KNO salt bridge was used as a reference electrode, and a Pt wire 1mm in diameter as a counter electrode.

A hydrophilic surface was needed because electrodes work in aqueous solution, and an even number of layer for the Pt substrate and an odd layer for the pyrrole-modified electrode were obtained. Afterwards, monolayer transfer electrodes were kept at 0 C - 5 C in a beaker, enveloped in a plastic bag to keep them wet.

RESULTS AND DISCUSSION

Figure 5 shows the cyclic voltammograms of the Pt electrodes coated with different layers of arachidic acid in 3M H_2SO_4 solution; Figure 6 shows the cyclic voltammograms of Pt

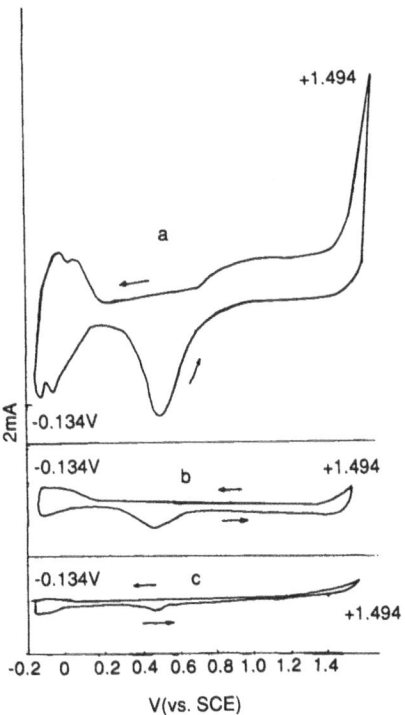

Figure 5. Cyclic voltammograms in 3M H SO for the Pt electrode coated with different layers of C LB film at 50mv.s.

    a. Bare Pt electrode
    b. Pt electrode with 2 layers C LB film
    c. Pt electrode with 12 layers C LB film

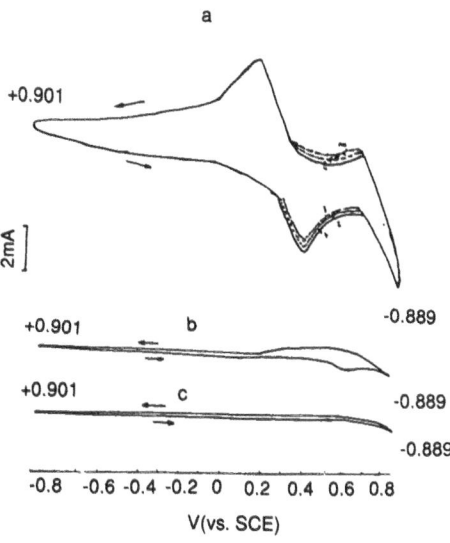

Figure 6. Cyclic voltammograms in 0.1M KCl for Pt electrode coated with different layers of C LB film at 50mv.s.

    a. Bare Pt electrode
    b. Pt electrode with 2 layers C LB film
    c. Pt electrode with 12 layers C LB film .cw10

electrodes coated with different layers of arachidic acid in H_2SO_4 : Figure 6 shows the cyclic voltammograms of Pt electrodes coated with different numbers of layers of arachidic acid in 0.1M KCl. Results in both cases indicate that arachidic acid LB films exhibit strong insulating ability and are stable in the electrolyte solution even in strong acid solutions like 3M H_2SO_4.

It was known that the LB film of arachidic acid has very high insulating ability because of the long hydrocarbon chain packed closely perpendicular to the surface of the solid substrate. However, it cannot block the electrode until the deposited layer is more than ten as shown in Figures 5 and 6. This might explain the holes in the LB film of alkyl acid[15]. Figure 7 shows the STM image of arachidic acid deposited on Pt plate[16] and gives direct evidence of a hole existence in the LB film. This fact told us that arachidate, the LB film, is a very good insulator for metallic electrodes despite the existence of many holes, implying that these holes are so small that the ions hardly penetrate or diffuse through them.

Figure 7. STM image of LB film of cadmium arachidate on platinum plate.

In order to improve the communication ability of LB films, electroactive compound (1),(2),(3) have been introduced into the Cd arachidate LB film. Monolayers of compound (1),(2),(3) with arachidic acid in a molar ratio of 1:5 were spread on an aqueous subphase, containing 3 x 10M CdCl at 18 C. For nitroazobenzene, the pH of subphase was kept at 4.0 using potassium hydrogenphthalate as a buffer. -A curves showed the possibility of mixed LB film formation.

After transferring these monolayers onto the platinum electrodes, cyclic voltammetric measurements were conducted. To our surprise, the mixed monolayers on the Pt electrode of these three compounds did not exhibit any measurable vectorial conductivity as of insulators. Figure 8 shows the cyclic voltammogram of such an electrode with the compound (I).

It was found that the electrical response of arachidate LB films with or without electroactive molecules appeared. Figure 9 shows the voltammograms of compounds (1) and arachidic acid mixed monolayers on the Pt electrode with coating polypyrrole. After modification with polypyrrole, the electrode response ability increased considerably.

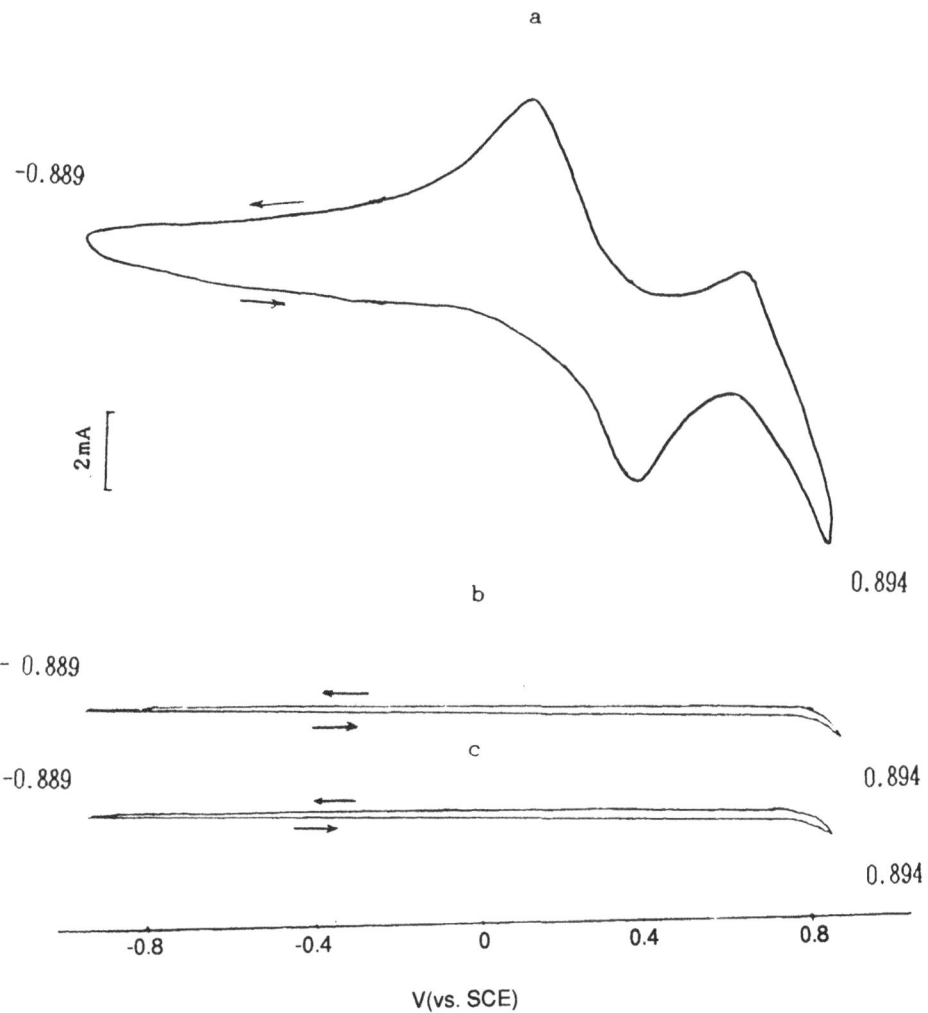

a

-0.889

2mA

b

- 0.889

-0.889

0.894

0.894

0.894

c

0.894

| -0.8 | -0.4 | 0 | 0.4 | 0.8 |

V(vs. SCE)

Figure 8. Cyclic voltammograms in 0.1M KCl for the Pt electrodeattached with C LB films with and without compound 1.
Scanrate = 50 mv.s.
   a. Bare Pt electrode
   b. Pt electrode with 11 layers C LB film
   c. Pt electrode with 11 layers mixed monolayers of I:C=1:5 .cw10

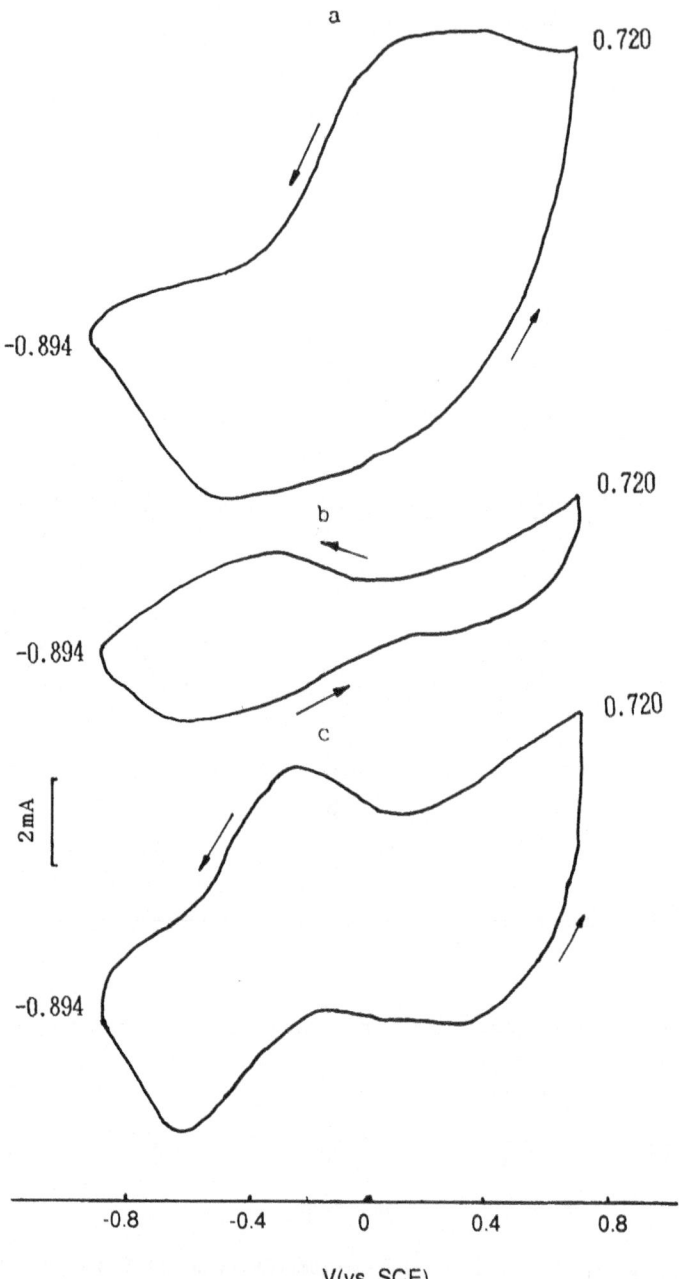

V(vs. SCE)

Figure 9. Cyclic voltammograms in 0.1M KCl for the polypyrrole modified electrode attached with compound 1 LB film. Scan rate = 50mv.s .
    a. Pt coated with polypyrrole
    b. Polypyrrole coated Pt with 11 layer C LB film.
    c. Polypyrrole coated Pt with 11 mixed layers of compound 1 and C (1:5).

     Although cadmium arachidate monolayers can markedly hinder the metal electrode from electrical response, it could not block the polypyrrole electrode even after 3 layers of deposition. When conductive molecules were introduced into the monolayers, the electrode response increased even more, and their voltammograms differentiate with special peaks characterized for each compound (Figure 10).

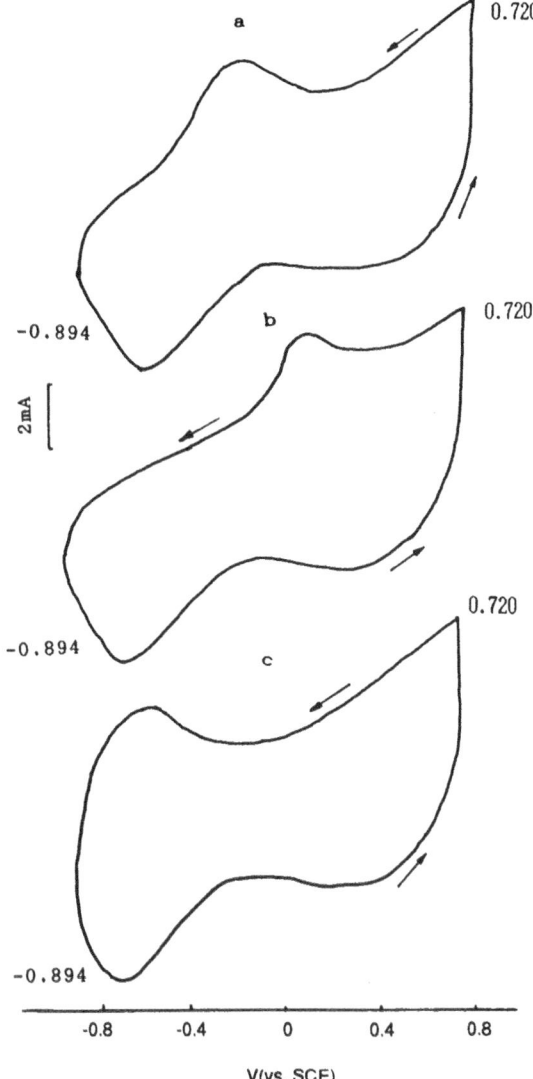

Figure 10. Cyclic volammograms in 0.1M KCl for the polypyrrole modified electrode attached with LB films of different compounds. Scan rate = 50 mv.s

    a. Polypyrrole coated Pt with 11 mixed monolayers of compound 1 and C

    b. Polypyrrole coated Pt with 11 mixed monolayers of compound 2 and C

    c. Polypyrrole coated Pt with 11 mixed monolayers of compound 3 and C

       Two reasons perhaps could explain why polypyrole can improve the electrical communication between metal electrodes and organic species. Firstly, the unevenness of the polypyrrole surface made the LB film deposition irregular, similar to the metal surface, enlarging the hole existing in LB films in facilitating the electrolyte ion transfer through the insulation film, that is, an ion transfer factor. Secondly, voltammograms show that there existed oxidation and reduction peaks characterized for each compound, according to the view that such electroactive compounds exhibit high electrical conductivity, only after they have been treated with oxidizing or reducing agents (doping)[9-11]. It means the polypyrrole can play the role of an oxidizing or reducing agent by using its flexible chain or tail to touch the electroactive molecules, by taking or giving an electron to the electroactive molecules, improving the communication ability between metallic electrode and organic molecules, that is, the electron transfer factor.

       A schematic representation of this mechanism is shown in Figure 11.

Though the result of the experiment is initial, it is also encouraging. We think this phenomenon could be of common significance in the design and manufacture of the molecular and biomolecular devices, and we are continuing the exploration of fundamental properties of these systems.

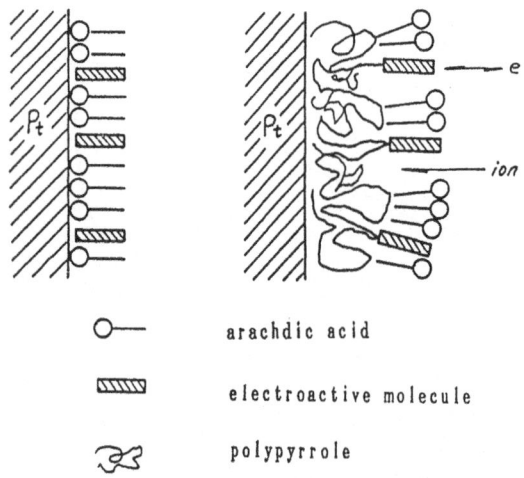

Figure 11. Improvement of the "communication" between metal electrode and organic film by polypyrrole.

CONCLUSION

1. Three electroactive molecules terthiophene, 2-formadehye-terthiophene, and 4-(N,N-dimethyamino)- 4'- nitroazobenzene have been introduced into an arachidic acid monolayer to make the LB film modify electrodes on a platinum plate.

2. These electroactive molecules did not assert themselves as a conductor unless the Pt working electrode was coated with an electroconductive polymer -- polypyrrole.

3. The mechanism of the ability of the polypyrrole for improving the "communication" ability between metallic electrodes and organic molecules has been discussed in terms of ionic transfer and electron transfer process.

ACKNOWLEDGEMENT

The authors wish to thank Professor Z.Q. Yao in Lanzhou Institute of Chem. Physics, and Professor Y.Q. Shen in our institute for their kindly offerings of the compounds. This work was supported by the National Natural Science Foundation of China (NSFC).

REFERENCES

1. J.D. Swalen, D.L. Allara, J.D. Andrade, E.A. Chandross, S. Garoff, J.Israelachvili, T.J. McCarthy, R. Murray, R.F. Pease, J.F. Rabott, K.J. Wynne, H. Yu, Molecular monolayers and films, Langmuir, vol.3, No.6, 932 (1987).

2. Y. Degani and A. Heller, Direct electrical communication between chemically modified enzymes and metal electrodes, 1. Electron transfer from glucose oxidase to metal electrodes via electron relays, bound covalently to the enzyme, J. Physical Chemistry, vol. 91, No.6, 1285 (1987).

3. M. Aizawa, S. Yabuki, H. Shinohara, Electrochemical preparation of conductive enzyme membrane, in: Proceeding of the 1st International Symposium on Electroorganic Synthesis, S. Torli ed.,Tokyo, Japan (1987).

4. M. Umana and J. Waller, Protein-modified electrodes. The glucose oxidase/polypyrrole system, Anal. Chem. vol. 58, 2929 (1986).

5. J.R. Li, M. Cai, T.F. Chen, and L. Jiang, Enzyme electrodes with conductive polymer membranes and Langmuir-Blodgett films, Thin Solid Films, 180 (1989) 205-210.

6. H. Kuhn, Molecular engineering - a begin and an endeavor, in Proceedings of international symposium on future electron devices. Nov.20-21, 1985, pp.1-6.

7. M. Sugi, Molecular engineering in Japan - A prospect of research on Langmuir-Blodgett films, Thin Solid Films, 152 (1987) 305.

8. A. Ulman, An Introduction to Ultrathin Organic Films: From Langmuir-Blodgett to Self-Assembly, Academic Press, 1991.

9. S. Roth, M. Filzmoser, Advanced Materials, vol. 2, No.8, (1990) 356-360.

10. H. Naarmann, Polymeric organic semiconductors, Naturwissenschaften, 1969, 56(6), 308-13.

11. C.K. Chiang, C.R. Fincher, Y.W. Park, A.J. Heeger, H. Shirakawa, E.J. Louis, S.C. Gau, A.G. Macdiarmid, Electrical conductivity in doped polyacetylene, Phys.Rev. Lett.39 (1977) 1098.

12. H. Kuhn, D. Mobius, and H. Bacher, in Physical Methods of Chemistry eds. A. Weissberger and B. Rossiter, vol.1, part 3B, p.577, Wiley, New York 1972.

13. H. Nakahara, J. Nakayama, M. Hoshino, and K. Fukuda, Langmuir-Blodgett films of oligo- and polythiophenes with well-defined structure. Thin Solid Films, 160 (1988) 87-97.

14. D.K. Lowde and J.O. Williams, The characterization of catalyst surfaces by cyclic voltammetry, Applications of Surface Science vol.1 (1978) 215-240.

15. N. Uyeda, T. Takenaka, K. Aoyama, M. Matsumoto, Y. Fujiyoshi, Holes in a steric acid monolayer observed by dark-field electron microscopy, Nature, vols. 327, 319 (1987).

16. Z.L. Ma, C.X. Zhu, S.J. Pang, Q.X. Chen, L. Jiang, Structure of LB film surface studied by STM, in Abstract of The 1990 China-Japan Bilateral Symposium on LB Films. Nov.4-8, 1990, Beijing, pp.33-34.

ENHANCEMENT EFFECTS OF SURFACTANTS IN FLAME ATOMIC ABSORPTION ANALYSIS

Daniel York Pharr

Chemistry Department
Virginia Military Institute
Lexington, VA 24450

INTRODUCTION

The use of surfactants in flame atomic absorption has been reported previously (1-7). Surfactants in dilute solutions below their critical micellar concentration (CMC) act as strong electrolytes. Above their CMC they exhibit unique solvent property changes such as apparent molar volumes, density, specific heat, electrical conductance, electromotive force, surface tension, vapor pressure, microfluidity or viscosity, and the temperature coefficient of solubility (9). It was the purpose of these studies to ascertain whether or not these changes in the solvent properties of water would effect atomic absorption signals. A review of the literature on this subject results in apparent inconsistencies in the final results and the theories that explain this phenomena. In the original papers some confusion might have resulted because the enhancement of signals had been reported three different ways. For ease in comparison these values have been recalculated as ratios of the signal studied compared to their aqueous solutions. For example a value of 1.10 could have previously been reported as a 10% enhancement or 110% of the aqueous signal.

HISTORICAL

In 1959 Foster and Hume used a total consumption burner and determined that variations in surface tension by the non-ionic surfactant Tergitol NPX had no significant effect in the analysis of sodium (1). It was believed that these solutions would behave like organic cosolvents which have lower surface tensions compared to water. Therefore more of the solution would be aspirated by the capillary tube and smaller droplets would be generated during the nebulization process, resulting in an increase in efficiency an enhancement in sensitivity. In 1960 Dean reported that small amounts of surfactants resulted in the enhancement of emission signals using a premix burner, but he recanted that statement in 1969 after reports by Lockyer in 1961 and Pungor in 1963 refuted those earlier findings (2). In 1974 Venable and Ballad reported on the effects of surfactants on the flame atomic absorption analysis of copper and nickel (3). Using sodium dodecyl sulfate (SDS) and tetradecylpyridinium bromide above their critical micellar concentrations (CMC) they found that SDS caused an enhancement in signal while tetradecylpyridinium bromide actually caused a decrease in the signal. Another effect noted was that SDS solutions prevented the formation of precipitates and stabilized colloidal dispersions over an extended pH range from 1 to 9 compared to solutions without SDS which had a pH range of 1 to 5.3. However, in their conclusions they warned against using surfactants because they might give erroneous results.

Kodama reported on the enhancement of chromium signals in the presence of SDS in flame atomic absorption spectroscopy (FAAS) using a premix burner (4). Later experiments focused on the droplet size distributions of water and aqueous micellar systems which exhibited proof that the surfactants produced a finer aerosol due to the surface tension lowering of these solutions above their CMC. Both SDS and dodecyltrimethylammonium chloride enhanced chromium (VI) absorbances up to a factor of 2.00. Other metals studied exhibited enhancement between 1.05 to 1.40 in the order: Fe>Ni>Mn>Cu>Zn. These results follow the decreasing order of analytical sensitivity and seem to indicate that the metals that form more stable solid particles in the flame are influenced by the size of the droplets (5). Armstrong studied the effects of SDS, ammonium dodecylsulfate and Aerosol OT, which all gave similar enhancements for copper flame atomic absorption spectrometry, while the cationic surfactant cetryltrimethylammonium bromide and the nonionic surfactant Triton X-100 either slightly depressed the signal or had no effect (2). He also reported enhancements due to the presence of calcium, chromium, manganese, nickel and magnesium, but actual enhancement figures were not given. It was concluded that the aerosol ionic redistribution theory of Borowiec et. al. could explain the enhancement effects that were reported for copper and chromium (10).

Taga reported the enhancement of flame absorption signals for titanium (IV) and vanadium (V) using SDS and cetyltrimethylammonium bromide (6). The surfactants enhanced the analyte signals 1.50 for titanium and 1.80 for vanadium compared to solutions containing only the analyte in water. A study by Pharr investigated micellar systems for fifteen metals using different cations for the anion surfactant dodecylsulfate. Ammonium, lithium, sodium, and potassium dodecylsulfates were investigated to determine if any salt or matrix effects would explain the enhancement of the absorbance signal of the analyte in the presence of SDS (7).

BACKGROUND

For flame atomic absorption spectrometry the flame acts as the sample cell. In order for the atoms to be present in the flame from a liquid solution they must go through several processes. The sample is first aspirated into the pneumatic nebulizer where the liquid is broken into droplets. This is often followed by one or more aerosol modifiers that break the droplets into finer droplets and remove the larger droplets so that only droplets of a specific size range enter the flame. Once in the flame the solvent evaporates, the analyte vaporizes and atomization follows. Several studies have been done to better understand the process of how the analyte is delivered into the flame for analysis by FAAS. A brief review will aid in the understanding of the behavior of surfactant solutions in FAAS.

Elements can be categorized by their thermochemical behavior into three groups (11). Group A includes elements that do not ionize significantly in flames and form little or no stable compounds. Their first ionization energies are high while the bond dissociation energy of their monoxides are small. This was the largest group that has been studied. It includes Cd, Co, Cu, Fe, Mn, Ni, Pb, Sb, and Zn. Copper in particular has been used as a reference in a study on the relative atomization efficiencies in FAAS as a function of temperature because it is believed to undergo complete atomization in the air/acetylene flame. Copper also has low dissociation energy molecules: CuOH at 2.6 eV ,CuH at 2.8 eV and CuO at 3.5 eV; therefore, even at low temperatures the atomization efficiencies of copper are considered to be equal to one (12). Group B consists of those elements that have low first ionization energies and do not form refractory compounds. Some of these easily ionized elements are used as spectral buffers. Many of these are usually studied by flame emission spectroscopy and they include three of the cations used with the dodecylsulfates, lithium, sodium and potassium in the study by Pharr (7). In this group Sr and Rb have been studied for micellar enhancement. Group C consists of elements that have a low ionization energy but do form refractory compounds. Aluminum, calcium, chromium and tin have been studied from this group.

LIGANDS AND FLAME CONDITIONS

An earlier study in 1972 on the effects of ligands on the analysis of cobalt by FAAS suggested that the type of ligand helps to determine flame conditions and the distribution of atoms in the flame (13). In burner height studies, oxygen and carbon coordinated complexes exhibit maximum absorbance in the upper portion of the flame when using an acetylene rich flame. This points to a delay in the dissociation of these complexes into the flame. In the enhancement studies of Armstrong, Kodama, Taga and Pharr, the changes in the optimum flame height for analytes in surfactant solutions compared with aqueous solutions did not exhibit large differences. Venable and Ballad studied burner height and looked for differences between an air/acetylene and nitrous oxide/acetylene flame for SDS compared to water and no effects on the enhancement were noted (3). The study by Armstrong shows that enhancement was not effected by any change in the reducing nature of the flame by the presence of SDS (2). A study by Kodama was conducted using sodium sulfate and SDS in combination so that when the total ionic strength was constant, the resulting curve peaked and then leveled off after the CMC. This showed that there was neither a salting in effect nor did the sodium present affect the results. Kodama reported that the enhancement effects of SDS with chromium (VI) and chromium (III) were identical (5). In order to discover if the organic matter present in SDS was having an effect, tris(hydroxymethyl)aminomethane (THAM) was added as a complexing agent and it was found that there were no changes in absorption compared to aqueous solutions (5). Indeed a study in 1981 by Cherry on the association constants of the cations of cesium, thallium, copper, manganese, cobalt, chromium and europium with dodecyl sulfate showed that they did not depend upon the charge of the metal ion (14).

DROPLETS IN THE FLAME

Hieftje and Malmstadt conducted a study in 1968 of what happens when a droplet travels through the flame using a droplet generator (15). As the droplets enter the flame the solvent evaporates from the droplet producing small salt or oxide particles. These particles then vaporize into atomic vapor. The presence of a droplet within a vapor cloud can cause chemical and thermal quenching of the excited atoms in the cloud, thereby reducing the total observed signal (15). Although desolvation usually depends upon the vapor pressure of the liquid in the droplet, at high vaporization rates it is usually heat transfer controlled.

Browner proposed an aerosol transport model for FAAS (16). Using several different spray chambers he reported that secondary impaction causes the formation of smaller droplets, and that tertiary drop impaction occurs in the Perkin-Elmer nebulizer/spray chamber with the paddle mixer. In a comparison of spray chambers and their drop size distributions using an empty chamber, one with the impact bead or one with the paddle mixer, the paddle generated smaller droplets with a much sharper cutoff below the important 2 µm size range (16).

Kodama's determination of droplet distributions exhibited that the effectiveness of the surfactant's improvement in sensitivity and reduction in interference for the determination of chromium were related to the production of finer aerosols resulting in a lowering of the surface tension (5). In that study of aerosols and droplet distributions for SDS and dodecyltrimethylammonium chloride (a cationic surfactant) compared to water, the surfactants exhibited fewer large size droplets greater than 6 µm. The creation of free atoms in the flame is controlled by the vaporization rate of the solid particles, which depends on flame temperature, their nature and particle size.

In a 1984 article by Browner and Farino a study of the effects of surface tension on aerosol properties in atomic spectroscopy, SDS and Tween-20 (a nonionic surfactant) were studied along with two organic solvents, a light machine oil and dioctyl phthalate (8). It was shown that the organic solvents provoked a shift in the tertiary aerosol to smaller droplet size, and the surfactants caused no such shift. They used a Perkin-Elmer 5500 AA/ICP but removed the burner/spray chamber and replaced it with the Varian nebulizer. Enhancement of copper with SDS was accomplished with 3.0 mM and 17 mM concentrations of SDS and 5000 ppm of copper analyte for the solvent properties study and 10 ppm for the transport efficiency study. These solutions apparently lead to a net reduction in transport efficiency when compared to water alone. Armstrong and others have reported decreases in absorbance enhancement for SDS concentrations that were above the CMC of SDS (2) and the high concentration of 5000 ppm copper may have interfered with the normal micelle formation and solubility of SDS. Browner maintained that in the time scale (μsec) of the nebulizer that the surfactants do not have time to orient themselves in order to reduce the surface tension (msec). Two years earlier Armstrong had maintained that there was indeed enough time for the formation of micelles in a nebulizer and used Einstein's relationship to prove this (2). The effects of the various solutions altered depending on the type of nebulizer that was used: Varian with impact bead, Perkin-Elmer bead or Perkin-Elmer mixer paddle. They believed that the mixer paddle would level any enhancement effects because of its superior droplet size distribution profile (8). Pharr, however, used a Perkin-Elmer mixer paddle nebulizer in his study and noted enhancement effects for over a dozen elements including copper (7). The surfactant systems in the micellar enhancement studies were used at the concentration that achieved the optimum enhancement of absorbance (2,5-7). This concentration was determined by selecting the highest absorbance value obtained when a low concentration of the metal ion remained constant while varying the concentration of the surfactant.

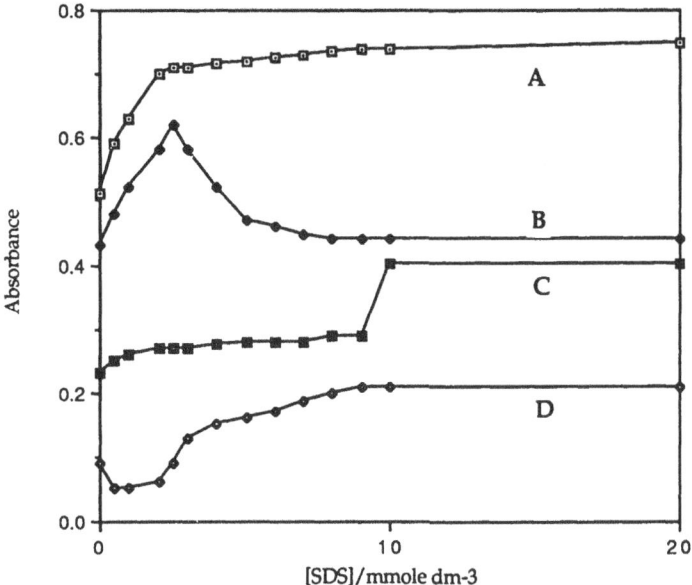

Figure 1. Patterns for surfactant optimization of enhancement.

Four general patterns have been observed. In Figure 1.A., for the cations of Co, Fe, Mn, Ni, Sn, Ti, V and Zn, the absorbance increased gradually and then leveled off after the CMC in the range from 6 to 20 mM SDS. This pattern supports the theory that enhancement was due to surface tension lowering and viscosity changes. In Figure 1.B. the absorbance exhibits an increase in the signal followed by a reduction to form a camel hump curve for the Al, Ca, Cd, Cu and Pb analytes. The optimum surfactant concentration varied with the different metals studied: Al-0.20 mM, Ca-0.30 mM, Cd-5.0 mM, Cu-10.0 mM and Pb-2.0 mM. An even different form was exhibited by antimony in Figure 1.C., a sudden peak at 10 mM SDS and then a plateau. For Rb and Sr in Figure 1.D. there was a unique decrease in the absorbance signal just below the CMC followed by a rise to a plateau above the CMC.

Dodecylsulfate surfactants achieved the best enhancement effects for most of the metals that were studied. The cationic surfactant, cetyltrimethylammonium bromide, a non-ionic surfactant, Triton X-100, and the anionic surfactants: ammonium dodecylsulfate (ADS), lithium dodecylsulfate (LDS), sodium dodecylsulfate (SDS) and potassium dodecylsulfate (KDS) have been studied with copper, chromium and lead (7). Both the cationic and nonionic surfactant gave lower absorbance values than water. The enhancement value of the micellar systems was obtained by comparison of the Beers Law slopes of the surfactant system correlated to water. For copper they demonstrated factors of 0.50 or less than those of the aqueous solutions. For chromium the values were 0.95 those of water, while for lead they were 0.07 for the cationic and 0.35 for the nonionic surfactant, Triton X-100 compared to the analyte in water (g).

The investigation of the relationship between aspiration rate and viscosity were consistent in the separate studies of both Armstrong and Pharr in that little variation was noted between the aqueous and the micellar solutions.

A pH study was also performed using the four dodecylsulfates (ADS, LDS, SDS and KDS) and the metal analytes of cadmium and copper with solutions at various pHs (0.35, 1.0, 2.0, 4.0, and 7.0). The absorbance values obtained were compared to those of water and SDS. The composition of the buffers supplied an acidic proton and other anions and cations. Both cadmium and copper exhibited similar behavior with a 1.10 to 1.13 enhancement for the acidic buffered surfactant systems compared to the unbuffered SDS surfactant system. The pH 7.0 buffered surfactant system gave similar enhanced values to the unbuffered SDS surfactant system.

Surfactant systems that had an enhancement of greater than 1.20 were considered to exhibit a large enhancement. The KDS system contained more elements in this category than any of the other surfactant systems. This included: Co, Cr, Cu, Fe, Mn, Ni, Rb, Sn and Sr. The LDS system contained the following elements with an enhancement value greater than 1.20 : Cd, Cr, Fe, Ni, Rb, Sn; for the ADS system: Fe, Ni and Sn, and for the SDS system only, Cr and Rb. Moderate enhancement was defined as those systems that had enhancement values between 1.10 and 1.20 compared to water. These included for ADS: Ca, Cd, Cu, Rb, Sb, and for LDS: Cu, and for SDS, Mn.

There were several metal analytes that actually had a signal depression in the presence of the surfactant system. Some were considered light with values between 0.95 and 0.99 those of water. These included for ADS: Zn, for SDS: Cd, Fe, Sb, Sr, and for KDS: Zn.

Others included for ADS: Al (0.89), Pb (0.43); for LDS: Mn (0.84), Pb (0.33), Sr (0.75); for SDS: Ca (0.85), and for KDS: Al (0.91), Pb (0.49).

THEORIES

There have been several models proposed to explain the enhancement of absorbance signals in FAAS with surfactants. The surfactant's ability to reduce surface tension at the CMC was first used to explain the enhancement (4,5). Surfactants such as SDS are known to lower surface tension, which results in the generation of smaller droplets during the aspiration and nebulization process. This increase in smaller droplets would result in a higher efficiency of the laminar flow burner so that a larger amount of the analyte that would go into the flame. In FAAS usually only 2 to 3 percent of the analyte reaches the burner flame. This theory suggests a pattern of absorbance versus concentration to increase at the CMC and then level off. This was the pattern that was seen for many of the metals studied including: Co, Fe, Mn, Ni, Sn, Ti, V and Zn. The theory also suggests that the observed absorbance would be independent upon the type of surfactant that was used (cationic, non-ionic or anionic) and independent of the counter ion that was used on the dodecylsulfate.

Taga supported this theory with a study of SDS enhancement for chromium, vanadium, titanium and magnesium (6). The cationic surfactant cetyltrimethylammonium bromide, CTAB, caused an enhancement for titanium of 1.80 and vanadium of 1.50 using a nitrous oxide flame. Vanadium absorbances and vanadium monoxide emissions in the presence of the CTAB both exhibited similar enhancements. They concluded that the enhancements were due to increased nebulization efficiency from smaller aerosol droplets as previously reported by Kodama.

The Aerosol Ionic Redistribution, AIR, theory was first used in the study of ocean sprays where it was found that a concentration enrichment occurs via the microscopic bubble bursting (10). The fractionization of aerosols according to size shows that the redistribution is size dependent. Enrichment depends upon ionic type, concentration ratio and significantly upon droplet size, with the smaller droplets generally exhibiting a greater enhancement factor. The enrichment study that was originally done with the Perkin-Elmer nebulizer was done, unfortunately, with the mixer paddles removed. The secondary fragmentation resulting from the paddles would have resulted in a further enrichment of the analyte. The stripping action that produces the smaller enriched droplets were in the 0.5-10 μm range. Loeb considered this redistribution in terms of an electrical double layer effect. As the smaller droplets are formed and enrichment is taking place, the smaller more mobile ions are drawn more quickly to the double layer. This would explain the selectivity that was seen in the enrichment of various ions that were studied. These AIR effects were found to be dependent upon nebulizer design and operation (2).

Armstrong used this theory to explain the enhancement of SDS on the absorbance of copper (2). A mechanism was proposed that allowed analyte transport to the hottest part of the flame by the interaction of the spectator ions of the anionic sulfate head group of the dodecyl sulfate with the cationic metal ion analyte. The enrichment occurs at the double layer on the outside surface of the large drops. As these drops divide, a stripping action occurs that effectively concentrates the metal cation into the smaller droplets that are nebulized and carried into the flame, resulting in an increase in the analyte signal. However, the dichromate anion was reported to exhibit an enhancement with SDS, which is contrary to this theory. With concentrations much above the CMC a decreased signal was observed because micelles in greater concentration throughout the bulk solution are competing for interaction sites with those involved in the surface stripping action and its resulting enhancement. This would be true for cations that were tightly bound to SDS like copper, manganese and nickel. This type of enrichment would not be observed in a total consumption burner. The major evidence given for this conclusion was the decreased amount of copper that was found when the nebulization waste was reanalyzed. This was believed to be a result of the decreased concentration of the analyte in the larger droplets that went down the drain. According to the AIR theory there was a stripping action that concentrated the metal cation into the smaller droplets which were carried into the flame. The larger droplets containing less analyte went down the drain and upon analysis of the waste solution the differences between these two solutions should be evident. In the study of copper absorbencies of original 10 ppm copper solution and the waste solutions from a liquid trap after nebulization, the waste solutions absorbencies were about 0.60 of the 10 ppm copper source solution. Armstrong has shown using the Einstein relationship that there would be time for the surfactant to migrate to the droplet surface during the nebulization process (2).

The decrease in nebulization waste was also studied in papers by Taga and Pharr by collecting and analyzing the waste from the analyte-surfactant system. Taga found in the study of vanadium and titanium that there were no appreciable differences between the absorption signals obtained originally or from the waste solutions (6). A series of solutions of the various dodecylsulfate surfactants was also analyzed by Pharr (7). The absorbance data of the waste solutions was compared to the corresponding absorbance value obtained by the solution that generated the aspirated waste. When a 2.0 ppm solution of cadmium in 5.0 mM SDS gave a signal of 0.264 absorbance units, the recovered waste was collected for one minute and then discarded. Then the portion collected after that was analyzed and gave an absorbance of 0.190 which was a -28% drop in the signal or a 0.72 ratio when compared to the original solution. This was what the AIR theory predicts, a decrease in waste analyte signal compared to the original analyte surfactant solution. For SDS the largest decreases were considered to be values less than 0.70 for the waste solution signals compared to their original solutions. This occurred with Co, Mn, Ni, Rb and Sb. A moderate decrease of 0.90 to 0.70 occurred with Cu (0.80), Cd (0.72) and Pb (0.71). A few metal analytes exhibited little or no effect, this included Cr, Sn, Sr, Ti, V all with values between 1.00 and 0.95 and Fe at 0.93. Surprisingly the waste recovery for Zn was greater than the bulk solution at 1.19 (7).

In using ammonium dodecylsulfate, ADS, lithium dodecylsulfate, LDS, and potassium dodecylsulfate, KDS, the results obtained were not always consistent with those of the sodium dodecylsulfate. Potassium dodecylsulfate waste recovery for four metals did give results very similar to the SDS study for Co, Mn, Ni and Rb. The studies with ADS and LDS did not exhibit large differences for Co and Ni. For Mn, the ADS, LDS and KDS systems all gave a much lower absorbances between the waste recovery and the generating analyte than did the SDS study. While for Rb the ADS, SDS and KDS systems all exhibited the same results, 0.42, LDS had a decrease of 0.81. These small inconsistencies with the same metal analyte but with different dodecylsulfates indicate other possible mechanisms might be operable even within the AIR theory.

The AIR theory was exemplified by the type of curve for the absorbance versus SDS concentration seen in Figure 1.B. for Al, Ca, Cd, Cu and Pb. For aluminum and lead there was a formation of cloudy solutions at high SDS concentrations. This solubility problem would explain the decrease in signal at the higher SDS concentrations for these two elements without using the air theory.

INTERFERENCES

Methods to reduce interferences have included the use of nitrous oxide/acetylene flame, the use of releasing agents and the control of particle size introduction into the flame. The generation of smaller droplets resulted in the least amount of interferences (17).

Salts such as ammonium chloride, potassium persulfate, potassium sulfate, sodium sulfite and sodium sulfate were used for the suppression of interferences in trace chromium analysis by Hurlbut and Chriswell in 1971 (18). They also reported that these salts caused an increased enhancement of the absorbance signal. The study of salt matrix effects have usually been confined to interference studies. The use of the four different cations (ammonium, lithium, sodium and potassium) with the dodecylsulfates was used by Pharr to determine if there was a cation metal ion contribution for the signal enhancement (7). The metal cation could cause enhancement by the prevention of the formation of refractory oxides in the flame or it could act as a releasing agent. In graphite furnace atomic absorption it has been shown that the anionic group can greatly effect the process when the metal goes from the atomization step to the vaporization step. Kodama reported on the effect of ten metal cationic interferences in the analysis of chromium (4,5). The metal interferences, when present as chloride salts, all gave positive interferences, but the nitrate salts exhibited depression effects for Fe, Ni, Co, Mn, Zn, Cu and Cd. These were especially severe for Fe, Co and Ni. This shows that the suppression of interferences takes place in the intermediate form of the salt, that is after evaporation of the solvent but before atomization. This was explained by the formation of the oxide from the nitrate salt while the chloride was formed from the chloride salt. With SDS and dodecyltrimethylammonium chloride the positive and the negative interferences were eliminated except for vanadium and magnesium whose values were 0.83. Interferences were more serious with a fuel rich flame compared to a fuel lean flame (5).

In 1980 Willis proposed that for FAAS both atomization efficiencies and chemical interferences behave according to droplet size distribution produced by the nebulizer. This size distribution influences transport efficiency, desolvation, vaporization of the remaining salt, and interferences (19). In 1985 a report by Browner on the influence of aerosol drop size on signals and interferences in FAAS showed that most interferences can be either totally eliminated or at least significantly reduced by the singular elimination of large drops from the aerosol. This was achieved by using a nebulizer with mixer paddles. However, an unfavorable decrease in the analytical signal often followed the decrease in the interference severity. This resulted in a net loss of sensitivity (17).

Pharr studied fifteen metal analytes with nineteen interferences (7). The concentration of the interfering cation had a concentration that was ten times that of the analyte. For example 20.0 ppm aluminum present in the analysis of 2.00 ppm chromium produced a value of only 0.89 that of the absorption signal. Similarly a 2.00 ppm chromium-SDS solution was analyzed with and without 20.00 ppm aluminum. The absorption signal exhibited only a 1.02 increase in the presence of aluminum which effectively eliminates aluminum as an interference compared to the decreased signal of 0.89 for the same system without SDS. The other twelve elements that caused interferences in chromium analysis were found, with the exception of magnesium, to produce a marked decrease in their interference in the presence of SDS. The interferences for iron, manganese, rubidium, tin, strontium and zinc were generally diminished in the SDS system compared to water for the analysis of chromium. For calcium analysis the interferences in the SDS system exhibited a marked increase in absorption signals while the nickel analysis absorbances in the presence of an interfering cation in the SDS system were generally depressed when compared to their aqueous solutions. The interfering cations of bismuth, cadmium, chromium, magnesium, lead and tin exhibited a decrease in interference for the analytes that were studied in the presence of SDS.

Calcium, cadmium and copper were studied with salts added to the solution instead of the dodecylsulfates (7). These salts had the same four cations (ammonium, lithium, sodium and potassium) that were on the dodecylsulfates and were prepared from twelve salts in six different concentrations, each with 5.00 ppm calcium, 0.80 ppm cadmium and 2.00 ppm copper. Many of these salts exhibited a signal enhancement for the analyte when compared to their aqueous solutions.

The calcium analyte in the presence of ammonium sulfate and lithium sulfate gave enhancements similar to those obtained with ammonium dodecylsulfate and lithium dodecylsulfate. In the presence of potassium sulfate the signal was enhanced 1.15 while in the presence of KDS it was less at 1.09. The sodium salts exhibited essentially the same absorbances as water, while in the presence of SDS they were only 0.86 that of water (a 14% signal decrease).

The study of cadmium resulted in absorbance enhancements with the various salts. The ammonium and lithium salts enhancement ratios of 1.34 and 1.31 were less than that obtained with ADS and LDS which were 1.86 and 1.93 respectively. The potassium salts gave values very similar to those obtained with KDS except that at concentrations of 20 mM and greater they were slightly higher at 1.15. Potassium sulfate caused greater enhancement at all concentrations. The sodium salts of sodium nitrate and trisodium phosphate exhibited enhancements similar to SDS, while sodium chloride and sodium sulfate had enhancements of 1.55 and 1.45 respectively compared to 1.36 for 5.0 mM SDS versus water.

For the copper solutions, the ammonium, lithium and potassium salts all exhibited enhancements similar to their respective dodecylsulfates. Potassium sulfate was slightly higher with an enhancement ratio of 1.54 compared to 1.38 for 5.0 mM KDS. The SDS value of 1.19 was less than the three sodium salts of chloride, nitrate and sulfate which exhibited 1.40 compared to water.

CONCLUSIONS

The use of the anionic surfactant dodecylsulfate resulted in an increased absorption signal and a masking of interferences for many of the metal ions studied. In most of the studies cationic and nonionic surfactants had no effect or a depressing effect on the absorbance signals. While the aerosol ionic redistribution (AIR) theory may be used to explain the enhancement effects for some of the analytes studied, not all of the metal analytes that have been studied have exhibited the predicted enhancement or the marked decrease of analyte in the nebulization waste. In varying the cationic head group of the surfactant, a significant difference was observed for many of the metal analytes studied (7). In conjunction with this difference a salt study showed that certain inorganic salts exhibited a similar enhancement behavior to the micellar systems. While this is not inconsistent with Borowiec's original AIR theory (10), it is not consistent with Armstrong's enrichment model of the anionic surfactant systems (2). Actually there would appear to be several competing chemical-physical relationships that may be existing in any one system. The dominance of one model for one system like chromium does not mean that all the analytes behave in a like manner. The presence of certain metals in a surfactant system may actually effect the total properties of that system. This was observed in the changes in viscosity and surface tension for the aluminum and lead analytes with SDS (7). The complex micellar systems which modify both the nebulization aerosols and the analyte atomization process are only now partially understood. Part of the difficulty may arise because each study used a different nebulizer sprayer system, as well as, changes in other experimental conditions. However, analytical applications of these systems and the models used to explain the aerosol transportation and atomization processes should give useful information towards a more complete understanding of flame-analyte interactions.

REFERENCES

(1) W.H. Foster and D.N. Hume, Anal. Chem.31:2028(1959).

(2) D.W. Armstrong,H. Kornahrens and K.D. Cook, Anal. Chem.54:1325(1982).

(3) R.L. Venable and R.V. Ballad, Anal. Chem.46:131(1974).

(4) M. Kodama,S. Shimizu,M. Sato and T. Tominaga, Anal. Letters,10: 591(1977).

(5) M. Kodama, and S. Miyagawa, Anal. Chem.52:2358(1980).

(6) M. Taga,Y. Takabatake, and H. Yoshida, Bunseki Kagaku.33:439(1984).

(7) D.Y. Pharr, H.E. Selnau, E.A. Pickral and R.L. Gordon, The Analyst.116:511(1991).

(8) J. Farino and R.F. Browner, Anal. Chem.56:2709(1984).

(9) W.L. Hinze, Use of surfactant and micellar systems in analytical chemistry, in: "Solution Chemistry of Surfactants,Vol. 1," K.L. Mittal ed.,Plenum,New York (1979).

(10) J.A. Borowiec,A.W. Boorn,J.H. Dillard,M.S. Cresser,R.F. Browner, and M.J. Matteson, Anal. Chem.52:1054(1980).

(11) B. Magyar, CRC Critical Rev.in Anal.Chem.17:145(1986).

(12) T. Takada, Spectrochimica Acta.41B:999(1986).

(13) K. Fujiwara, H. Haraguchi, and K. Fuwa, Anal. Chem.44:1895(1972).

(14) H. Ziemiecki and W.R. Cherry, J. Am. Chem. Soc.103:4479(1981).

(15) G.M. Hiefte and H.V. Malmstadt, Anal. Chem.40:1860(1968).

(16) R.F. Browner, A.W. Boorn, and D.D. Smith, Anal. Chem.54:1411(1982).

(17) D.D. Smith and R.F. Browner, Anal. Chem.56:2702(1984).

(18) J.A. Hurlbut, and C.D. Chriswell, Anal. Chem.43:465(1971).

(19) J.W. Novak Jr. and R.F. Browner, Anal. Chem.52:287(1980).

THE EFFECT OF CATIONIC ELECTROLYTES ON
THE ELECTROSTATIC FORCE BETWEEN TWO
DISSIMILAR IONIZABLE SURFACES

You-Im Chang,

Dept. of Chemical Engineering, Tunghai University

Taichung, Taiwan 40704

INTRODUCTION

It is widely recognized that the electrostatic interaction is one of the most important forces (including long-range, short-range and hydrodynamic forces) in promoting or inhibiting adhesion of the colloidal particles in many technological processes such as filtration, paper-making, wastewater clarification and clay migration in oil reservoir. Usually, the effects of electrostatic force on colloidal adhesion follow the rule given by the DLVO theory [1]. The basic principle of the DLVO theory is that the total interaction energy between two interacting colloidal particles is the sum of the electrostatic repulsive energy and the van der Waals attractive energy. The magnitude of the electrostatic energy depends on the thickness of the electrical double layer and the surface potentials of the interacting particles. In calculating the electrostatic force, it always assumes that the two interacting surfaces are fixed either with constant potential or with constant charge [2,3]. However, in dealing with the adhesion of colloidal particles bearing with ionizable surface groups, it was pointed out that, for the purpose of minimizing the total interaction free energy, the particles will regulate their surface charges or potentials so that the interacting surfaces can maintain at ionic

Advances in the Applications of Membrane-Mimetic Chemistry
Edited by T.F. Yen *et al.*, Plenum Press, New York, 1994

equilibrium with the bulk solution. Hence, neither the particle potential nor the particle charge remains constant during the period of adhesion [4,5,6,7,8,9]. Recently, based on the above charge regulation model, the author [10,11] examined the effect of the presence of multivalent cations on the electrostatic force between a particle bearing with ionogenic groups and a collector with either constant potential or constant charge. It was interesting to find that the presence of cations in the suspension medium reduces the electrostatic repulsion force between particle and collector surface only if the separation distance between them is greater than some critical value. If the separation distance is smaller than this critical value, the repulsive force is greater than that if the cations are absent. This is interpreted as the requirement of continuous reequilibration of the ionogenic groups on the particle surface, and the screening out of those overloaded cations in the interaction region as the separation distance decreases [10,11].

However, in the author's previous study, efforts were always put on the analysis of the effect of cations on the electrostatic force between a particle bearing with ionogenic groups and a collector with either constant potential or constant charge. No attempt as yet has been undertaken aimed at investigating this effect on the interaction between two dissimilar surfaces, bearing with different surface charges and different surface potentials of the same sign or the opposite sign. Hence the main purpose of this paper is, with the aid of solving nonlinear charge regulation model, to present a number of numerical examples to illustrate the effect of cationic electrolyte on the electrostatic force between two dissimilar surfaces.

THEORETICAL FORMULISM

Our formulism is analogous to the model of Ninham and Parsegian [4] in which the electrostatic potential between two approaching surfaces is regulated by equilibria of the surface bearing ionizable groups. The detailed formulation can be found in Ninham and Parsegian [4], Prieve and Ruckenstein [5,6], Chan et al [7,8] and Healy et al [9]. In the present paper, we only intend to retain the final dimensionless results. Consider that there are two dissimilar particles approach one another, and assume that the approach rate of the particle is sufficiently slow that electrochemical equilibrium is maintained at all times during adhesion [4]. For one-dimensional rectangular coordinate, the nonlinear Poisson-Boltzmann equation can be described as:

$$\frac{d^2\phi}{dH^2} = \frac{1}{2}\left[e^\phi - (1-\eta)\cdot e^{-\phi} - \eta\cdot e^{-q\phi}\right] \tag{1}$$

The boundary conditions on the two particles are:

$$\frac{d\phi}{dH}\bigg|_{H=\kappa l} = \frac{\kappa}{2n}\cdot\left(\frac{[H^+]_{s1}}{K_{b1}+[H^+]_{s1}}S_{b1}^{-1} - \frac{K_{a1}}{K_{a1}+[H^+]_{s1}}S_{a1}^{-1}\right) \qquad (2)$$

with $[H^+]_{s1} = [H^+]_r\cdot\exp(-\phi_1)$

and

$$\frac{d\phi}{dH}\bigg|_{H=0} = -\frac{\kappa}{2n}\cdot\left(\frac{[H^+]_{s2}}{K_{b2}+[H^+]_{s2}}S_{b2}^{-1} - \frac{K_{a2}}{K_{a2}+[H^+]_{s2}}S_{a2}^{-1}\right) \qquad (3)$$

with $[H^+]_{s2} = [H^+]_r\cdot\exp(-\phi_2)$

and the electrostatic interaction force can be written as:

$$F = -\left(\frac{d\phi}{dH}\right)^2 + \left[(e^\phi - 1) + (1-\eta)\cdot(e^\phi - 1) + \frac{\eta}{q}\cdot(e^{-q\phi} - 1)\right] \qquad (4)$$

Hence, the force, F, acting between two approaching surfaces is composed of two contributions: the Maxwell stress (first term on the right hand side of Eq. 4) and the osmotic pressure (the remaining terms on the right hand side of Eq.4) caused by the presence of cations in the suspension medium.

The definition of the parameters shown in the above equations are given at the end of this article.

CALCULATION EXAMPLES

In the following calculations, as shown in Table 1, characteristics of two interacting particles are obtained from the work of Prieve and Ruckenstein[6]. Two types of surfaces are employed in the following illustrative examples. Type "A" surface is defined as one having a single strongly acidic or strongly basic site, which remains fully dissociated for any degree of double layer interaction. Type "B" surface is defined as one having equal numbers of a single kind of acidic site and a single kind of basic site, with a significant buffer capacity. Therefore, during the period of adhesion, the type "A" surface remains nearly constant charge, whereas the type "B" surface maintains nearly constant potential. In the following calculations, the interaction surfaces are referenced by type and surface potential. For example, B+10 surface is the amphoteric surface having a surface potential of +10 mv when it is infintely far from any other surface. The pH value of the suspension medium is set at 7.0.

Table 1. The surface characteristics of interacting surfaces illustrated in the numerical analyses.

Type	$\Psi_{s\infty}$ (mv)	pK_a	pK_b	S_a^{-1} (groups/cm^2)	S_b^{-1} (groups/cm^2)
A	-20	2	-	2.860×10^{12}	0
	-10	2	-	1.403×10^{12}	0
	+10	-	10	0	1.403×10^{12}
	+20	-	10	0	2.860×10^{12}
B	-20	3.297	10	1.404×10^{17}	1.404×10^{17}
	-10	3.639	10	4.637×10^{16}	4.637×10^{16}
	+10	3.639	10	4.637×10^{16}	4.637×10^{16}
	+20	3.297	10	1.404×10^{17}	1.404×10^{16}

CASE 1. INTERACTION OF TWO IDENTICAL TYPE SURFACES HAVING POTENTIAL OF DIFFERENT MAGNITUDE, BUT THE SAME SIGN

The variation of the electrostatic force per unit area between A-20/A-10 surfaces as a function of the separation distance is shown in Fig. 1a. When there is no divalent cation present in the suspension medium (q=1 and η=0.0, shown by the solid line in Fig.1a), the electrostatic force is always repulsive (positive sign for F in Fig 1a), which means the osmotic pressure term in Eq.4 overwhelm the Maxwell stress during the adhesion period, and this repulsive force increases with the decrease of the separation distance. The same variation tendency is observed when there is a small amount of divalent cation (q=2 and η=0.1, shown by the dashed line in Fig.1a) present in the suspension medium. In Fig. 1a, as we expect, the presence of divalent cation in the suspension medium has the effect of decreasing the electrostatic repulsion force during the whole adhesion period.

In Fig.1b, the electrostatic force for B-20/B-10 interaction is illustrated. For the situation where there is no divalent cation present in the suspension medium, the repulsion force reaches a maximun value as the separation distance decreases. As $\kappa\ell\rightarrow$ 0, the electrostatic force changes from repulsive to attractive, causing the two interacting particles adhere one another irreversibly. Such attraction for two surfaces with potentials of the same sign with a separation distance $\kappa\ell<1.0$, can be explained by the effect that, in order to maintain the electroneutrality condition, the charge density on surface undergoes a sign reversal and both surface charge densities tend to become large in absolute magnitude and opposite in sign, and therefore an attractive force is induced (For details, see the work of Prieve and Rickenstein in ref. 6). Analogous to the author's previous work[10,11], it is worth noting that, compared with

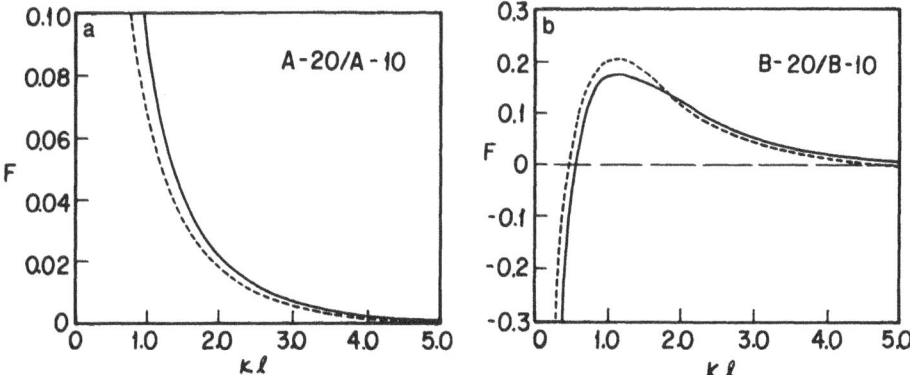

Figure 1. (a) The variation of electrostatic force versus separation distance for A-20/A-10 interaction. (b) The variation of electrostatic force versus separation distance for B-20/B-10 interaction. The solid lines stand for q=1 and η=0.0, the dashed lines stand for q=2 and η=0.1.

the curve at which q=1 and η=0.0, with small amount of divalent cation present (shown by the dashed line in Fig. 1b), the repulsion force will be reduced only when the separation distance is greater than a critical value at $\kappa\ell=1.75$. When the separation distance is smaller than this critical value, the presence of divalent cation causes a higher repulsion force than in its absence. The explanation of the existence of this critical separation distance is that, as the separation distance decreases, the dissociation degree of surface sites on the particle decreases and the divalent cations in the interaction region gradually becomes overloaded in neutralizing the negative charge on the particle surface. Finally, when the distance is smaller than the critical value shown in Fig. 1b, the osmotic pressure required to screen out these overloaded cations begins to raise the repulsion force from that of the case with no divalent cation present. Hence, the pressure of divalent cation in the suspension medium dose not always have the effect of decreasing the electrostatic force during the whole adhesion period.

CASE 2. INTERACTION OF TWO DIFFERENT TYPE SURFACES HAVING POTENTIAL OF DIFFERENT MAGNITUDE, BUT THE SAME SIGN

The electrostatic force for B-20/A-10 interaction and A-20/B-10 interaction is shown in Fig. 2a and Fig.2b, respectively. When there is no divalent cation, analogous to the curve shown in Fig. 1b, the curves in Fig. 2a and Fig. 2b reach a maximum value as the separation distance decreases. At small separation, unlike B-20/A-10 interaction, the repulsion force for A-20/B-10 interaction becomes attractive. The sign of

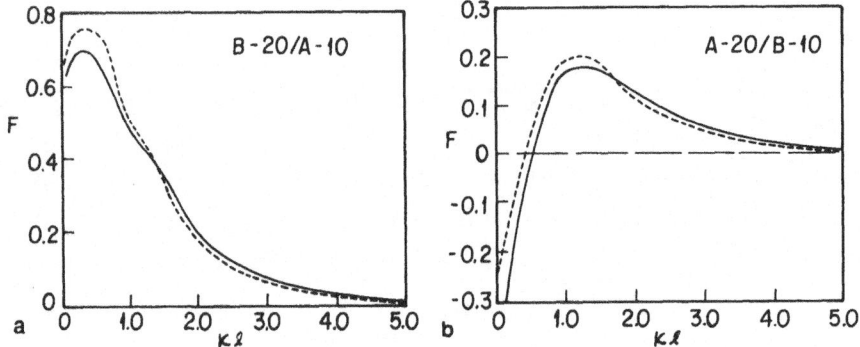

Figure 2. (a) The variation of electrostatic force versus separation distance for B-20/A-10 interaction. (b) The variation of electrostatic force versus separation distance for A-20/B-10 interaction. The solid lines stand for q=1 and η=0.0, the dashed lines stand for q=2 and η=0.1.

the force at $\kappa\ell \to 0$ for this type of interaction is determined solely by the difference between the absolute magnitude of type B surface potential and that of type A surface potential at large separation (see Eq. [10] in ref.6). Considering the dashed lines shown in Fig.2a and Fig.2b, again, the presence of divalent cations in the suspension medium does not always reduce the repulsion force in the period of adhesion. The critical separation distance for B-20/A-10 interaction and A-20/B-10 interaction occurs at $\kappa\ell$= 1.43 and $\kappa\ell$ =1.62, respectively.

CASE 3. INTERACTION OF TWO IDENTICAL TYPE SURFACES HAVING POTENTIAL OF THE SAME MAGNITUDE, BUT OPPOSITE IN SIGN:

As shown in Fig. 3a and Fig.3b, the electrostatic forces for both A+20/A-20 interaction and B+20/B-20 interaction are attractive during the whole adhesion period. More importantly, it is interesting to find that the presence of divalent cations in the suspension medium has the effect of decreasing these attractive forces shown in Fig.3a and Fig.3b. Compared with the curve at which no divalent cation is present, the increased osmotic pressure required to screen out these divalent cations accumulate near the oppositely charge surface can be applied to interpret this reduction of electrostatic attraction force.

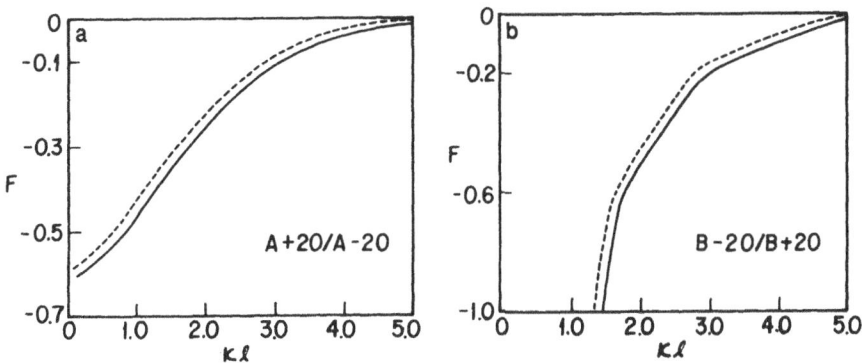

Figure 3. (a) The variation of electrostatic force versus separation distance for A+20/A-10 interaction. (b) The variation of electrostatic force versus separation distance for B+20/B-10 interaction. The solid lines stand for q=1 and η=0.0, the dashed lines stand for q=2 and η=0.1.

CONCLUSION

From the above numerical analyses, with the model of charge regulation, it can be confirmed that the influence of cationic electrolyte on the electrostatic force is dependent on the type of two interacting surfaces and the separation distance between them. The divalent cation can decrease the electrostatic repulsion force for A-20/A-10 interaction, and also can decrease the electrostatic attraction force for the interactions of A+20/A-20 and B+20/B-20. For the interactions of B-20/B-10, B-20/A-10 and A-20/B-10, the critical distance occurs at $\kappa\ell=1.75$, $\kappa\ell=1.43$ and $\kappa\ell=1.62$, respectively. As shown in Fig. 1a, Fig.2a and Fig.2b, the divalent cation decreases the repulsion force only when the distance is greater than the critical value. An opposite effect of divalent cation is observed when the distance is smaller than the critical value.

APPENDIX

English characters

F   dimensionless interaction force per unit area per nkT.

H   dimensionless separation distance between two interacting particles. $H = \kappa x$.

$[H^+]_r$ the hydrogen ion concertration in the suspension medium (mol/l).

$[H^+]_{si}$ the hydrogen ion concertration on the ith particle surface (mol/l).

k   Boltzmann's constant (erg/K).

K_{ai} dissociation equilibrium constant for acid groups on the particle surface (mol/l).

K_{bi} dissociation equilibrium constant for base groups on the particle surface (mol/l).

n   ionic strength in the suspension medium (mol/l).

q   valence for cationic electrolyte.

S_{ai} the reciprocal of acidic groups density on the particle surface (cm^2/groups).

S_{bi} the reciprocal of basic groups density on the particle surface (cm^2/groups).

T   absolute temperature (K)

x   normal distance from particle surface (cm).

Greek characters

η the fraction of cationic electrolyte in the suspension medium. $0 \leq \eta \leq 1$.

κ reciprocal Debye length (cm^{-1}).

ϕ dimensionless interaction potential between two interacting particles. $\phi = e\psi/kT$.

ϕ_i dimensionless potential on the ith particle surface.

ℓ shortest distance between two interacting surfaces (cm).

$\psi_{s\infty}$ the surface potential of ith particle at infinty separation distance (mv).

REFERENCE

1. E.J.W. Verwey and J.Th.G. Overbeek, "Theory of the stability of lyophobic colloids , " Elserier pub., Amsterdam (1948).
2. R. Hogg., T.W. Healy, and D.W. Fuerstenaue, Mutual coagulation of colloidal dispersions, *Trans. Faraday Soc.* 62:1638 (1966).
3. G.R. Wiese , R.O. James , and T.W. Healy, Discreteness of charge and solvation

effects of cation adsorption at the oxide/water interfaces, *Discuss. Farday Soc.* 52: 302 (1971).

4. B.W. Ninham and V.A. Parsegian, Electrostatic potential between surfaces bearing ionizable groups in ionic equilibrium with physiologic saline solution, *J. Thero. Biol.* 31: 405 (1971).

5. D.C. Prieve and E. Ruckenstein, Role of surface chemistry in particle deposition, *J. Colloid and Interface Sci.* 60:337 (1977).

6. D.C. Prieve and E. Ruckenstein, The double-layer interaction between dissimilar ionizable surfaces and its effect on the rate of deposition, *J. Colloid and Interface Sci.* 63:317 (1978).

7. D. Chan., J.W. Perram, L.R. White, and T.W. Healy, Regulation of surface potential at amphoteric surfaces during particle-particle interaction. *J.C.S. Faraday I.* 71 :1046 (1975).

8. D. Chan., T.W. Healy, and L.R. White, Electrical double layer interactions under regulation by surface ionization equilibria-Dissimilar amphoteric surfaces. *J.C.S. Faraday I.* 72:2844 (1976).

9. T.W. Healy, D. Chen, and L.R. White, Colloidal behavior of materials with ionizable group surface, *Pure Appl. Chem.* 52:1207 (1980).

10. Y.I. Chang, The effect of cationic electrolytes on the electrostatic behavior of cellular surface with ionizable groups, *J. Thero. Biol.* 139:561 (1989).

11. Y.I. Chang, Divalent cations can increase cell-substrate repulsion at small distances, *Colloids and Surfaces.* 41:245 (1989).

CHARACTERIZATION OF COLLOIDAL AGGREGATES

Eric Y. Sheu

Texaco R&D, P.O. Box 509, Beacon, New York 12508

ABSTRACT

Characterization of the self-assembled colloidal solutions is introduced. The main focus is on the comparison between the capability of the macroscopic techniques in which the experimental probes have length scale comparable to the measured sample, and the microscopic techniques where the probes are comparable to the colloidal size. Viscosity measurement is used as an example for the macroscopic technique, and small angle X-ray and neutron scattering for the microscopic technique. Three systems: sodium dodecyl sulfate micellar solutions, bis(2-ethyl)hexylsulfosuccinate/water/decane (AOT) three component microemulsions, and asphaltenes colloids in organic solvents, are used as exapmles for demonstration.

I. Introduction

Colloidal dispersions are recognized by containing suspended supramolecules of ~ 100 to 10,000 Å in size [1]. These particles may be colloidal particles (such as polyballs), macromolecules (such as polymers), or self-assembled aggregates (such as micelles, reversed micelles, or microemulsion droplets). Many books, chapters, and review articles [1-20] have been written to discuss these systems in great details. Throughout these reviews, many experimental techniques applicable for characterizing colloidal dispersions have been discussed. These techniques can in principle be categorized into the macroscopic and the microscopic measurements. Some examples of the macroscopic measurements are viscosity, viscoelestisity, surface tension, dielectric relaxation, densitometry, and scanning calorimetry

(DSC) measurements, etc. The probes in a macroscopic measurement usually has a length scale much larger than the colloidal size, while in a microscopic measurement the probe is considerably smaller, and usually comparable to the colloidal size. The microscopic techniques include laser light scattering, forced Rayleigh scattering, nuclear magnetic resonance (NMR), fourier transfer infrared spectroscopy (FTIR), electric spin resonance (ESR), neutron and X-ray scattering, scanning electron microscopy (SEM), transmission electron microscopy (TEM), and gel permeation chromatography (GPC), and many others.

Frequently, the quantities measured from a macroscopic technique directly represent the properties of the system. For example, a DSC spectrum directly exhibits the thermal history of the system, including melting, glass transition, and other phase transition temperatures. However, the molecular details that result in these observed macroscopic properties are usually not provided. In order to obtain the microscopic details from a macroscopic measurement, sophisticated theories are often needed. If the theories are adequate, one may be able to characterize the molecular details through a macroscopic measurement. The microscopic measurements, on the other hand, provide certain microcharacteristics of the system. The data obtained from such a measurement usually contain information on the molecular length scale, with the resolution limited by the resolution of the probe. Unfortunately, most of the microscopic measurements do not provide the desired physical quantities directly, thereby, often require models to rationalize the observed data. The advantage, however, is that the model can be established from a microscopic point of view. This allows one to characterize the microscopic properties more accurately. In the characterization of a self-assembled colloidal solution, for example, the models are often established based on the statistical mechanical theories. Therefore, the detailed information like intermolecular correlation, particle size distribution, and the pair correlation function can be obtained.

In this chapter, I will discuss how to obtain the microscopic information from a macroscopic measurement, and how to develop an appropriate model for analyzing the microscopic data, so that maximum amount of information can be obtained. This chapter is organized as follows. In section II, the viscosity measurement will be used as an example for the macroscopic techniques, and small angle neutron scattering for the microscopic techniques. I will discuss the theories commonly used for analyzing the viscosity data. Also, the statistical mechanical theory used for treating the small angle neutron scattering data will be discussed. In section III, three examples will be used for demonstration. They are sodium dodecyl sulfate micelles in aqueous solutions, bis(ethyl)hexylsulfosuccinate/water/decane (AOT) three component microemulsion, and asphaltenes in organic solvents which is a natural reversed micellar system. Both macroscopic and microscopic techniques used for investigating these systems will be reviewed. A brief discussion will be given in section IV with a conclusion and some future perspective about this subject.

II. Theory of Viscosity and Small Angle Scattering

A. Viscosity

Viscosity is a measure of the response of a solution upon shear force. It is defined as the ratio between the shear stress and the shear rate. The shear stress is calculated from the shearing force, acting on the shearing surface, to which the sample adheres, and the shear rate is defined as the quotient of the speed difference of the shearing surfaces and their distance. A solution is classified as a Newtonian fluid, if the viscosity is independent of shear rate, otherwise, it is called a non-Newtonian fluid. For many pure liquids, the viscosity responses are Newtonian, unless the shear rate is extremely high. However, when solutes, like polymer, are added, it may become Non-newtonian even at very low concentration and low shear rate. Since the systems to be discussed in this chapter are primarily Newtonian, the theories to be described below will be mainly for zero shear viscosity.

The first theoretical work for viscosity was done by Einstein at the beginning of this century. Einstein derived the shear viscosity for a Newtonian fluid containing hard spheres in the dilute concentration regime. The result he obtained was

$$\eta(\phi) = \eta_o (1 + 2.5\phi) \tag{1}$$

where η is the solution viscosity, ϕ the volume fraction of the solute, and η_o the solvent viscosity. The slope of 2.5 was found to be for hard sphere only. If we rewrite Eq.(1) as

$$\eta(\phi) = \eta_o(1 + [\eta]_\phi \phi) \tag{2}$$

then $[\eta]_\phi$ can be expressed as

$$[\eta]_\phi = \frac{\eta/\eta_o - 1}{\Phi} \tag{3}$$

where $[\eta]_\phi = 2.5$ in the Einstein equation. Since both Eq.(1) and (2) are primarily for low volume fractions, one can say that as $\phi \to 0$, $[\eta]_\phi$ will essentially represent the viscosity effect of a "single" particle. We thus defined $[\eta]$ as the intrinsic viscosity representing the viscosity effect of a single particle,

$$[\eta] = \lim_{\phi \to 0} [\eta]_\phi = \lim_{\phi \to 0} \frac{\eta/\eta_o - 1}{\Phi} \tag{4}$$

Since $[\eta]$ represents the effect of a single particle, it is expected to strongly depend on the shape of the particle. The advantage of this dependence is that one may be able to determine the particle shape from a simple viscosity measurement. In the past, many formula have been derived or empirically given to describe $[\eta]$ for various particle shapes. First, Simha [21] derived $[\eta]$ for an elongated particle with axial ratio p greater than 15,

$$[\eta] = \frac{14}{15} + \frac{p^2}{5}[\frac{1}{3(\ln 2p - \lambda)} + \frac{1}{\ln 2p - \lambda + 1}] \tag{5}$$

where $\lambda = 1.5$ for ellipsoids and 1.8 for rods. In order to cover a wider p range, Kuhn and Kuhn [22] derived the following equation,

$$[\eta] = 2.5 + \frac{32}{15\pi}(\frac{1}{p} - 1) - 0.628\frac{(1/p) - 1}{(1/p) - 0.075} \qquad 0 < p < 1$$

$$= 2.5 + 0.4075(p - 1)^{1.508} \qquad 1 < p < 15$$

$$= 1.6 + \frac{p^2}{5}[\frac{1}{3(\ln 2p - 1.5)} + \frac{1}{\ln 2p - 0.5}] \qquad 15 < p \tag{6}$$

In addition to the elongated particles, Riseman and Ullman [23] derived an equation for a dumbbell particle, assuming that the Brownian motion is overwhelmed,

$$[\eta] = \frac{\pi c L^2 N_o}{25M}[1 + \frac{9}{20}\frac{a}{L} + \frac{9}{40}(\frac{a}{L})^2] \tag{7}$$

where c is the concentration, L the distance between the centers of mass of the two spheres forming the dumbbell, a the radius of the dumbbell (assuming two spheres have equal size), and M the molecular weight. For flexible molecules (such as polymers) Houwink [24], Kuhn [25], and Mark [26] postulated a two-parameter relation for the intrinsic viscosity

$$[\eta] = KM^\alpha \tag{8}$$

where K and α depend on the nature of the solvent, the particle material, and the temperature, but generally independent of the concentration and molecular weight of the polymer.

In self-assembled systems, the formation of prolate ellipsoidal particles is commonly observed [27]. For these systems, Jeffery [28] and Eisenschitz [29] derived an expression for the intrinsic viscosity, provided the systems are ideal (i.e., the interparticle interactions are negligible). Unfortunately, the expression given in both papers were misprinted. The correct expression was recently given by Kohler and Strnad [30]. This equation covers the entire p range (for P > 1), and converges to the Simha equation and the Kuhn's formula within their applied p ranges. This equation has the following form,

$$[\eta] = \frac{4\beta^2}{15} \left\{ \frac{14}{p^2(4p^2-10+3\alpha)} + \frac{3\beta}{p^2(p^2+1)[(2p^2-1)\alpha-2]} + \right.$$

$$\left. \frac{6}{(P^2+1)(2p^2+4-3p^2\alpha)} \frac{4p^2+2-(4p^2-1)\alpha}{p^2(4p^2-10+3\alpha)[(2p^2+1)\alpha-6]} \right\} \qquad (9)$$

Since this equation is valid for p>1, the axial ratio of an ellipsoid can be determined by analyzing the intrinsic viscosity data using this equation. However, it is not as trivial, when dealing with a self-assembled system. In a self-assembled system, the true meaning of intrinsic viscosity becomes obscure. This is because the shape of the particles changes as a function of concentration. Under this condition, only the shape factor $[\eta]_\phi$ can be experimentally obtained using Eq.(3). The shape factor represents the particle shape at the concentration where data is taken (instead of at the zero concentration limit), assuming the interparticle interactions are negligible. Because only $[\eta]_\phi$ can be obtained, the data analysis has to be based on either the $[\eta]_\phi$ or the relative viscosity data. In addition to the problem of concentration dependent particle shapes, the self-assembled systems often exhibit polydispersity, which further complicates the resulting η and $[\eta]_\phi$. Fortunately, the problem of polydispersity can be solved, at least for a hard sphere system. Roscoe [31] derived the relative viscosity η_r for a polydispersed hard sphere system and obtained

$$\eta_r = \frac{\eta}{\eta_o} = (1 + \phi)^{2.5} \qquad (10)$$

It is noted from this equation that neither the polydispersity nor the particle size distribution parameters are explicitly included. This means that no polydispersity or size distribution parameters can be determined, when using this equation to analyze the data. Kohler and Strnad [30] recently developed an equation for analyzing the $[\eta]_\phi$ data, with the polydispersity incorporated. They used this equation to analyze a self-assembled system and obtained the average particle axial ratio as a function of concentration, as well as the standard work for micellar formation. The results obtained were reasonably accurate. The formula Kohler and Strnad derived is good only for $p > 6$.

Another problem in dealing with the viscosity of a self-assembled system is that the surfaces of the particles may not be smooth. This makes the particles drag the surrounding solvent molecules under shear, when viscosity is measured. Due to this solvation, the hydrodynamic volume of the particles are larger than the their true volumes. This solvation effect also modifies the ϕ dependence of the viscosity, as well as $[\eta]$. Since both polydispersity and the solvation effects modify $[\eta]$, in a way competing with the effect of the particle shape [3], one has to be very careful in analyzing the [η] data for a self-assembled system, in order to differentiate these effects.

To differentiate the effect of solvation from that of the particle shape is not a easy task. It may be easier if the particle are hard spheres. For a dilute hard sphere system, Pal and Rhodes [32] derived a simple relation

$$\eta_r = (1 - K\phi)^{-2.5} \tag{11}$$

where K is an adjustable parameter, representing the degree of solvation. Although this equation was developed for the dilute concentration regime, it was often found to be applicable even for volume fraction much higher than the Einstein region (i.e., for $\phi \gg 0.1$) [32]. Unfortunately, the K value does not directly represent the solvation quantitatively. Moreover, the K value may become very inaccurate when the particles are not true hard spheres. Krieger [33] gave a formula similar to Eq.(11),

$$\eta_r = (1 + \frac{\phi}{\phi_o})^{-\frac{1}{v}} \tag{12}$$

where v is an adjustable parameter, often found to be near 0.4 for spherical systems [34]. Eq.(12) is a two parameters formula. ϕ_o was described by Krieger as a scaled volume. It may, in some way, represent the solvated volume, but was not explicitly described by Krieger. To quantify the degree of solvation, the Eiler's equation is a better alternative. The Eiler's equation reads

$$\frac{(\eta_r^{1/2} - 1)}{\phi} = \frac{[\eta]}{2} + \frac{1}{\phi_m}(\eta_r^{1/2} - 1) \tag{13}$$

This equation contains ϕ_m which is the maximum packing volume fraction of the particles. For many solution systems, the nearest neighbor structure factor was found to be describable by a face center cubic (fcc) crystal structure [35,36]. If this is the case, then the maximum packing volume fraction would be 0.74. Due to the effect of solvation, ϕ_m extracted using Eq.(13) is expected to be lower than 0.74. For ionic systems, it can be even lower, because of the long range Coulombic interactions. If we attribute the difference between the ϕ_m value extracted using Eiler analysis and 0.74, to the effect of solvation, then the degree of solvation can be estimated as

$$S = \frac{\phi_o}{\phi_m} - 1 \tag{14}$$

where ϕ_o is 0.74 for fcc structure, 0.68 for random structure, and 0.53 for simple cubic structure. Because the fcc structure presumed by Prins [35] and Bahe [36] to construct the structure factor for solution systems is only accurate to the first order approximation, the solvation estimated is expected to be no better than the first order of approximation. A more accurate method for estimation of the solvation is to apply the contrast variation technique (to be described later in this chapter) in a small angle neutron scattering experiment.

The next question to be asked is: how is the solvation formed? Is it by solvent entrapment within a temporarily agglomerated particle cluster, or by the direct swelling of the particles? This can be answered by a theory recently developed by Tsenoglou [37]. This so called scaling theory is able to identify the formation of the solvated volume, if it is due to the entrapment of the solvent molecules between the temporarily agglomerated particles. It, however, can not provide any information, if the solvation is through other mechanisms. The scaling theory concludes that if the solvation is due to the solvent entrapment between agglomerated particles, the K extracted from Pal and Rhodes equation (see Eq.(10)) will be volume fraction dependent, according to

$$\frac{1}{K} = 1 - \frac{\phi}{\phi_m}(1-\phi_m) \tag{15}$$

Another problem associated with the viscosity property of a self-assembled system is: can a theory developed based on a hard sphere system be used? This is because most of the colloidal particles in self-assembled systems are not true hard spheres, but theories were often developed for a hard sphere system. It is thus necessary to check how much the system viscosity differs from a true hard sphere system. In the dilute concentration regime, the Einstein equation is used for this purpose. One can either compare the relative viscosity or the intrinsic viscosity. If the particle are "soft", the Einstein equation should be replaced by the Taylor equation [38],

$$\eta_r = 1 + 2.5(\frac{\eta'+\frac{2\eta}{5}}{\eta'+\eta})\Phi \tag{16}$$

where η' is the viscosity of the suspended particles. If the concentration goes beyond the Einstein region and above the percolation concentration (roughly 0.16 volume fraction [39]), the Campbell and Forgacs [40] can be used for comparison. This equation was derived based on the percolation phenomenon of a hard sphere system. When the concentration is higher than the percolation threshold, the particles begin to "connect" with each other, and the "connectivity" extends from a molecular length scale to a macroscopic scale. The Campbell and Forgacs equation is only applicable for $\phi > \phi_c$, the threshold volume fraction of percolation. It has the following form,

$$\eta_r = \exp[\frac{\phi_m-\phi_c}{\phi_m-\phi}] - 1 \tag{17}$$

To apply this equation for comparison, one plots the measured relative viscosity data as a function of volume fraction, together with the simulated η_r using Eq.(17) (the ϕ_m value can be obtained from the Eiler analysis). The advantage of this comparison is that if there exists a concentration range where the relative viscosity of the system behaves like a hard sphere system, then, the theories developed based on a hard sphere system may be adopted for this concentration range. This point will be demonstrated in the asphaltenes reversed micellar system later in this chapter.

The final question to ask is: if all the discussed effects (solvation, polydispersity, concentration dependent particle shape, and non-hard sphere) are difficult to differentiate, then, can one represent the combination effect by a parameter, which not only serves as an indicator but also has some physical meaning? The answer is "possible". It is possible, when the particle shape does not show appreciable effect (i.e., the particles are not too different from a sphere), or when its rheological behavior is similar to a hard sphere system within a certain concentration range. In this case, the effects of solvation, polydispersity and non-hard sphere may be combined into a potential function between particles, with the particles taken as hard spheres. This potential can be theoretically recognized as the interactions between particles. The viscosity for such a system has been derived by Grimson and Barker [41] assuming a very general potential form. In their equation, two parameters were involved. One is the coupling constant representing the strength of the interaction, the other quantifies the dependence of the potential on the interparticle distance,

$$\eta_r = (1+2.5\phi) + \Lambda(\frac{\phi}{\phi_m})[1 - (\frac{\phi}{\phi_m})^{1/3}]^{-n} \tag{18}$$

where Λ is the coupling constant and n is related to the interparticle potential as

$$V(r) = \Lambda[\frac{\sigma}{r-\sigma}]^n \tag{19}$$

with σ being the particle diameter. Again, ϕ_m in Eq.(18) can be obtained from the Eiler's analysis. When using Eq.(18) to fit the viscosity data, the extracted Λ represents how much the deviation (compare to a noninteracting hard sphere system) has been made by the combination effect. The n parameter, on the other hand, indicates how much and how far this combination effect will rheologically "disturb" the neighboring particles.

Before concluding this subsection, it is worth noting that the recent theory developed by Kohler and strnad [30] for self-assembled system has been extended by Coello et al [42]. These authors extended the theory for treating the low axial ratio cases (the original Kohler-Strnad theory is applicable for micelles with axial ratios greater than 6, while the Coello's theory is good for 2.8 and up). This is to say that the standard work for forming an elongated micelle and the subsequent micellar growth can now be estimated by a simple zero shear viscosity measurement, as long as the axial ratio is greater than 2.8. However, one must keep in mind that viscosity measurement is a macroscopic technique, the results and the analysis methods are better verified by at least one independent microscopic technique.

B. Small Angle Scattering

Small angle scattering technique makes use of a fundamental concept that the interaction between a probe (such as neutron or X-ray) and a colloidal particle should reflect the physical characteristics of the system. One of the important properties of a colloidal systems is the colloidal structure, which can be determined by a scattering measurement. The way it determines the colloidal structure is through the following argument. In a scattering process, the interaction between the probe and the particle depends on the particle structure and the interparticle structure. Because this interaction in turn determines the spatial scattering pattern of the scattered wave, the scattering pattern (it is the differential cross section per unit volume of the system) should carry the structural information of the colloids. Thus, the particle structure and the interparticle structure can in principle be determined through analysis of the scattering pattern. This pattern is usually defined in a space characterized by the Bragg wave vector,

$$\vec{Q} = \vec{k}_i - \vec{k}_f \tag{20}$$

where $\vec{k}_i$ and $\vec{k}_f$ are the momentum transfer of the incident and the scattered wave respectively. When the scattering angle is small, the interactions between the radiation and the scatterers are predominately elastic or quasielastic scattering. In this case, $|\vec{k}_i| = |\vec{k}_f| = 2\pi/\lambda$, where λ is the wavelength of the radiation. Consequently, the scattering vector Q has a magnitude of

$$Q = |\vec{Q}| = \frac{4\pi}{\lambda} n \, \sin(\frac{\theta}{2}) \tag{21}$$

where n is the index of refraction of the system and θ is the scattering angle. The index of refraction for laser light is 1.33 for water, and about 1.4 for some saturated alkyl chain oil. For X-ray and neutron, n is very close to unity. This is why n is often omitted in the formulation of X-ray or neutron scattering.

The scattering intensity as a function of angle θ, and thus Q, provides the structural information of the scatterers in the spatial space through the Fourier transform of the intensity spectrum [43]. The resolution of the experiment in θ naturally propagates to the spatial space. Thus, the maximum resolution in the spatial space depends on the maximum Q value achievable in an experiment. As a general rule, if Q_m is the maximum achievable Q value, the resolution will be

$$\Delta R = \frac{\pi}{Q_m} \tag{22}$$

As mentioned earlier, the pattern of the scattered wave in the Q space represents the differential scattering cross section, provided the energy transfer between the incident wave and the scattered wave is negligible. This is usually true when the scattering angle is small (or when the incident wave length is long see Eq.(21)). Under this condition, it is plausible to apply the static approximation concept. The differential cross section $d\Sigma(Q)/d\Omega$ (or the scattering intensity function) in this case can be written as,

$$\frac{d\Sigma}{d\Omega} = \frac{1}{V} < |\sum_{\alpha=1}^{N} \sum_{\alpha'=1}^{N} b_\alpha \, b_{\alpha'} \exp[i\vec{Q}\cdot(\vec{r}_\alpha - \vec{r}_{\alpha'})]| > \qquad (23)$$

where b_α is the bound scattering length of the α-th nucleus in the scatterer, and r_α is the position vector of the α-th nucleus. The bracket represents an average over all possible configurations. For simplicity, Eq.(23) can be rewritten as

$$\frac{d\Sigma}{d\Omega} = \frac{1}{V} < |\sum_{\alpha=1}^{N} b_\alpha \exp(i\vec{Q}\cdot\vec{r}_\alpha)|^2 > \qquad (24)$$

When the scattering angle is small, the bound state scattering length b_α depends on the species of the nuclei and their corresponding spin states, it is thus more advantageous to separate b_α into the coherent part and the incoherent part. The incoherent part is Q independent. Experimentally, it contributes to a constant background, and is generally subtracted before data analysis proceeds. Mathematically, these two parts can be expressed as

$$(\frac{d\Sigma}{d\Omega})^{inc} = \frac{1}{V} \sum_{\alpha=1}^{N} (b_\alpha^{inc})^2 \qquad (25)$$

and

$$(\frac{d\Sigma}{d\Omega})^{coh} = \frac{1}{V} < |\sum_{\alpha=1}^{N} b_\alpha^{coh} \exp(i\vec{Q}\cdot\vec{r}_\alpha)|^2 > \qquad (26)$$

In what follow, $d\Sigma/d\Omega$ will be used for coherent scattering cross section, unless it is denoted.

For most colloidal systems, the colloidal particles are in a relatively well defined structure distinguishable from the surrounding solvent molecules. It is thus convenient to mathematically treat each colloid as a scattering center. From this view point, the scattering volume can then be divided into M cells with the center of mass of the colloid being the center of the cell. Eq.(26) in this case becomes

$$(\frac{d\Sigma}{d\Omega}) = \frac{1}{V} < |\sum_{i=1}^{M} \exp(i\vec{Q}\cdot\vec{r}_i) \sum_{cell j} b_{ij} \exp(i\vec{Q}\cdot\vec{x}_{ij})|^2 > \qquad (27)$$

where x_{ij} is the position vector for the j-th nucleus in the i-th cell. In Eq.(27), the second summation is evaluated for the i-th particle. Since it is an intraparticle term, we thus define the form factor as

$$F_i(\vec{Q}) = \sum_{cell\,j} b_{ij} \exp(i\vec{Q}\cdot\vec{x}_{ij}) \tag{28}$$

At this point we introduce a scattering length density for the i-th particle $\rho_i(r)$, defined as

$$\rho_i(\vec{r}) = \sum_j b_{ij}\delta(\vec{r}-\vec{x}_{ij}) \tag{29}$$

and assume the solvent molecules are much smaller than the colloids, so that the solvent can be treated as a continuous media with a constant scattering length density ρ_s. With the $\rho_i(r)$ of Eq.(29) and the assumption for ρ_s, one can write $F_i(Q)$ as

$$F_i(\vec{Q}) = \int_{particle\,i} d^3r\ [\rho_i(\vec{r})-\rho_s]\exp(i\vec{Q}\cdot\vec{r}) + \int_{cell\,i} d^3\ \rho_s\ \exp(i\vec{Q}\cdot\vec{r}) \tag{30}$$

The second term of Eq.(30) is approximately a delta function at Q = 0. Thus, for Q > 0 we have

$$F_i(\vec{Q}) = \int_{particle\,i} d^3r\ [\rho_i(\vec{r})-\rho_s]\exp(i\vec{Q}\cdot\vec{r}) \tag{31}$$

Obviously, $F_i(Q)$ depends only on the structure of the colloids, if the particles are uniform. Substitute Eq.(28) and (31) into Eq.(27), one immediately obtains, assuming the particles are monodisperse,

$$\frac{d\Sigma}{d\Omega}(Q) = \frac{N_p}{V}|F(Q)|^2\frac{1}{N_p}<\sum_{i=1}^{N_p}\sum_{i'=1}^{N_p} \exp[i\vec{Q}\cdot(\vec{r}_i-\vec{r}_{i'})]> \tag{32}$$

Define $P(Q)=|F(Q)|^2$, $n_p=$ number density of the particles$= N_p/V$, and the interparticle correlation factor (known as interparticle structure factor) as

$$S(Q) = \frac{1}{N_p}<\sum_{i=1}^{N_p}\sum_{i'=1}^{N_p} \exp[i\vec{Q}\cdot(\vec{r}_i-\vec{r}_{i'})]> \tag{33}$$

then, the differential cross section $d\Sigma(Q)/d\Omega$ can be simplified as

$$\frac{d\Sigma}{d\Omega}(Q) = n_p P(Q)S(Q) \tag{34}$$

In the case of polydisperse systems, Eq.(34) becomes

$$\frac{d\Sigma}{d\Omega}(Q) = \frac{N}{V}<P(Q)>\tilde{S}(Q) \tag{35}$$

where

$$<P(Q)> = <|F(Q)|^2> = \sum_{p=1}^{n} \frac{N_p}{N}|F_p(Q)|^2, \quad (N = \sum_{p=1}^{n} N_p) \tag{36}$$

and $\tilde{S}(Q)$ is an effective one-component interparticle structure factor,

$$\tilde{S}(Q) = 1 + \beta(Q)[S(Q)-1], \quad \beta(Q) = \frac{|<F(\bar{Q})>^2|}{<|F(\bar{Q})|^2>} \tag{37}$$

From Eq.(32) to (37) one can see that P(Q) is governed by the structure of the particle and S(Q) by the interparticle correlations. For various particle geometries, P(Q) has analytical expressions [43] and can be easily computed. For example, P(Q) for a spherical particle of radius R is

$$P(Q) = |F(Q)|^2, \quad F(Q) = V(\bar{\rho}-\rho_s)\frac{3j_1(QR)}{QR} \tag{38}$$

where ρ is the average scattering length density of the colloid, and $j_1(x)$ is the spherical Bessel function of the first kind. For a triaxial ellipsoid of semiaxes a, b, and c, the form factor P(Q) can be rigorously derived to be

$$P(Q) = \int_0^1 dx \int_0^1 \phi^2 \left\{ Q[a^2\cos^2(\frac{1}{2}\pi x) + \right.$$

$$\left. b^2\sin^2(\frac{1}{2}\pi x)(1-y^2) + c^2y^2]^{\frac{1}{2}} \right\} dy \tag{39}$$

For a hollow cylinder of height H, outer radius R_1 and inner radius R_2, P(Q) has the following form

$$P(Q) = \int_0^1 \Phi[Q, \sqrt{R_1(1-\mu^2)}, \sqrt{R_2(1-\mu^2)}] \frac{\sin(\frac{QH\mu}{2})}{QH\mu} dx$$

$$\Phi(Q, R_1, R_2) = \frac{1}{1-(\frac{R_2}{R_1})^2}[2J_1(QR_1) - (\frac{R_2}{R_1})^2 2J_1(QR_2)] \qquad (40)$$

For other geometries the reader can refer to reference [44].

Compare to P(Q), it is much more difficult to compute S(Q) because the basic interparticle correlation function can not be presumed or determined trivially. For many colloidal systems, S(Q) is calculated through the statistical mechanical theory using the Ornstein-Zernike equation [45]. Several colloidal structures factors were determined using this theory [46-49]. The S(Q) so computed can be very accurate. The details for calculation of S(Q) will be described in the following section.

III. Characterization of Colloidal Aggregates - Examples

A. Sodium Dodecyl Sulfate Micellar Solutions

Sodium dodecyl sulfate (SDS) is an anionic surfactant consisting of a $NaSO_3$ hydrophilic head and a linear alkyl chain with twelve carbons. As SDS is dissolved in water with concentration exceeding ~ 8 mM (the critical micelle concentration (CMC) [50]), the SDS molecules start to self-assemble into micelles. Since the counterion of a SDS molecule is Na, the micelle will be negatively charged, as Na dissociates from the micelle. The interactions between SDS micelles thus include not only the steric interaction but also the electrostatic interaction, which is strong and long-ranged [51]. Because the interactions are long ranged, the potential between two micelles would involve the effect from the third micelle. As a result, the potential form can be much more complicated than a simple Coulombic potential. In spite of the complicated interparticle interaction, the viscosity measurements for SDS micellar systems (SDS in aqueous or in brine solutions) show that the systems remain Newtonian for shear rate up to about 5000 s^{-1}. Fig. 1 is the shear viscosity measurements for SDS at 0.1 M NaCl aqueous solutions as a function of shear rate. For all measured shear rates, the viscosity increases monotonically as a function of concentration.

Since the micellar structure may vary as a function of concentration (due to the micellar growth), the measured viscosity can provide the shape factor data only (see Eq.(3)), instead of intrinsic viscosity data (see Eq.(4)). Under this situation, our data analysis has to rely on either the relative viscosity data or the shape factor data.

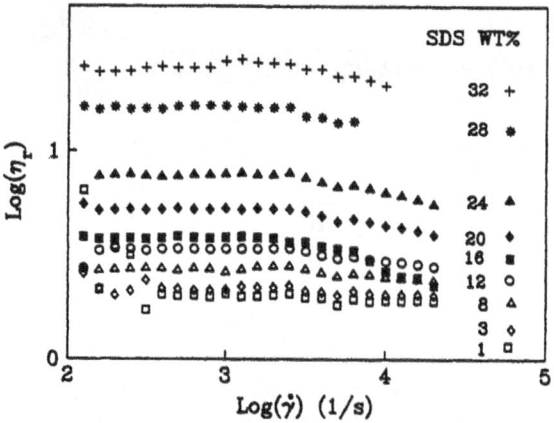

Fig. 1. Shear viscosity for SDS micelles in 0.1 M NaCl brine solutions.

Fig. 1 shows two distinctive regions, one is the Newtonian region and the other the shear thinning region. Before performing data analysis, we first calculated the relative viscosity as a function of concentration for both regions (Fig. 2 and 3). Since the shear thinning region involves the flow birefringence, the following data analysis will focus only on the Newtonian region.

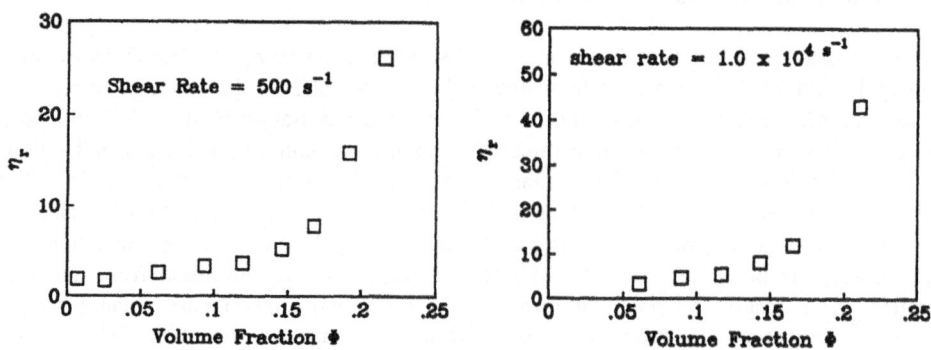

Fig. 2 Relative viscosity for 500 s⁻¹ shear rate.   Fig. 3 Relative viscosity for 1x10⁴ s⁻¹ shear rate.

The first analysis performed was the Pal and Rhode analysis. Fig. 4 shows the plot according to Eq.(9). Two striking features were observed: (1) the curve exhibits a break point at approximately 0.13 volume fraction, and (2) the zero volume fraction limit does not converge to unity as it should, according to Pal and Rhodes' equation. If one carefully examines the curve, it is not difficult to find that the data points for concentration above the break point can be extrapolated back to unity at the zero concentration, but the low concentration data can not. This is probably due to the long-ranged interactions between SDS

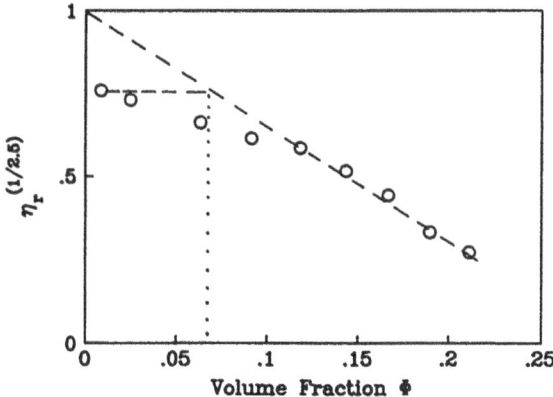

Fig. 4. Pal and Rhodes analysis using Eq.(9).

micelles at the dilute concentration regime, where the counterions dissociation fraction are higher [52]. As concentration increases the counterions dissociation considerably decreases, particularly, when there is structural transition [53,54]. The break point is likely where the structure transforms from a spherical-like to a cylindrical-like structure. Since there are more counterions condensed onto an elongated particle than a sphere [53,54], the elongated micelles are expected to have less effective surface charges. This means that the intermicellar interactions would become shorter-ranged. On the contrary, the interactions in the dilute regime are longer-ranged. This makes the inter-micellar distance always greater than a certain distance, usually much greater than 2R (R is the micellar radius). It is equivalent to having an interaction volume for each micelle. If this interaction volume is approximately the rheological volume of the micelle, (i.e., the effective volume of the micelle as it moves under shear force) then, this interaction volume can be regarded as the "solvated" volume (or the scaled volume, ϕ_o, in Krieger's equation, Eq.(12)). Assuming this argument is correct, then, the effective volume fraction for the lowest volume fraction shown in Fig. 4 ($\phi = 0.01$) is equivalent to about 0.06, if one estimates by making a straight line between the break point and 1 at ϕ=0. Then, takes the volume fraction on this line where the y-axis value is equal to that of the 0.01 data point (see Fig. 4). In this way, one can actually evaluate the interactions between micelles by fitting an adjustable volume fraction parameter in the Pal and Rhodes equation. In addition, if the interaction parameter, such as the Debye screening length, can be incorporated into the equation, some quantitative results may be obtained. However, not such works have been addressed.

So far, the information obtained from the Pal and Rhodes analysis is qualitative. An interesting phenomenon to investigate from these data is how the micellar structure evolves as a function of concentration. This can be obtained from the shape factor data, but requires an appropriate model. The model to be presented here assumes micelles to grow from a spherical-like shape into an ellipsoidal-like shape with a major radius a and a minor radius b. With this presumed micellar shape, one can analyze the experimental shape factor data as a function of micellar concentration using Eq.(9). Fig. 5 shows how the a/b evolves as a function of micellar concentration using this analysis. One has to be cautious here that the

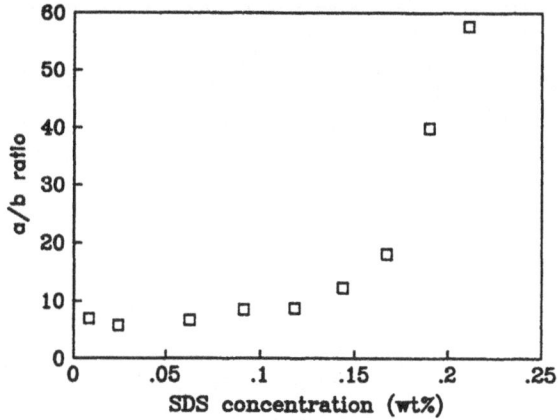

Fig. 5. a/b ratio as a function of micelle concentration.

a/b ratios so obtained are based on the assumption that the intermicellar interaction is negligible, which is actually not true, especially, in the dilute regime. We will compare these data with the results obtained from the small angle neutron scattering measurements later.

As one can see, the viscosity data can provide some molecular information, if the analysis is appropriate. However, it is always necessary to verify the results by a microscopic measurement. In the following I will present some small angle neutron scattering data for SDS system to compare with the viscosity results.

Before describing the small angle scattering studies for the SDS systems, let us look at the phase of the SDS/water system. Ekwall [55] showed a series of phase diagrams for the

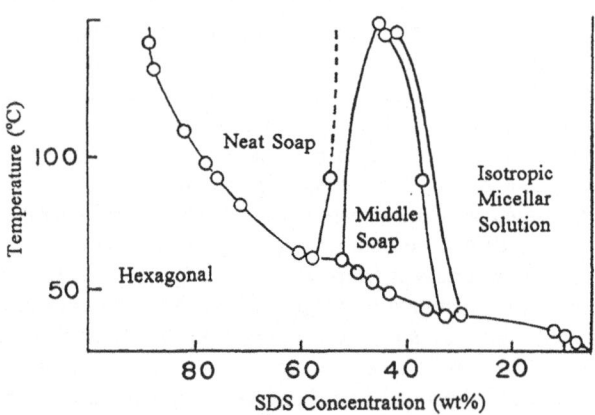

Fig. 6. Phase diagram for SDS micelle as a function of temperature.

SDS related systems. In the isotropic phase region, SDS surfactants often form globular micelles, until concentration is fairly high where the hexagonal tubular phase is observed (see Fig. 6). A series of papers using small angle neuron scattering (SANS) technique to characterize the SDS micellar solutions were reported previously by the author and his coworkers [56-60]. In the following I shall summarize these results.

The SANS experiments for the SDS systems reported here were performed on the H-9B biological spectrometer at the High Flux Beam Reactor (HFBR) at Brookhaven National Laboratory. Fig. 7 shows the diagram of this spectrometer [61]. The HFBR heavy water reactor has a maximum power of 60 MW. The neutrons produced have an average energy of approximately 1 MeV with a Maxwellian distribution in energy. Since the SDS micelles are of the colloidal sizes, it is necessary to increase the wavelength of the neutron, in order to serve as a proper probe for these systems, and to maintain the scattering in the elastic or the quasielastic energy range (see Eq.(21)). To achieve this, the "hot" neutrons, after being moderated by the heavy water, are further "cooled" by a cold source containing ~ 1.5 liter of liquid hydrogen at 1.5 K (cooled by liquid helium). After passing this cold source, the "cold" neutrons (the wave length spans from ~1 to ~20 Å) are guided to the monochromator by a nickel coated guide pipe. The monochromator was made by multi-beryllium layers. By adjusting the Bragg angle of the monochromator, one can select the neutrons of the desired wave length. The selected neutrons are then collimated by a series of pin holes to define the beam and flux, before impinging onto the sample cell. The wavelength spread $\Delta\lambda/\lambda$ at the sample position for the H-9 spectrometer was about 6%. Because the energy spread is low, incorporation of the resolution function in the data analysis is not necessary, which makes the data analysis much easier. The scattered neutrons were recorded by a two dimensional detector, containing 128 x 128 pixels. Each pixel is a He^3 gas proportional counter. The total sensitive area is 64 cm x 64 cm. The detector can be moved back and forth to adjust the covered Q range. It can also be moved side way, in case high momentum transfer is needed (see Fig. 7). This particular spectrometer can cover a Q range from 0.008 to 0.3 Å^{-1}

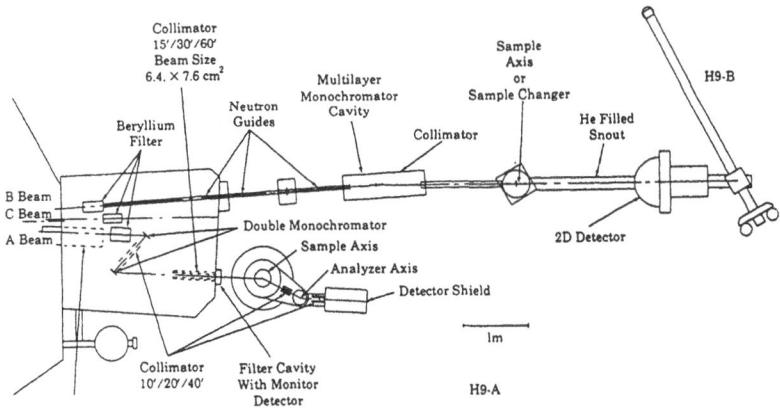

Fig. 7. SANS spectrometer (H-9B) at the HFBR, Brookhaven National Laboratory.

by moving the detector back and forth. By moving it to the side way, the maximum Q can be further increased to ~ 0.6 Å^{-1}. The sample cell used was a quartz circular cell with 1 or 2 mm path length, depending on the sample transmission. The temperature was controlled by a water bath temperature controller with ± 0.2 °C accuracy.

Fig. 8 to 10 show some typical SANS spectra for SDS micellar in aqueous or brine solutions. Due to the strong electrostatic intermicellar interactions, a pronounced peak is always observed. Fig. 11 shows the extracted S(Q) for Fig. 8 and 9.

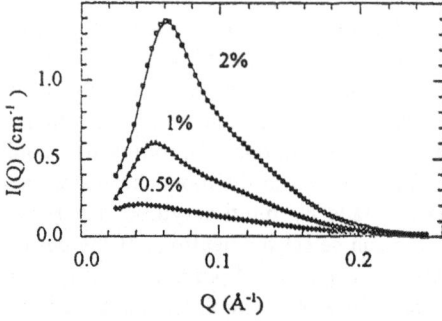

Fig. 8. I(Q) for SDS/water micellar solutions at 0.5, 1, and 2 wt%.

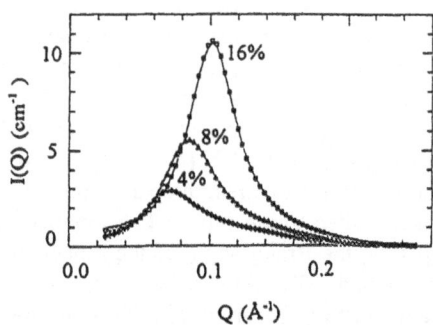

Fig. 9. I(Q) for SDS/water micellar solutions at 4, 8, and 16 wt %.

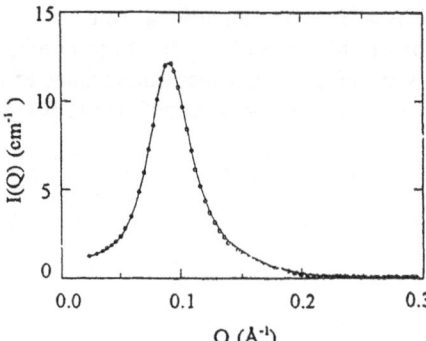

Fig. 10. I(Q) for SDS/water micellar solution at 20 wt%.

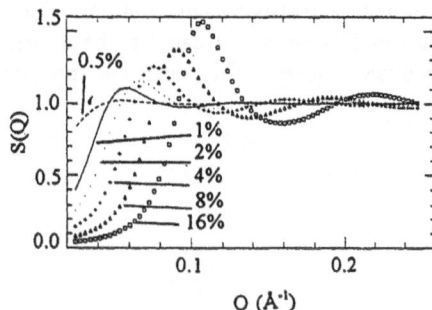

Fig. 11. Extracted S(Q) from Fig. 8 and 9.

As mentioned earlier, a microscopic measurement usually does not provide the desired information directly. In the case of a SANS measurement, the scattering intensity distribution function I(Q) represents the differential cross section. The structure of the micelles as well as how they grow as a function of concentration are not given directly, but are carried in I(Q). In order to determine the structure and the intermicellar interactions, one need to use or develop an appropriate model to analyze I(Q). A simple model often used for colloidal solutions assumes that the micellar size is monodispersed with a structure describable by an ellipsoid. If the system is noninteracting, then, there is only one parameter (the semimajor axis of the ellipsoid) needed to fit the I(Q) data (the semiminor radius is assumed to be the fully stretched length of the surfactant tail). In the SDS case, however, the interaction is strong and long ranged. The interparticle structure factor S(Q) needs to be taken into account. S(Q) can be calculated by solving the Ornstein-Zernike (OZ) equation. For an one-component macroion system (for example, in a micellar system, the micelles are the macroions and the counterions are assumed to be point size) the OZ equation reads [45]

$$h(r) = c(r) + n_o \int_{all\ space} c(r')h(|\vec{r}-\vec{r}'|)d^3r' \tag{41}$$

where h(r) is the total correlation function between macroions, c(r) the direct correlation function, and n_o the number density. If one knows the total correlation function h(r), the pair correlation function g(r) can be obtained by

$$g(r) - 1 = h(r) \tag{42}$$

Once g(r) is obtained, the structure factor S(Q) is simply the Fourier transform of g(r),

$$S(Q) = 1 + \int_0^\infty dr\ 4\pi\ r^2\ n_o\ [g(r)-1]\ \frac{\sin Qr}{Qr} \tag{43}$$

Since g(r)-1 represents the number of micelles at distance (r,r+dr) given one at origin, S(Q) can be regarded as a summation over different neighboring shell of thickness dr. At Q=0, S(Q) represents the inverse of the osmotic compressibility, which can be determined experimentally by laser light scattering.

It is obvious that one needs to solve the OZ equation, in order to get h(r). Because the OZ equation is a coupled equation, one way to solve it is to presume a closure [62,63]. There are several closures being used for solving the OZ equation. The hard sphere closure [64], the Percus Yevick closure [65], the mean spherical approximation (MSA) [66], the Roger-Young closure [67], and the hypernetted chain closure (HNC) [62] are the most commonly used closures. The hard sphere closure is a simple one, and has been very useful for colloidal systems like polystyrene ball systems. The MSA closure is the linearized version of the HNC closure. It takes the advantage that h(r) can be obtained analytically for several interparticle potentials. This makes it a very handy closure for experimentalists. MSA method usually gives fairly accurate results, not only for the S(Q) and g(r), but also for the subsequently derived thermodynamic parameters, such as excess internal energy, and the excess enthalpy [68]. The advantage of Roger-Young closure, on the other hand, is in its successful prediction of the contribution of particle polydispersity at S(0). This makes it a favored closure for evaluation of the osmotic compressibility. Finally, the HNC closure gives the most satisfactory S(Q) for charge systems. However, it is a nonlinear theory, and requires numerical iterations, which may be time consuming, since it may converge slowly. Recent advancement in this calculation has made the convergent rate much faster, and becomes more experimentally practical [69]. Since this chapter mainly discusses the experimental aspects, I will only focus on the MSA closure.

The MSA closure assumes $c(r)$ to be the potential between two particles for $r > \sigma$ ($\sigma =$ the diameter of the macroion) and $h(r) = -1$ for $r < \sigma$. This is

$$c(r) = -\frac{u(r)}{k_B T}, \quad r > \sigma$$

$$h(r) = -1, \quad r < \sigma \tag{44}$$

where k_B is the Boltzmann constant, and T the absolute temperature. Since the pioneering work of Hayter and Penfold [47] it has been a customary in SANS literature to use a so called DLVO (Derjaguin-Landau-Verway-Overbeek) double layer repulsive potential as the intermacroion potential function in MSA. This potential has a form:

$$V_R(r) = \frac{Z_o^2 e^2}{\epsilon(1+\frac{\kappa\sigma}{2})^2} \frac{e^{-\kappa(r-\sigma)}}{r} \tag{45}$$

where Z_o is the macroion charge number, e the electronic charge, ϵ the dielectric constant of the solvent, and κ the Debye screening constant. In the case of ionic micellar solution without added salt, κ is completely determined by the degree of counterion dissociation. Thus, Z_o and κ are related via the charge neutrality. In this case, the contact potential ($r=\sigma$) $Z_o^2 e^2/[\epsilon\sigma(1+\kappa\sigma/2)^2]$ is coupled to the decay constant of the potential function. Using Eq.(45) as the direct correlation function in the MSA ansatz, an analytical solution of the OZ equation can be obtained [66].

The solid lines in Fig. 8 to 10 show the fits of SDS small angle neutron scattering data. The form factor P(Q) was taken to be prolate ellipsoid of major radius a and minor radius b (b was taken as the full stretched tail length of a SDS surfactant). As one can see, the fits are excellent. Table I lists the parameter values extracted from these fits. In view of Table I, one realizes that the extracted parameters from a SANS spectrum , including the interaction parameters and the structural parameters, are fairly adequate for characterizing a micellar solution. However, one has to bare in mind that the accuracy of the extracted parameter values depends on how accurate the models are (the structural model for P(Q) and the

Table I. Extracted parameter values from Fig. 8 to Fig. 11 (T = 40 °C).

Conc. (wt%)	$<N_w>$ [1]	Poly.(%) [2]	α [3]	a/b
0.5	52.2	21	0.24	1.3
1.0	56.6	23	0.23	1.4
2.0	68.7	33	0.25	1.6
4.0	85.4	43	0.24	2.0
8.0	98.8	48	0.19	2.4
16.0	116.0	54	0.14	2.8
20.0	128.0	59	0.13	3.4

1. $<N_w>$ = weight averaged aggregation number. 2. Poly.=polydispersity.
3. α=fractional charge=Z_o /$<N_w>$.

interaction model for S(Q)). In the case of charged micellar systems, the Z_o extracted, using the DLVO potential and the MSA ansatz, has to be questioned, while the other parameters listed in Table I are more or less accurate. In the following I will explain why the extracted Z_o is not necessary accurate. First of all, the scattering spectra exhibit a pronounced peak, which results from the coupling of P(Q) and S(Q). The height of this peak is largely determined by the intermicellar correlation, while the width of the peak as well as the low Q data have something to do with the particle form factor. This is to say that P(Q) and S(Q) are somewhat independent. One thus may model P(Q) and S(Q) independently until the best fitting on the so called absolute intensity scale is obtained (i.e., to achieve the best agreement between the experimental and the computed differential cross section). In the absolute intensity calculation, one computes P(Q) with its prefactor (see Eq.(31) and (34)) calculated according to the micelle-solvent contrast, which is rigorous [70]. But, in the calculation of S(Q) there is a prefactor Z_o which is often taken as an adjustable parameter. This is because the effective surface charges of a micelle is usually not the full charge (i.e., all the counterions are well dissociated). Some counterions may physisorb to the micelles and form a "dressed" shell. In fact, many experimental and theoretical results [71-74] indicated that the effective double layer charge for a micelle is about 25% of the full charge. In addition to the effect of the dressed charges, the Z_o value is renormalized during the fitting, in order to accommodate the approximation of the MSA ansatz. This is to say that the extracted Z_o value is not even the effective surface charges. To compare the Z_o values extracted from the MSA analysis with those from the HNC analysis, the readers can refer the paper by Bratko et al [75].   What I like to point out here is that the parameters values extracted, using the scheme described above, are accurate except the Z_o value which is somewhat renormalized [76].

The analysis shown so far are based on an one-component system, while the SDS systems consist of micelles, counterions, and possibly coions. These other components should also be explicitly accounted in the calculation of S(Q) as well. For such a multicomponent system, the Ornstein-Zernike equation in Q space can be written as

$$\tilde{h}_{ij}(Q) = \tilde{c}_{ij}(Q) + \sum_k n_k \tilde{C}_{ik}(Q) \tilde{h}_{kj}(Q) \tag{46}$$

where $\tilde{h}_{ij}(Q)$ is the total correlation between particle i and j, $\tilde{c}_{ij}(Q)$ is the direct correlation between them. The second term is an indirect term, summing over the third particle k. To calculate the structure factor from Eq.(46) directly would be too much involved in computation, because of the many body effect. To handle this situation, Beresford-Smith and Chan [77] came up with an idea.   They showed that it is possible to contract a multicomponent system into an effective one component system and obtain an effective direct correlation function. This effective direct correlation function can then be used as the MSA ansatz to solve the one-component OZ equation as described above. To contract the multicomponent system into an effective one component system, one still has to write down the basic direct correlation function between particles. Since the natural potential between two charged particles is the Coulomb potential, the direct correlation function between particle i and j can be expressed as

$$c_{ij}(r) = \frac{-\beta Z_i Z_j e^2}{\epsilon r_{ij}}, \qquad r_{ij} > \sigma_{ij}(=\frac{\sigma_i + \sigma_j}{2}) \tag{47}$$

where ϵ is the solvent dielectric constant, and σ_i and σ_j are the diameters of the particle i and j respectively. Then, we define the Fourier transform as

$$h_{ij}(Q) = \int_{all\ space} d^3r\ e^{-i\vec{Q}\cdot\vec{r}}\ h_{ij}(r) = \int_0^\infty dr\ 4\pi r^2 \frac{\sin QR}{Qr} h_{ij}(r) \tag{48}$$

By denoting the macroion as 0, and the other components with much smaller sizes as 1, 2, etc, the effective one-component OZ equation can be obtained by contracting Eq.(47) [69,76], namely,

$$\tilde{h}_{00}(Q) = \tilde{c}_{00}^{eff}(Q) + n_o\ \tilde{c}_{00}^{eff}(Q)\tilde{h}_{00}(Q) \tag{49}$$

where

$$\tilde{c}_{00}^{\text{eff}}(Q) = \tilde{c}_{00}(Q) + \bar{c}_0^T \; [\bar{\bar{I}} - \bar{\bar{c}}]^{-1} \; \bar{c}_0 \tag{50}$$

represents the effective one-component direct correlation function in Q-space for the macroion. $\bar{c}_0$ is a column matrix having elements

$$(\bar{c}_0)_i = \sqrt{n_0} \; \tilde{c}_{i0}(Q), \qquad i=1,2,3,... \tag{51}$$

$\bar{\bar{I}}$ in Eq.(50) represents the two dimensional identity and $\bar{\bar{c}}$ the two dimensional matrix with elements

$$(\bar{\bar{c}})_{ij} = \sqrt{n_i n_j} \; \tilde{c}_{ij}(Q), \qquad i=1,2,3,... \tag{52}$$

Now, the problem is to derive the effective direct correlation function $c_{00}^{\text{eff}}(r)$ in the spatial domain. To do it, we first "guess" that the effective one-component direct correlation function is a modification of the natural potential between charged particles. Under this hypothesis the direct correlation function between particle i and j can be written as

$$c_{ij}(r) = -\beta \frac{Z_i Z_j e^2}{\epsilon r} + c_{ij}^s(r), \qquad i,j=0,1,2,... \tag{53}$$

where the second term takes into account the deviation from the Coulomb form, due to the multicomponent effect. If we neglect the size effect for the small ions, and only concentrate on the direct correlation between macroions, then

$$c_{00}^s(r) = 0, \quad \text{for} \quad r > \sigma_0$$

$$c_{oi}^s(r) = 0, \quad \text{for} \quad r > \sigma_0$$

$$c_{ij}^s(r) = 0, \quad \text{for} \quad r > 0 \tag{54}$$

Using these conditions, one can perform Fourier transform for Eq.(53) and then substitute into Eq.(50) to yield

$$\tilde{c}_{00}^{eff}(Q) = \tilde{c}_{00}^{s}(Q) + \sum_{i=1}(\tilde{c}_{0i}^{s})^2 - \frac{(\tau_0 + \sum\limits_{i=1}\tau_i \tilde{c}_{0i}^{s})^2}{Q^2 + \kappa^2} \tag{55}$$

with

$$\tau_i = \left[\frac{4\pi\beta e^2 n_i Z_i^2}{\epsilon}\right], \quad i = 1,2,3,\dots \tag{56}$$

being Q-independent, thus do not cause difficulty in the subsequent inverse Fourier transformation of $\tilde{c}_{00}^{eff}(Q)$ to obtain $c_{00}^{eff}(r)$. The effective one-component direct correlation function in r-space can be obtained by performing an inverse Fourier transform of Eq.(55). Instead of going into the derivation details, I will only write down the effective one-component direct correlation function so derived,

$$c_{00}^{eff}(r) = -\frac{\beta Z_0^2 e^2}{\epsilon} X^2 \frac{e^{\kappa r}}{r}, \quad r > \sigma_0$$

$$X = \cosh(\frac{\kappa\sigma_0}{2}) + U[\frac{\kappa\sigma_0}{2}\cosh(\frac{\kappa\sigma_0}{2} - \sinh(\frac{\kappa\sigma_0}{2})]$$

$$U = \frac{\mu}{(\frac{\kappa\sigma_0}{2})^3} - \frac{\xi}{(\frac{\kappa\sigma_0}{2})} \tag{57}$$

where κ is the Debye screening constant. μ and ξ in Eq.(57) are

$$\mu = \frac{3\eta}{(1-\eta)}$$

$$\xi = \frac{(\Gamma\sigma_0 + \mu)}{(1 + \Gamma\sigma_0 + \mu)} \tag{58}$$

The parameter Γ in Eq.(58) can be obtained by solving the following equations,

$$4\Gamma^2 = \kappa^2 + \frac{\tau_0^2}{(1+\Gamma\sigma_0+\mu)^2}$$

$$\tau_0^2 = \frac{24\beta Z_0 e^2 \eta}{\epsilon\sigma_0} \tag{59}$$

When thi system is in the dilute limit (i.e., $n_0 \to 0$), X in Eq.(57) becomes

$$\lim_{n_0 \to 0} X = \frac{\dfrac{e^{\kappa\sigma_0}}{2}}{1+\dfrac{\kappa\sigma_0}{2}} \tag{60}$$

The effective one component direct correlation function at this limit reduces to

$$\lim_{n_0 \to 0} \tilde{c}_{00}^{eff}(r) = -\beta \frac{Z_0^2 e^2}{\epsilon\sigma_0(1+\dfrac{\kappa\sigma_0}{2})^2} \frac{e^{-\kappa\sigma_0(\frac{r}{\sigma_0}-1)}}{\dfrac{r}{\sigma_0}} \tag{61}$$

which is precisely the DLVO potential. This point was in fact mentioned by Medina-Noyola and McQuarrie [78]. They pointed out that this DLVO form is simply due to the mean force potential. This is why the DLVO potential was successful in analyzing the SANS data [79-83]

From the above derivation, the DLVO potential is apparently a natural form for the direct correlation function of a multicomponent system in the dilute limit. In the concentrated regime, the form only differs from the DLVO potential by a different prefactor, thus, not affecting the quality of the data fitting. However, the micellar charge (see Eq.(45)) extracted will not be not accurate, if the original DLVO formed is used. Instead of the DLVO potential, Eq.(57) is a more rigorous form for multicomponent cases.

Other information obtainable from a SANS measurement is the polydispersity, the micellization energy as well as the energy gained by the system, when a surfactant molecule is added to an existing micelle, which provides information about the tendency of the micellar growth. To extracted these thermodynamic related parameters, the analysis is much more involved, and an appropriate polydispersity should be used. For the detailed analysis procedure, readers can refer the paper by the author [84].

In comparing the results obtained from the SANS studies with that obtained from the viscosity measurements, the first observed similarity was that the micellar shapes were found to be ellipsoidal in both cases. However, the a/b ratios obtained from two techniques were not quantitatively consistent. This is mainly due the fact that the interparticle interactions are not included in the viscosity analysis (see Table I).

B. AOT/water/decane microemulsion

Microemulsions have been widely applied in industries, from painting, food agents, to drug delivery systems [85-88]. Many applications solely depend on the phases and the structures of the microemulsions. To determine the phases and the structures, rheology and viscosity measurements are often applied [89-97]. Recently, the scattering techniques has become a main theme for these systems [6,8,9,10].

Among the available microemulsion systems, AOT/water/n-alkyl chain system is the most studied systems [89-102]. This is partly due to its wide thermodynamically stable phase region (see Fig. 12), and partly due to its relatively simple sample preparation procedure. A generally used notation for a systematic study of this system across the phase diagram is to maintain the molecular ratio (w) of water to AOT (w=[water]/[AOT]) and vary the volume fraction of oil. According to the phase diagram (Fig. 12), the system remains thermodynamically stable and optically transparent for w up to about 50. A well studied case was w=40.8, which corresponds to 3 gm of AOT to 5 cm^3 water. In many articles it is called the 3/5 system. Recently, many experimental works were devoted to this system. The main conclusion drawn from these studies was that this particular system forms water-in-oil microemulsion droplets for AOT+water volume fraction up to ~0.8 at room temperature. The average droplet size remains fairly constant as a function of oil volume fraction. However, it has a relatively large polydispersity (about 30%) and the polydispersity decreases slightly as a function of both AOT+water volume fraction and temperature [94]. These characteristics differ from the micellar systems where the micellar sizes often grow with increasing concentration, and decreases drastically as a function of temperature. The thermodynamic origin of these rather unusual phenomena was reviewed by Borkovec [103] in a recent paper. I will not go into the details of it.

Similar to the previous subsection, the main purpose here is to compare what one can obtain, using macroscopic measurements and microscopic measurements. And, the viscosity measurement is again used as an example for the macroscopic measurement. Many viscosity measurements have been performed for studying the AOT microemulsion systems [89-94]. Information obtained from these studies includes the structure, the droplet-droplet interaction, and the percolation phenomenon. As we mentioned in the previous section, the polydispersity was not explicitly incorporated into the Roscoe's equation, thus, no polydispersity was derived from these viscosity measurements. However, an important contribution from the viscosity measurement is the identification of the percolation phenomenon, which was defined through the conductivity measurement [98]. As the percolation occurs, the water droplets somewhat "connected" to each other, and the charges in the water pools hop to the neighboring droplets, thus, abruptly increasing the conductivity. In a viscosity measurement, the percolation means the "connectivity" of water droplets, and the "connected" droplets form a macroscopic network. Once the network is formed, the viscosity should in principle vary abruptly. However, the droplets in this case are water pools, which has a viscosity of 1 centipoise (versus 0.93 of oil like decane). Since the droplets are not hard spheres, the relative viscosity observed is expected to be modified according to the Taylor equation (Eq.(16)). Apparently, this low "viscosity contrast" between the droplets and the solvent has reduced the abrupt viscosity change at the percolation concentration to an inconspicuous level. As a result, the percolation phenomenon was not observed for water pool case. Recently, Boned and Peyreleases [89] replaced water with glycerol, which creates enough "viscosity contrast", and observed the percolation phenomenon clearly. The percolation concentration obtained agreed very well with the conductivity measurement. This is an important result, because it shows that one can use the viscosity measurement to study the percolation phenomenon for a nonconducting system, which can not be achieved by conductivity measurements. Other contribution from viscosity study for microemulsion systems was the determination of the interparticle potential. Mallamace [93] examined the AOT system using Grimson and Barker's theory (Eq.(18)) and found that the most appropriate potential between AOT/water droplets is an attractive one with $r^{-1.8}$ dependence. Quantitatively, it may not be accurate, it however, qualitatively agrees with an earlier work by Huang et al [99] using a square well potential, from the fact the potential obtained from both experiments was short-ranged and attractive. In addition to the droplet region, viscosity can also be used to determine the structural transition. Fig. 13 and Fig. 14 show the viscosity behavior for a high AOT/water volume fraction 3/5 system (AOT/water volume fraction = 0.85). The scattering data have shown that there exists a structural transition for temperature going from 13 to 21 °C (see Fig.

15). It is from a randomly connected lamellar structure to a well defined lamellar structure [104]. The transition is graduate, with a relative large two-phase coexistent region (from about 14 to about 20 °C). A viscosity measurement can not determine the structure, it however, can detect the transition temperature fairly accurately. From Fig. 13 and 14 one can see that the system starts to undergo some transition for T > 15 °C, and go into the new phase for T ≥ 21 °C, because the system becomes non-Newtonian.

Fig. 13. Viscosity for ϕ=0.85 as a function of shear rate.

Fig. 14. Viscosity as a function of temperature at ϕ=0.85

Fig. 15. Small angle X-ray scattering for 3/5 AOT/water/decane system.

Obviously, the scattering techniques are much better techniques for the structural determination. For the 3/5 AOT/water/decane system in the droplet region, they have unambiguously determined the droplet structure and the polydispersity (by SANS) [94]. It also identified the glass-like behavior at the high concentration range [105] (by laser light scattering). As for the phase transition studies, Fig. 15 shows a series of small angle X-ray scattering measurements, again for the 3/5 system at 0.76 AOT+water volume fraction. The structural transition between droplet phase and the lamellar phase is clearly demonstrated. By using the paracrystal model [104], one can analyze the structure for the lamellar phase and extracted the structural parameters such as lamellar thickness, and the number of lamellar sheets in each stack. Table II shows the extracted parameters using this model.

Table II. Extracted parameters from the paracrystal model fits to the AOT/water/decane systems.

	w=15	w=25	w=40.8
$<d>^1$ (Å)	39.9	48.2	60.7
g^2	0.081	0.075	0.097
M	14±1	14±1	15±1

1. $<d>$=average interlayer spacing. 2. g=The Hosemann's factor. It should be less than 0.18 for the bilayer lamellar structure. 3. M=number of correlated layers.

As a conclusion for this subsection, viscosity is an useful technique for microemulsion study, particularly, for detecting the percolation phenomenon. In order to extract the molecular information, such as the inter-droplet interactions, theories are needed, and the results obtained may only be qualitatively significant. It is better to combine the viscosity results with microscopic techniques, such as the small angle scattering measurements, in order to better characterize the microemulsion systems.

C. Petroleum Surfactant Systems

Unlike the previous two subsections, the petroleum surfactant systems are real systems derived from the crude oils (or from the vacuum residues), while the micellar systems and the microemulsion systems are made of well defined surfactants. Among petroleum materials, there exists a particular class of component called asphaltenes. They can be derived either from the crude oil, or from the residual material in the vacuum tower after vacuum distillation. A conventional definition of asphaltenes is based on the solubility. Asphaltenes are commonly realized as the heptane insoluble fraction, representing the heavy ended part. The general procedure for extracting asphaltenes from the crude oil or from the vacuum residue is by adding 40 cm^3 of heptane to 1 gm of the crude (or vacuum residue) and stirred overnight at room temperature. The insoluble fraction was then filtered and dry to make asphaltenes.

The main question about asphaltenes is why this material hinders the refining yields. From the recent works [106-109], it is becoming clear that asphaltenes form colloidal-like particle via self-association, and the hinderance of the refining yield is probably due to the formation of these associates. A possible scenario is that the associated particles are treated as an unit during the refining process. As a results, the internal material is not refined at all. The technological objective for studying asphaltenes is to develop a method to "break" theses associates. The first step toward achieving this objective is to understand the association. A rational procedure for understanding this system is to: (1) prove that the self-association indeed occurs, (2) characterize the structure of the associates, and (3) explain the observed phenomena with a thermodynamic theory, so that the type of association can be identified, and its strength can be evaluated.

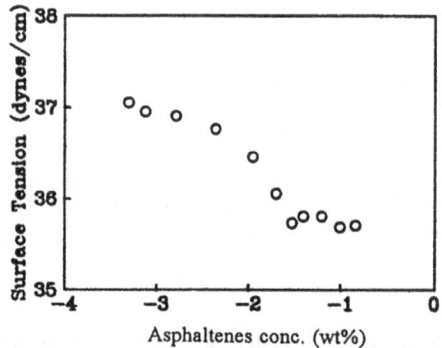

Fig. 16. Surface tension for asphaltenes in Pyridine.

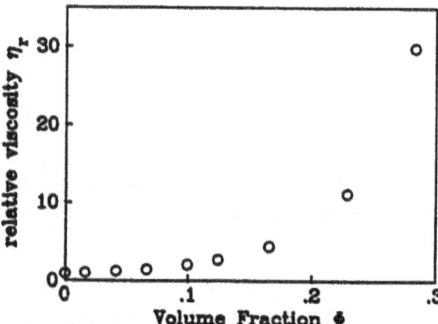

Fig. 17. The relative viscosity for asphaltenes in toluene.

Fig. 16 shows the surface tension data for the vacuum residue asphaltenes. The association (or reverse micellization) was clearly observed (the existence of a breaking point), signifying the onset of the asphaltenes self-association. Once the association is known, the next step is to characterize the structure of the associates. Here the macroscopic measurements (viscosity) and microscopic measurement (small angle neutron scattering) are combined. The results are given in the following.

Fig. 17 shows the relative viscosity of the asphaltenes in toluene as a function of asphaltene volume fraction. At the first glance, this curve is nothing but a viscosity curve commonly observed for any solution systems. For volume fraction lower than 0.1, it behaves like a Einstein fluid (the relative viscosity linearly proportional to asphaltene concentration), and deviates from it for volume fraction greater than 0.1. One may ask: what information can we obtained from such a simple curve? The answer is: it can provide as much information as possible, as long as there are theories to handle the data. Actually, one may be surprised that even some molecular scale information can be determined from such a simple measurement. And, information so obtained can be fairly accurate.

The first data analysis performed for Fig. 17 data is again the Pal and Rhodes analysis (Eq.(11)). The result is shown in Fig. 18. The curve shows linearity with the exponent being 2.5, as expected by Eq.(11) for a hard sphere system. We thus think that the particles are "hydrodynamically" spherical, but not necessary geometrically spherical, due to the multicomponent packing (asphaltenes may contain components of more than one million chemical structures). The slope (or the k solvation constant) obtained from this analysis differs from unity, indicating the asphaltene colloids are solvated. Since Pal and Rhodes' analysis does not provide the degree of solvation quantitatively, Eiler's formula (Eq.(13)) was

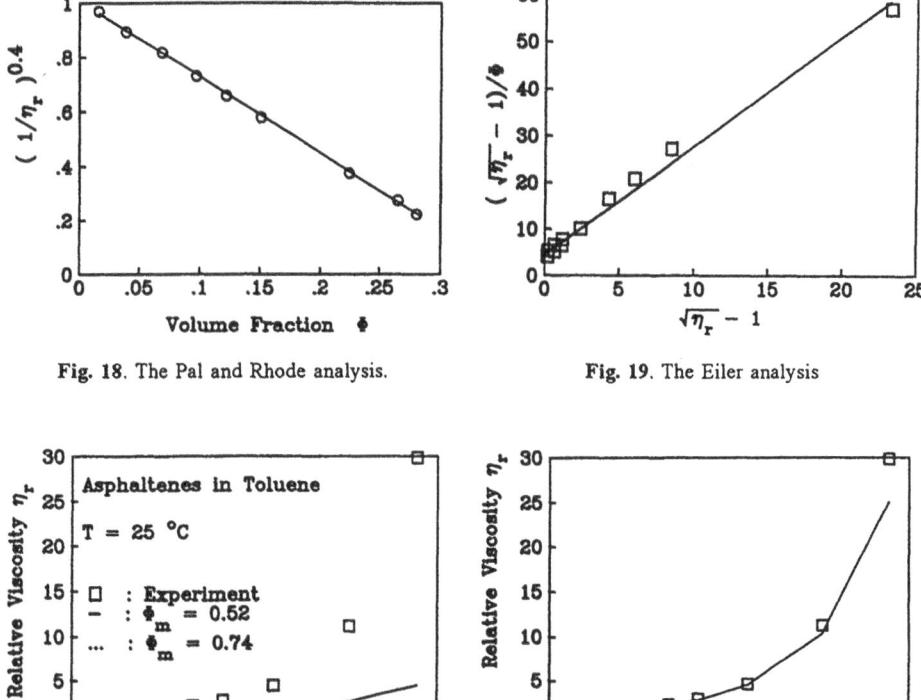

Fig. 18. The Pal and Rhode analysis.

Fig. 19. The Eiler analysis

Fig. 20.Tsenoglou's scaling theory.

Fig. 21. The Grimson and Barker's analysis.

applied for evaluating the degree of solvation. Fig. 19 shows such an analysis. The maximum packing fraction obtained was ϕ_m=0.43. The solvated particle volume was estimated to be 1.72 times of the dry volume (Eq.(14)). The next question to ask is: how is this solvation formed, by solvent entrapment through temporary agglomeration of the asphaltene colloids, or by swelling the asphaltene colloids? To answer this question, the scaling theory (Eq.(15)) developed by Tsenoglou [37] was used to simulate the relative viscosity, assuming temporary agglomeration is the case.

Fig. 20 shows the experimental data and the simulated curves for two extreme ϕ_m values (0.74 represents the face center cubic packing and 0.53 the simple cubic structure). For both cases, the simulated curves and the experimental curve differ substantially. This indicates that the solvation mechanism is not through solvent entrapment. In a way, it also suggests that the temporary agglomerations do not occur. The other possible mechanism is by swelling the asphaltene colloids. I will discuss this point later. Since the solvation is equivalent to creating a hydrodynamic field, which will results in building an interparticle potential (see section II). We thus applied the Grimson and Barker's two fluid theory (Eq.(18)) to evaluate this potential. Fig. 21 shows the fitting. The coupling constant extracted was 0.6, and n was 2.0 similar to the Column potential.

Other analysis performed was the comparison with a true hard sphere system, using Campbell and Forgacs' equation (Eq.(17). Fig. 22 shows the comparison. The viscosity behavior is very similar to a hard sphere system for $0.16 < \phi_m < 0.32$. Although the data and the theory seem to agree, even for $0.06 < \phi_m < 0.16$, we did not intend to draw any conclusion for this volume fraction range, since it is out of the Campbell and Forgacs' theoretical range.

As one can see from the above analyses, a simple viscosity measurement can provide so much information, even for complicated systems like asphaltene solutions. However, as I always emphasize, any molecular information extracted from a macroscopic measurement is better confirmed by an independent microscopic technique. In our case, the particle was found to be spherical, solvated, with a Coulomb-like potential from the viscosity measurement. We shall apply the small angle neutron scattering measurement to verify these results.

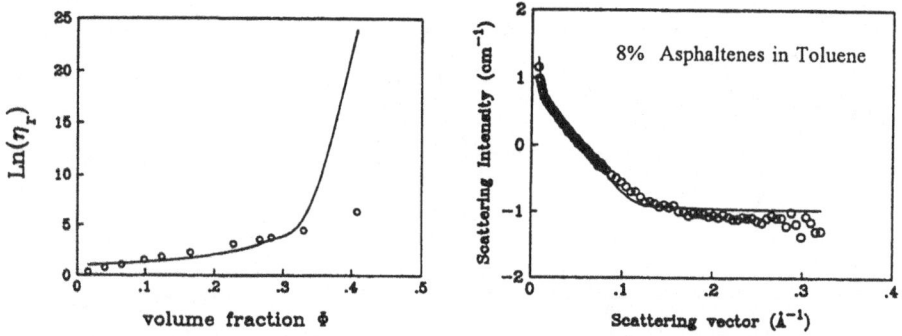

Fig. 22. Campbell and Forgacs' analysis.   Fig. 23. A typical SANS spectrum for asphaltene solution.

The small angle neutron scattering was performed for asphaltenes in deuterated toluene/pyridine (deuterated solvent is for enhancing the contrast between the asphaltene colloids and the solvent [70]). Fig. 23 shows a typical SANS data. To analyze the SANS data in the dilute regime, we neglected the interparticle interaction and followed the method developed by Sheu [43] to determine the polydispersity. The solidline in Fig. 23 represents the fit, using the spherical structure for the colloid, and the Schultz distribution for the particle size. The fitting was excellent on the absolute scale, indicating that the colloidal structure was spherical, the same as our viscosity result. In addition to the structure determination, the SANS measurements also identified the particle size distribution to be like the Schultz distribution function. Fig. 24 shows the particle size distributions for three different temperatures [110]. The width of the distribution becomes narrower as temperature increases, suggesting that the larger particles start to dissociate at higher temperature. As for estimation of the solvation, we determined the contrast between the asphaltene colloids and the solvent. The contrast obtained was much less than that between the dry asphaltenes and the solvent, indicating a substantial amount of solvation. Also, this results indicates that the solvation is through swelling the colloidal particles. We estimated the swelling from the extracted contrast and found that the solvated volume was about 1.85 times of the dry volume, agrees fairly well with the viscosity result (1.72). A theoretical calculation by Gokhman [111] concluded that the swelling is between 1.7 and 2.0, again agrees with our results.

Other information obtainable from SANS is the surface roughness. Fig. 25 shows the plot for evaluating the surface area per unit volume of the sample (S/V) according to the equation,

$$\frac{S}{V} = \frac{\pi \cdot \lim_{Q \to \infty}[Q^4 I(Q)]}{G}$$

(62)

$$G = invariant = \int_0^\infty Q^2 \, I(Q)dQ$$

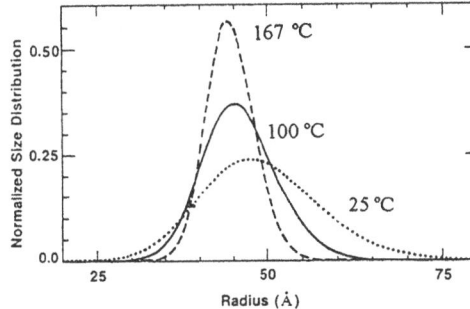

Fig. 24. Size distribution for three temperatures.

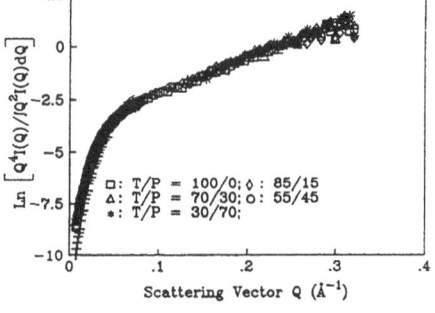

Fig. 25. Surface analysis (see text).

Since Eq.(62) is true only when the surfaces are smooth. A rough surface usually does not converge to a constant value at $Q \to \infty$ for $I(Q)Q^4$. This is precisely the case (see Fig. 25) for the asphaltene solutions, since asphaltene colloids are expected to have very rough surfaces.

The asphaltene system sets a good example of how one can characterize a complex system using macroscopic measurements like viscosity, and how much information can one obtains from these measurements. Since asphaltene system is a real system, the viscosity study demonstrated that a simple macroscopic measurement can provide reasonably accurate information, even for a real system.

IV. Conclusion

In many industrial research laboratories, the systems are much more complicated than the model systems, and most of the experimental techniques for practical uses are macroscopic type of measurements. The message I like to deliver here is that these macroscopic measurements can still give fairly good molecular picture of the systems, if one uses right analysis methods. Of cause, any analysis based on the macroscopic data needs to be verified by some microscopic measurements. Once the analysis methods are justified by the microscopic techniques, the simple laboratory scale (or even the bench top scale) instruments can be nearly as powerful as the large scale sophisticated instrument, like small angle scattering.

V. Acknowledgement

I am indebted to many of my colleagues in carrying out much of the studies reported in this chapter. In particular, I have been greatly benefited from many collaborations with S. H. Chen, D. A. Storm., M. M. De Tar, C. F. Wu, L. Blum, D. Bratko, J. S. Huang, M. Kotlarchyk, J. S. Lin, M. Capel, and J. C. Sung. Also, I have been benefited from the discussion with R. Strey, R. Klein, J. Samseth, B. Carvalho, and K. Mortenson. The SDS scattering data reported here were taken from the H-9 SANS spectrometer at Brookhaven National Laboratory, where D. Schneider and B. P. Scheonborn generously granted beam time and provided invaluable technical assistant. The AOT/water/decane small angle X-ray data were taken from the 10 m small angle X-ray spectrometer at the Oak Ridge National Laboratory, with the beam time allowance and assistance from J. S. Lin. This facility was supported in part by the Division of Materials Sciences, U.S. Department of Energy, under contract number DEAC05-84OR21400, with Martin Marietta Energy Systems, Inc. Finally, the SANS data for asphaltenes solutions were taken from the Intense Pulsed Neutron Source (IPNS) under the technical assistance from P. Thiyagarajan and D. Wozniak. IPNS is operated under the auspices of the US Department of Energy, BES-Materials Sciences, under contract number W-31-109-ENG-38, to whom thanks are extended for the use of their facilities.

VI. References

1. C. S. Hirtzel and R. Rajagopalan, *Colloidal Phenomena: Advanced Topics*, Noyes Publications, Park Ridge, New Jersey (1985).
2. K. Shinoda, T. Nakagawa, B. Tamamushi, and T. Isemura, *Colloidal Surfactants*, Academic Press, New York (1963).

3. P. C. Hieminz, *Principle of Colloid and Surface Chemistry*, Marcel Dekker, New York, (1977).

4. M. J. Rosen, *Surfactants and Interfacial Phenomena*, 2nd ed., John Wiley & Sons, New York (1988).

5. R. Buscall,T. Corner, and J. F. Stageman, *Polymer Colloids*, Elserier Applied Science Publishers, New York (1985).

6. S. H. Chen, J. S. Huang, and P. Tartaglia, *Structure and Dynamics of Strongly Interacting Colloids and Supramolecular Aggregates in Solution*, Kluwer Academic Publishers, Netherlands (1992).

7. E. J. Verwey and J. Th. G. Overbeek, *Theory of the Stability of Lyophobic Colloids*, Elsevier, New York (1948).

8. J. Meunier, D. Langevin, and N. Boccara, *Physics of Amphiphilic Layers*, Springer-Verlag, New york (1987).

9. S. H. Chen and R. Rajagopalan, *Micellar Solutions and Microemulsions*, Spring-Verlag, New York (1990).

10. S. Safran an N. A. Clark, *Physics of Complex and Supermolecular Fluids*, Wiley, New York (1987).

11. A. W. Adamson, *Physical Chemistry of Surfaces*, 4th ed., John Wiley and Sons, New York (1982).

12. H. R. Kruyt, *Colloid Science*, Elsevier Publishing Company, New York (1952).

13. E. Dickinson and G. Stainsby, *Colloids in Food*, Applied Science Publishers, London, U.K. (1982).

14. R. J. Good and R. R. Stromber, *Surface and Colloid Science, Vol. 11, Experimental Methods*, Plenum, New York (1979).

15. S. Frieberg, *Microemulsion Theory and Practice*, ed. by L. M. Prince, pp.133, Academic Press, New York (1977).

16. A. A. Caljé, W. G. M. Agterof, and A. Vrij, in *Micellization, Solubilization, and Microemulsions*, ed. by K. L. Mittal, Vol. 2, Plenum Press, New York, (1979).

17. J. Th. G, Overbeek, in *Colloid Science*, H. R. Kruyt (Ed.), Elsevier, Amsterdam, Vol I, Chap. 6 (1952).

18. D. S. Cannell, in *Physics of Amphiphiles: Micelles, Vesicles, and Microemulsions*, ed. V. Degiorgio and M. Corti, North-Holland, Amsterdam (1985).

19. B. Lindman, O. Söderman, H. Wennerström: in *Surfactant Solutions*, ed. by R. Zana, Marcel Dekker, New York (1987).

20. S. H. Chen, *Ann. Rev. Phys. Chem.*, **37**, 351-399 (1986).

21. R. Simha, *J. Phys. Chem.* **44**, 25 (1940); *J. Chem. Phys.* **13**, 188 (1945).

22. W. Kuhn and H. Kuhn, *Helv. Chim. Acta*, **28**, 97 (1945).

23. J. Riseman and R. Ullman, *J. Chem. Phys.* **19**, 578 (1951).

24. R. Houwink, *J. Prakt. Chem.*, **157**, 15 (1940).

25. W. Kuhn, *Kolloid-Z.*, **62**, 260 (1933).

26. H. Mark, *Der Feste Körper.* Hirzel, Leipzig, (1938).

27. Y. S. Chao, E. Y. Sheu, and S. H. Chen, *J. Phys. Chem.*, **89**, 4862 (1985); E. Y. Sheu, C. F. Wu, and S. H. Chen, *J. Phys. Chem.*, **90**, 4179 (1986).

28. G. B. Jeffery, *Proc. R. Soc. London, A.* **102**, 163 (1923).

29. R. Z. Eisenschitz, *Phys. Chem. A*, **163**, 133 (1933).

30. H. H. Kohler and J. Strnad, *J. Phys. Chem.*, **94**, 7628-7634 (1990).

31. R. Roscoe, *Br. J. Appl. Phys.* **3**, 267 (1952).

32. R. Pal and E. Rhodes, *J. Rheology, 33*, 1021 (1989).

33. I. M. Krieger, *Adv. Colloid Sci, 3*, 111 (1972).

34. C. Cametti, P. Codastefano, G. D'Srrigo, P. Tartaglia, J. Rouch, and S. H. Chen, *Phys. Rev. A., 42*, 3421 (1990).

35. J. A. Prins and H. Peterson, *Physica, 3*, 147 (1936).

36. L. W. Bahe, *Phys. Chem.*, 76, 1062 (1972).

37. C. J. Tsenoglou, *Rheology, 34*, 15 (1990).

38. G. I. Taylor, *Proc. Roy. Soc. A138*, 41 (1932).

39. P. G. de Genne, *J. Phys. (Paris) 40*, 783 (1979).

40. G. A. Campbell and G. Forgacs, *Phys. Rev. A., 41*, 8 (1990).

41. M. J. Grimson and G. C. Barker, *Europhys. Lett., 3*, 511 (1987).

42. A. Coello, J. V. Tato, and E. R. Núñez, *J. Phys. Chem, 96,* 1478 (1992).

43. E. Y. Sheu, *Phys. Rev. A., 45*, 2428 (1992).

44. L. A. Feigin and D. I. Svergun, *Structure Analysis By Small Angle X-ray and Neutron Scattering,* Plenum Press, New York (1987).

45. L. S. Ornstein and F. Zernike, *Proc. Akad. Sci., 17*, 793 (1914).

46. J. K. Percus and G. J. Yeciv, *Phys. Rev., 110*, 1 (1958).

47. J. B. Hayter and J. Penfold, *J. Chem. Soc. Faraday Trans. I, 77*, 1851 (1981).

48. R. J. Baxter, *J. Chem. Phys., 52*, 4559 (1970).

49. L. Blum and J. S. Høye, *J. Phys. Chem., 81*, 1311 (1977).

50. D. Bendedouch and S. H. Chen, *J. Phys. Chem., 87*, 1473 (1983).

51. J. N. Israelachvili, D. J. Mitchell, and B. W. Ninham, *J. Chem. Soc., Faraday Trans. II, 72*, 1525 (1976).

52. E. Y. Sheu, C. F. Wu, and S. H. Chen, *J. Phys. Chem., 90*, 4179 (1886).

53. G. V. Ramanathan, *J. Chem. Phys. 88*, 3887 (1988).

54. G. Manning, Q. *Rev. Biophys. 11,* 179 (1978).

55. P. Ekwall, in *Advances in Liquid Crystal, Vol. 1* edited by G. H. Brown, Academic Press, New York (1975).

56. Y. S. Chao, E. Y. Sheu, and S. H. Chen, *J. Phys. Chem., 89*, 4395 (1985).

57. Y. S. Chao, E. Y. Sheu, and S. H. Chen, *J. Phys. Chem., 89*, 4862 (1985).

58. E. Y. Sheu, C. F. Wu, S. H. Chen, and L. Blum, *Phys. Rev. A, 32,* 3807 (1985).

59. S. H. Chen, E. Y. Sheu, J. Kalus, and H. Hoffmann, *J. Appl. Cryst. 21*, 751 (1988).

60. S. H. Chen and E. Y. Sheu, *Makromo. de Chimie, 15*, 275 (1988).

61. D. K. Schneider and B. P. Scheonborn, in *Neutron in Biology,* ed. by B. P. Scheoborn, pp. 119, Plenum, New York (1984).

62. J. P. Hansen, and I. R. McDonald, *Theory of Simple Liquids*, Academic Press, London (1976).

63. R. Klein, in *Structure and Dynamics of Strongly Interacting Colloids and Supramolecular Aggregates in Solution,* ed. S. H. Chen, J. S. Huang, and P. Tartaglia, Kluwer Academic Publishers, Netherlands (1992).

64. E. Thiele, *J. Chem. Phys. 39*, 474 (1963).

65. J. K. Percus and G. Yevick, *Phys. Rev. 136*, B290 (1964).

66. E. Waisman, J. L. Lebowitz, *J. Chem. Phys., 56*, 3086 (1972); L. Blum, *Mol. Phys. 30,* 1529 (1975); J. S. Høye, L. Blum, *Stat. Phys. 19*, 317 (1978).

67. F. J. Roger and D. A. Young, *Phys. Rev. 30A*, 999 (1984).

68. G. Senatore, and L. Blum, *J. Phys. Chem., 89*, 2676 (1985).

69. L. Belloni, *J. Chem. Phys. 85,* 519 (1986).

70. V. F. Sears, *Thermal Neutron Scattering lengths and Cross Section For Condensed Matter Research,* Report of Chalk River Nuclear Laboratories (1984).

71. S. S. Berr, *J. Phys. Chem., 91,* 4760 (1987).

72. N. Chang, and E. W. Kaler, *J. Phys. Chem., 88,* 6348 (1985).

73. H. Wennerström and B. Lindman, *Micelles Physical Chemistry of Surfactant Association, Physics Report, 52,* 1, (1979).

74. L. Cantù, M. Corti, and V. Digiorgio, *EuroPhys. Lett., 2,* 673 (1986).

75. D. Bratko, E. Y. Sheu, and S. H. Chen, *Phys. Rev. A, 35,* 4359 (1987).

76. S. H. Chen and E. Y. Sheu, in *Micellar Solutions and Microemulsions,* ed. S. H. Chen and R. Rajagopalan, Spring-Verlag, New York (1990).

77. B. Beresford-Smith and D. Y. C. Chan, Chem *Phys. Lett., 92,* 474 (1982).

78. M. Medina-Neyola and D. A. McQuarrie, *J. Chem. Phys. 73,* 6279 (1980).

79. D. Bendedouch and S. H. Chen, *J. Phys. Chem., 88,* 648 (1984).

80. J. P. Hansen and J. B. Hayter, *Mol. Phys. 42,* 109 (1981).

81. N. J. Chang, and E. W. Kaler, *Langmuir 2,* 184 (1986).

82. R. Triolo, J. B. Hayter, L. Magid, and J. S. Johnson, Jr., *J. Chem. Phys. 79,* 1977 (1983).

83. B. Cabane, R. Duplessiz, and T. Zemb, *Surfactant in Solution, Vol. 1,* pp373, ed., by K. L. Mittle, and B. Lindman, Plenum Press, New York (1984).

84. E. Y. Sheu and S. H. Chen, *J. Phys. Chem., 92,* 4466 (1988).

85. P. Becher, ed., *Encyclopedia of Emulsion Technology,* Marcel Dekker, New York, (1981).

86. J. L. Bennett, Jr., and M. F. Wulfinghoff, *Emulsions,* Chemical Publishing Co., New York, (1968).

87. S. Berkman and G. Egloff, *Emulsions and Forms,* Reinhold, New York, (1961).

88. A. T. Florence and D. Attwood, *Physicochemical Principles of Pharmacy,* 2nd ed., Chapman and Hall, New York (1981).

89. Z. Saidi, C. Mathew, J. Peyrelasse, and C. Boned, *Phys. Rev. A., 42,* 872 (1990).

90. J. Peyrelasse, and C. Boned, *Phys. Rev. A., 41,* 938 (1990).

91. F. Mallamace, N. Micali, C. Vasi, and G. D'Arrigo, *Phys. Rev. A., 43,* 5710 (1991).

92. A. D'Aprano, G. D'Arrigo, M. Goffredi, A. Pararelli, and V. Turco Liveri, *J. Chem. Phys., 95,* 1304 (1991).

93. D. Majolino, F. Mallamace, and N. Micali, *Solid State Comm., 74,* 465 (1990).

94. J. S. Huang and M. W. Kim, *Phys. Rev. Lett., 47,* 1462 (1981); M. Kotlarchyk, S. H. Chen, J. S. Huang, and M. W. Kim, *Phys. Rev. A.29,* 2054 (1984); J. S. Huang, J. *Surface Sci, Tech., 5,* 83 (1989); J. S. Huang and M. Kotlarchyk, *Phys Rev. Lett. 57,* 2587 (1986); S. H. Chen and S. L. Chang, and R. Strey, *J. Chem. Phys. 93,* 1907 (1990); M. W. Kim and J. S. Huang, *Phys. Rev. B 26,* 2703 (1982).

95. D. O. Shar, ed., *Macro- and Micro Emulsions - Theory and Applications, ACS Symposium Series, 272,* Americal Chemical Society, Washington D.C., (1985).

96. C. Cametti, P. Codastefano, A. Di Biasio, P. Tartaglia, and S. H. Chen *Phys. Rev. A., 40,* 1962 (1989).

97. C. Boned, J. Peyrelasse, and M. Moha-Ouchane, *J. Phys. Chem., 90,* 634 (1986).

98. S. Bhattacharya, J. P. Stokes, M. W. Kim, and J. S. Huang, *Phys. Rev. Lett., 55,* 1884 (1985).

99. J. S. Huang, *J. Chem. Phys. 82,* 480 (1985).

100. R. Strey, W. Jahn, G. Prte, and P. Bassereau, *Langmuir, 6,* 1635 (1990).

101. J. M. Tingey, J. L. Fulton, and R. D. Smith, *J. Phys. Chem., 94,* 1997 (1990).

102. M. Skouri, J. Marignan, and R. May, *Colloid Polym. Sci., 269,* 929 (1991).

103. M Borkovec, *J. Chem. Phys.* *91*, 6268 (1989).

104. E. Y. Sheu, M. M. De Tar, M. Kotlarchyk, J. S. Lin, M. Capel, and D. A Storm, in *Structure and Dynamics of Strongly Interacting Colloids and Supramolecular Aggregates in Solution,* pp 419, ed. S. H. Chen, J. S. Huang, and P. Tartaglia, Kluwer Academic Publishers, Netherlands (1992); M. Kotlarchyk, E. Y. Sheu, and M. Capel, *Phys. Rev. A. 46*, 928 (1992).

105. E. Y. Sheu, S. H. Chen, J. S. Huang, and J. S. Sung, *Phys. Rev. A., 39,* 5867 (1989).

106. E. Y. Sheu, M. M. De Tar, D. A. Storm, and S. J. DeCanio, *Fuel, 71,* 299 (1992).

107. E. Y. Sheu, D. A. Storm, and M. M. De Tar, *J. Noncryst. Solid, 131-133,* 341 (1991).

108. E. Y. Sheu, M. M. De Tar, and D. A. Storm, *Macromolecular Reports, A28* (Suppl. II), 159 (1991).

109. S. I. Anderson and K. S. Birdi, *J. Coll. Int. Sci., 142,* 497 (1991).

110. E. Y. Sheu, K. S. Liang, S. K. Sinha, and R. E. Overfield, *J. Coll. Int. Sci., 153,* 399 (1992).

111. L. M. Gokhman, *ACS Preprint, Div. Petro. Chem., 35(3),* 308 (1990).

POLYMERIZABLE PHOSPHOLIPIDS:
VERSATILE BUILDING BLOCKS FOR NOVEL BIOMATERIALS

Alok Singh and Joel M. Schnur

Center for Bio/Molecular Science and Engineering
Naval Research Laboratory, Washington, D.C. 20375

1. INTRODUCTION

Phospholipids fulfill a number of important functions of basic cell membranes. Synthetic phospholipids, in particular phosphatidylcholines, have received considerable attention because of their ability to produce a variety of morphologies including liposomes or vesicles. These vesicular structures have been used as models for biological membranes[1-3] to understand *in vitro* membrane structure and functions with an intention to enhance their applicability. In vesicles the attention is focussed on the both the central cavity as well as the outer surface.[4] During the past 40 years a large number of papers have been published on synthetic and natural phospholipids. It is clearly reflected from the reviews[1,5,6] that a large number of those papers have focused on the potential of these microstructures for applications in the areas of encapsulation, controlled release, biosensors, enzyme immobilization, and functional protein reincorporation.

A large number of reports have successfully demonstrated the proof of principle experiments for such applications outlined in the preceding paragraph. Yet few, if any, of these are currently in commercial use. One of the reasons for this is the lack of stability of lipid based microstructures. This "stability problem" has been realized since the beginning stages of research on liposomes. Earlier strategies for improved stability have included the incorporation of proteins, sugars, and cholesterol. These incorporation strategies have had only limited success. The main reason being the perturbation of the organized assemblies by the incorporated substances use in the stabilization process. The strategy that seems to have the most potential for wide spread application is the modification of the phospholipids to enable them to form stabilized microstructures via polymerization.[7-10] However, this approach is also not free from complications. The modification of the lipid by the incorporation of a polymerizable moiety can inhibit the formation of the microstructure that was supposed to be stabilized.[11] Interestingly, these structural modifications can also lead to new structures that also are of technological importance.[12-14]

This article reviews the role of polymerizable phospholipids on the properties of the self-assemblies formed from these lipids, stabilization of microstructures, and the potential for technological applications.

2. DESIGNING A POLYMERIZABLE PHOSPHOLIPID

Since the first appearance of reports describing the concept of polymerized vesicles,[7-10] polymerizable lipids have attracted considerable attention from different disciplines. This concept led to numerous literature reports on the design and synthesis of novel polymerizable phospholipids. A phospholipid molecule consists of three structural components; a chiral glycerol backbone, a phosphate ester linked to glycerol, and hydrocarbon chains also linked to glycerol. A stereospecific numbering system (*sn*) is used to define the position of acyl chains. In natural phospholipids phosphate ester occupies *sn-3* position. The nature of the linkage as well as the linker provides ample opportunities for modifications in a lipid molecule. Structurally, there are four possible sites available for the attachment of polymerizable moieties in a phospholipid molecule. The head group region offers one site and the hydrocarbon region offers other three sites; at the chain terminus, in the mid chain, and near the glycerol backbone. As the field has progressed, the focus on polymerizable lipids has shifted from stabilization of the microstructures to specific applications based on polymerization (e.g. induced phase separation, biodegradable carriers, morphology transformations, controlled release etc). Most of these topics are briefly addressed in the reviews that have appeared on the subject during the last decade.[4,11,13,15-19]

The nature of polymerizable moieties and their placement in the lipid molecule are likely to influence the physical properties of the vesicles or other microstructures formed from the polymerizable lipids. The polymerization in the hydrophilic region of the molecule may change the properties of the head groups,[20] while the polymerization at the chain terminus may cause inter-layer cross-linking. Mid-region cross-linking also can produce membrane with interesting properties including perturbation of the bilayer structure.[21] Increased understanding about the effects of molecular geometry upon microstructure formation may allow us to tailor the properties of biologically derived membranes to meet a specific need by carefully selecting the polymerizable groups, and their placement in phospholipid molecules.

The polymerizable moieties incorporated in phospholipids are vinyl, methacryloyl, diacetylenic, styryl, acetylenic, and dienoyl (sorbate). Other monomers used in cross-linking vesicles are sulfhydryl and disulfides in the acyl region and substituted amino acids in the head group region of the phospholipids. Both of these constrain the neighboring molecule by a condensation reaction in order to provide overall stabilization.

3. POLYMERIZABLE FUNCTIONALITIES AND POLYMER CHARACTERIZATION

Polymerization of monomer phospholipids in microstructure assemblies has been used for the purpose of stabilization of microstructures. Characterization of polymers produced by crosslinking phospholipid monomers in lipid-assemblies is inadequate from polymer chemists' perspective. The following discussion covers phospholipids and excludes results reported on polymerizable, non-phospholipid surfactants. Diacetylenic phospholipids have been discussed separately in section-5.

Methacrylate Lipids

The methacryloyl group has been linked to the phospholipids via an ester linkage. This functionality has been linked to both hydrophilic[20-26] and lipophilic[27-30] sites and did not adversely affect lipid assemblies. Other noteworthy features of methacryloyl moiety are; polymerization in bilayer membranes by thermal as well as photochemical means,[7,27,28] and temperature independent polymerization.

A broad phase transition temperature (Tm) in vesicles from methacrylate lipid **1b** ($T_m = 11°C$)[27,28] has been observed, which is caused by polymerization. Broadening indicates that overall hydrocarbon chain packing in the vesicles remains undisturbed.

The most convenient method to routinely follow the course of polymerization is by thin layer chromatography. Polymer spot does not move on silica plate when developed with lipid solvent system; chloroform, methanol, water (65:25:4). For quantitative evaluation of polymerization efficiency NMR was used as a simple monitoring tool. Polymerization causes disappearance of vinyl methyl protons in NMR spectrum.

Typically, one hour UV irradiation is needed complete the polymerization.[27,28] There is no report available on the degree of polymerization on the methacrylate phosphatidylcholine (PC). However, the number-average degree of polymerization is reported on the methacrylate ammonium surfactant: 600 for radical catalyzed polymer and 135 for UV polymerized vesicles.[31]

Photopolymerization in vesicles of phospholipid containing methacrylate chain only in sn-1 position in vesicles is found slower than that of the lipid which occupies methacrylate moiety in the acyl chain at sn-2 position.[27] No apparent explanation is reported for this behavior. There are two reports in the literature in which one methyl group on choline has been replaced with acrylate[32] and methacrylate[27] moiety. The former lipid produced a photo-polymer of molecular weight of 98,000 as determined by gel permeation chromatography. In the latter case complete polymerization is observed by TLC but no polymer analysis is carried out.

Phospholipids in monolayers have been polymerized by 254 nm UV irradiation for an hour. Polymerization of methacrylate moieties placed on the headgroup of phospholipids and glycerides is studied in monolayers at air-water interface.[20,24,26]

Polymerization in vesicles has also been initiated by heating the dispersion with radical initiator 2,2'-azoisobutyronitrile, AIBN, at 60°C for 20 hours[24].

Thiol and Disulfide Lipids

Thiol chemistry is quite abundant in biological systems. Protein fragments are held together via a disulfide linkage. Upon reduction protein fragments containing sulfhydryl groups (-SH) are produced which are linked to solid surfaces.[33] The reversible thiol-disulfide redox cycle is extended to lipid molecules to achieve polymerization and depolymerization in the phospholipid vesicles. Placement of the -SH group is being reported both near the head group and at the end of the fatty acyl chain.[34-36] Thiol lipids, 2a, did not exhibit any chain melting transition (T_m) even in a lipid having 16 carbons in its acyl chains. On the other hand, lipid 2b, (-SH group next to ester linkage) with similar alkyl chain exhibits a transition at 22°C independent of polymerization.

Sulfhydryl dimerization stabilizes the vesicles by producing linear polymers. The chemistry of the sulfhydryl or thiol group is straight forward. Upon oxidation thiol dimerizes to give disulfide by involving a thiol group from the neighboring lipid chain. To achieve rugged cross-linked polymers, the -SH group is replaced by cyclic disulfide moiety (lipoyloxy,-CH[S]-CH_2-CH_2[S]).[37] The lipoyloxy group is placed at the terminus

of the fatty acyl chain. The polymerization is monitored by TLC and UV spectroscopy. Disappearance of absorption at 333 nm due to five membered cyclic disulfide indicated polymerization reaction. The presence of extensively crosslinked polymer is demonstrated by its insolubility in chloroform or chloroform/methanol (1:1).

In two chain amphiphiles the dimerization of thiols leads to a polymer whose molecular weight is dependent on the size of the vesicle. Vesicles from a thiol lipid are polymerized by ultra-violet irradiation (30-60 minutes) or heat in the presence of H_2O_2 at 40°C (3 hours).[34,38] Hydrogen peroxide oxidation is pH sensitive; pH 8.5 being optimal (95% oxidation). Tri-n-butylphosphine is used to reverse the reaction in vesicles (depolymerization). Based on the estimation of thiol content in the vesicles the extent of the polymerization for lipid **2a** (n=10,15) and **2b** is calculated to be 17, 25, and 20 lipids per polymer chain, respectively.[34] Ring opening polymerization of lipoic acid is induced by UV exposure, heating or using DTT.

Dienoate Lipids

Dienoate (or sorbate) functionalities are incorporated in the hydrocarbon region of phospholipids. The placement of this group is accomplished at two positions; at chain terminus and near the head group. A mid-chain placement is also possible but to date no such lipid has been reported. This could be attributed to the unavailability of synthetic precursors. The dienoate phospholipid dispersions do not exhibit chain melting transition after polymerization.

Near the glycerol backbone ester linkage is used[20,39-57] and at the acyl chain terminus both ether[58] and ester linkages[45,59] are used to incorporate diene functionality.

Dienoate lipids have been polymerized thermally as well as by photo-irradiation. Photopolymerization is achieved by exposing vesicle dispersions to 254 nm UV light, at 50°C.[43-45] Thermally, lipid vesicles are polymerized utilizing both hydrophilic (azobis(2-amidinopropane) dihydrochloride, AAPD) and lipophilic (azobis isobutyronitrile, AIBN) free-radical initiators.[47,48,54] Polymerization in **3a** vesicles by lipophilic or hydrophilic initiator converted 50% lipid into polymer. This effect is attributed to the fact that the diene group on the acyl chain at sn-2 position is accessible to aqueous medium. This conclusion is supported by an experiment in which 1-palmitoyl, 2-(2,4-octadecadienoyl)-sn-glycero-3-phosphocholine vesicles were polymerized by hydrophilic radical initiator and not by lipophilic radical initiator.[47] Additionally, the polymerization of this lipid is favored above the phase transition temperature.[54]

The polymerization has been checked by observing disappearance of diene peak in 255-260 nm range.[20,44-46,49,59,60] Some have also reported the disappearance of chemical shifts due to diene carbons at 118.4, 128.2, and 145.2-146.0 ppm in ^{13}C NMR.[46,52,55] Insolubility of photopolymerized, freeze-dried vesicles in methylene chloride is related to the presence of oligomer or polymers.[60] The degree of polymerization as estimated by gel permeation chromatography was 26.[55] Polymer obtained after acetolysis of polymerized vesicles is checked by GPC using chloroform or tetrahydrofuran as solvent. The average molecular weight of UV polymerized vesicles is around 5000.[48]

Styryl Lipids

Styryl monomers provide improved stability to vesicles. Styrene is linked to the acyl chain through a ketone,[61-66] ether or amide[61] bond. The vesicular structures were obtained only if the styryl group was introduced to a single chain; styryl group in both chains disrupt the vesicle formation.[62] The second acyl chain in these lipids is usually linked via an ether linkage. Polymerized vesicles from these lipids withstand freeze drying and can be re-dispersed in organic solvents without loosing their structural integrity.[66]

The styryl group in phospholipid vesicles is polymerized employing ultra violet irradiation at 50°C as well as free radical initiators, 2,2' azobis-[2-(imidazolin-2yl)propane]dihydrochloride (VA-44) and 2,2'-azobis (4-methoxy-2,4-dimethyl) valeronitrile (V-70).[67] The course of polymerization is monitored by analyzing UV absorption and ^{13}C NMR spectra.[63,65,66] Disappearance of 265 nm peak (due to vinyl group) and chemical shifts at 116.8 and 136.1 ppm indicates the completion of polymerization. The polymerization in this lipid is suggested to proceed at an intermolecular level via sn-1 chain. The photo-polymer in the vesicles is analyzed after acetolysis or methanolysis of lipid chains. GPC analysis indicated the number-average degree of polymerization of the methanolyzed polymer to be 400.[66] Reduction in time for complete polymerization from 6 h at 45°C to 3.5 hours at 60°C indicated that polymerization extent is temperature dependent.

The polymer produced in lipid $\underline{5}$ vesicles provided mechanically strength to the structures, which is demonstrated by the formation of channels in polymerized vesicles with the aid of non-polymerizable phospholipids.[64]

Other Polymerizable Lipids

In addition to the polymerizable groups described in the preceding sections, many less frequently used polymerizable groups have also been reported for the stabilization of vesicles. Some of these have additional influence on the microscopic properties of bilayer assemblies.

In an intriguing system, the polymerizable moiety "itaconate ($-OCO-CH_2-C(=CH_2)-COO-$)" is used in the stabilization of vesicles.[41] Polymerization of the itaconate moiety is achieved by UV photo-initiation. Radical initiated polymerization of itaconate is slow in free form, but in assembled structures it is rapid. Isocyano moiety constitutes a potentially useful system in which crosslinking is reported to proceed parallel to the vesicle surface.[68,69] This system also provides an example of transbilayer crosslinking.

To crosslink the neighboring lipid chain or protein molecule present in phospholipid bilayers, photo-cleavable carbene precursors consisting of omega-(2)-diazo-3,3,3,-trifluoropropionyloxy ($-OCOCN_2 CF_3$),[70-72] diazirinophenoxy ($-O-C_6H_4-CHN_2$), 2-nitro-4-azidophenoxy ($-O-C_6H_4(NO_2)N_3$), and m-azidophenoxy[73] groups are introduced. In another report, an α, β unsaturated keto group is used as a polymerizable moiety to stabilize the vesicles by photo polymerization.[74] The morphological integrity of the vesicles prepared from this phospholipid did not prevail during photo-polymerization. Closed tubular structures are obtained.

Polymerizable Groups in the Polar Headgroup Region

Phospholipids containing a polymerizable moiety in the polar head-group region have not been pursued as actively as their acyl counterpart. This class of lipids are good for crosslinking to solid substrate to form stable monolayers. Polymerization in vesicles is not efficient because of the lack of neighboring monomer for polymerization. The phospholipid analogues bearing the methacrylate moiety usually lack a phosphate or choline group.[20,24,25] Phosphocholine analogues equipped with the methacrylate moiety have been reported by Kusumi et al.[25] and Ohkatsu et al.[32] Headgroup polymerizable methacrylate shows an increase in the phase transition temperature.[23,24] Photopolymerization at 60°C produced a polymer of 98,000 molecular weight.

A novel polymer network in an organized system has been produced in which counterions are polymerizable moieties. Counterions create a thin protective shield on the vesicles upon polymerization. The polymerizable counterions are used with phosphate or ammonium surfactants. The polymerizable moieties used are methacrylate, 4-vinylpyridine, and styrene sulfonate.[75,76] 1,2-dimyristoyl phosphatidic acid consisting of ammonium methacrylate or dimethylammonium diethyldimethacrylate counterion produces vesicles.[77]

Monomer counterions in organized bilayers have been polymerized by UV irradiation (methacrylate ions),[76-79] addition of protic acid (vinylpyridine ions),[75] and the addition of photo initiator (EtO-C(S)-S-CH$_2$CO-(bis undecyl)-L-glutamate)(styrene sulfonate ions).[76] Selective polymerization initiation of cholinemethacrylate counterion is achieved by using water soluble radical initiators[77] without affecting methacrylate counterion.[79] This opens up a new way to use vesicles as template for polymerization.

$$CH_2\text{-O-R } (sn\text{-}1)$$
$$|$$
$$CH\text{-O-R' } (sn\text{-}2)$$
$$| \quad\quad O$$
$$| \quad\quad \parallel$$
$$CH_2\text{-O-P-O-R''}$$
$$|$$
$$O_-$$

1. Methacrylate lipids   (R" = -CH_2-CH_2-NMe_3 or choline)

  a. R=R'= OC-$(CH_2)_{11}$-OC(O)-C(Me)=CH_2 (methacryloyl)

  b. R= palmitoyl R'= methacryloyl

  c. R= methacryloyl R'= palmitoyl

2. Sulfhydryl lipids (R" = choline)

  a. R=R'= OC$(CH_2)_n$-SH (n= 10,15)

  b. R=R'= OC-CH(SH)-$(CH_2)_{13}$-CH_3

  c. R=R'= OC-$(CH_2)_{11}$-O-C(O)-$(CH_2)_4$-CH[S]-CH_2-CH_2S
  ⎿_____⏌

  d. R= lipoyl R'= palmitoyl

3. Dienoate lipids (R" = choline)

  a. R=R'= OC-CH=CH-CH=CH-$(CH_2)_n$-CH_3 (n= 10,12)

  b. R= OC$(CH_2)_{16\sim18}$-CH_3 R"= as in **a** (n= 12)

  c. R (sn-1)= R" (sn-3)= as in **a** (n=6,8,10,12)

4. Diacetylenic lipids (R" = choline)

  a. R=R'= OC$(CH_2)_8$-C≡C-C≡C-$(CH_2)_n$-CH_3 (n= 9,11,12)

  b. R= OC$(CH_2)_m$-CH_3 (m= 14,16) R'= as in **a** (n=9,13)

  c. R=R'= as in **a** (n= 9) R"= -$(CH_2)_n$-OH (n= 2,3,4)

5 Styryl lipids (R" = choline)

  a. R= OC$(CH_2)_8$-C(O)-4-vinylbenzene R'= -$(CH_2)_{17}$-CH_3

Structure of polymerizable phospholipids

Figure-1

4. POLYMERIZABLE PHOSPHOLIPIDS AND TECHNOLOGICAL APPLICATIONS

As described earlier, studies on "conventional" vesicles, under carefully controlled laboratory conditions have demonstrated their technological potential. The major goal for the development of polymerizable lipids was to produce stabilized vesicles to solve an important problem in "ruggedness". Conventional vesicles are not sufficiently stable against chemical, biochemical, and mechanical stress for most technological applications.

The increase in number of articles published each year over the last two decades indicates that the topic of polymerized vesicles is one of strong scientific interest. A perusal of the articles show that this increased level of scientific activity has also extended the list of realistic applications. Their usefulness may apply from the biomedical field[49,66,80-82] to microelectronic applications.[83,84] The commercial promise for polymerizable lipid systems is reflected by the large number of patents issued. The applications described in the patents ranged from preparation of polymerized vesicles,[85-89] encapsulating materials for use in food, cosmetics and pharmaceuticals,[90-96] controlled release of bioactive materials,[97] site-specific targeting by magnetic vesicles,[121] controlled release to enhance the oil recovery,[99] enzyme encapsulation[91] or immobilization,[100] and biocompatible monolayers surfaces,[95,96,104-107] immobilized enzymes,[100] antigen analysis,[108] and preparation of flexible membranes mimicking the skin.[32]

Stabilized Vesicles

Incorporation of polymerizable moieties in a phospholipid or any other surfactant (anionic or cationic) molecule has resulted in the stabilization of the self-assembled structures while preserving their morphological features. The criteria for determining the stability of polymerized vesicles depends on the applications in mind and thus varies from one research group to another. The stabilized vesicles should demonstrate at least one of the following three properties to be considered as technologically attractive; a) prolonged shelf-life, b) physical and mechanical stability, and c) chemical stabilization with respect to biological or other nearby fluids. It should be emphasized here that stability can occur from physical interactions as well as chemical crosslinking. Thus, extensive polymerization alone is not necessarily a technological necessity.

The stability of vesicles can be assessed by monitoring the absence of any precipitated material in the polymerized dispersions. Polymerized dispersions of thiol lipids are stable up to 6 days at 25°C. Polymerized vesicles containing polystyryl backbone are stable for a month as demonstrated by no-change in the particle size of the dispersion.[61,62] Similar results are reported for polymerized methacrylate derived lipid assemblies.[101,102]

An easy method to test vesicle stability is by addition of a perturbant such as ethanol,[27,60,102] salt[9] or detergents[24,34,60,102] to polymerized and monomer vesicles and measuring any changes in their turbidities. A constant turbidity indicates vesicle stability.

Mechanical stability of the vesicles is tested by one of the following techniques; subjecting polymerized vesicles to heating and cooling cycle; addition of detergent to disrupt vesicles filled with fluorescent dye[24] or radioactive marker,[102] mild sonication[38] and washing the vesicles with solvent.[64] Tests for stability have become harsher as the field matures. Thus, stability of vesicles with crosslinked disulfide polymer backbone is now evaluated by addition of SDS and heating up to 60°C.[103] Similarly, photo-polymerized and freeze-dried giant vesicles from a mixture of **5** and cholesterol upon washing with chloroform provides rugged, skeletonized frame structure from the polymerized vesicles.[64]

Encapsulation and Controlled Release

Polymerizable lipids offer opportunities for developing controlled release systems with a potential for applications in the area of improved microencapsulation, drug delivery, catalysis, sensors, micro-repair of surfaces (biological and non-biological), etc. Stability is directly associated with the polymerization efficiency independent of the type of monomer. An impact of the nature of the monomer and its placement in the lipid molecule on the properties of polymerized vesicles, which may lead to novel applications, is remained to be seen.

Methacrylate phospholipids (**1a-c**) are used in entrapment and leak-rate studies and their performance is compared with non-polymerized dispersions.[27,28,38] Vesicles from lipids containing polymerizable moiety in sn-1 or sn-2 position retained ~53% of encapsulant after 8 hours of dialysis. Dialysis against NaCl and 20% ethanol (v/v) also revealed sustained superiority of polymerized systems over non-polymerized ones. Vesicles from **1a** were tested for their suitability in coating biomedical devices which may come in contact with blood.[101] Polymerized liposomes have a modest effect on platelet aggregation, inhibit antifungal growth in vitro when filled with amphotericin B, and provide stable, reproducible, pharmaceutically acceptable liposome preparation.[80,81]

Polymerized vesicles from thiol lipids[34] provide a biodegradable polymer system. Polymerized disulfide vesicles have larger captured volume as compared to conventional vesicles and extremely slow release rate of encapsulant. Highly crosslinked vesicles prepared from dilopoyl lipid **2c** are leaky to glucose but retain sucrose.[37] In vesicles from **2d** palmitoyl group significantly increased the lipid packing efficiency and afforded tighter membranes causing decreased permeability of markers in the non polymerized vesicles. The mechanism of decreased permeability in the case of mixed PC (**4a,b**) is attributed to the polymer boundaries in the membrane which serve as primary avenue for release of sucrose.[37] Thus, permeability decreases with a decrease in the effective number of polymer boundaries by increasing ratio of dilopoyl to monolipoyl PC .

Highly polymerized vesicles from **3a** released about 20% glucose in first few hours of testing; it then it took a week for the rest of the glucose to leak.[43,44] The results were attributed to two populations of lipids in polymerized vesicles.

A novel approach for the controlled release of markers is pursued by polymerizing the hydrophilic core of a vesicle to yield microparticles encapsulated by vesicles.[40]

Self-quenching fluorescent dye, carboxyfluorescein (CF) makes a non-radioactive but sensitive alternative marker to study the leak behavior of the vesicles. The vesicles from **3a (n=12)** released all the trapped CF in ~50 hours at 20°C (above Tm) while upon polymerization no detectable release was observed and the addition of 30% ethanol caused only 15% release.

Lipid **3c** provided large unilamellar vesicles (35 - 85 nm dia).[49] The shorter chain (C=12) gave the largest size (85 nm) vesicles. Suitability of these vesicles in the entrapment of large molecules such as proteins or enzymes is suggested.

Mixed lipid system containing polymerizable and non-polymerizable lipids constitute a good system to develop triggering mechanisms for marker release. Vesicles from a mixture of **3a** with egg phosphatidylethanolamine (PE), dioleoyl(DO) PE or DOPC are photopolymerized causing PE to phase separate.[59] The nonlamellar phase due to PE causes an increase in membrane permeability. Stimuli-responsive polymerized vesicles are produced by mixing **3a** with dimyristoyl PC, dipalmitoyl PC, DPPE, or phosphatidylserine (PS).[109] In **3a** and DPPC/DMPC vesicles a rapid release of contents is achieved at the chain melting transition temperatures (Tm) of non polymerizable lipids. Similarly, vesicles were prepared to release their contents upon changing the pH or Ca^{++} ions level.

Surface Modification, Enzyme Immobilization and Biosensors

Polymerizable lipids have been successfully used in the covalent enzyme immobilization on the surface of vesicles (34,63,68) or incorporation into the lipid bilayers.[57,59,110-112] The enzymes are covalently attached to a surface via an amino or carboxylic group. An approach is reported in which the head group of methacrylate phosphatidylcholine is modified to enable vesicles to link to the amino group of chymotrypsin via schiff base.[29] Other enzymes immobilized on polybutadiene vesicle surfaces were phospholipase A_2,[56] trypsin and soybean trypsin inhibitor.[51]

Polymerized matrices in vesicle bilayers provide a chemically and mechanically stable environment to membrane proteins. Bacteriorhodopsin is one of the proteins most studied in reincorporation experiments involving polymerizable lipids. Earlier experiments described the insertion of the protein in the lipid bilayers of vesicles by co-sonication followed by polymerization.[113] A net inward proton flow in the presence of visible light proved the stability and vectorial orientation of the protein in lipid bilayers. Later, the procedure was improved by inserting rhodopsin directly into the pre-polymerized layers from a mixture of **5a** and DOPC,[57] and **4a(n=9)** & DNPC.[110] Utilizing mechanical stability from polymerizable lipids and the flexibility of conventional lipids, asymmetric monolayers are deposited on glass patch electrodes to incorporate ion-channels in the membranes.[83,111,112] Thus, alamethicin[83] and acetylcholine[111] are incorporated in bilayer made by mixing **4a(n=9)** with asolectin and Cholesterol to fabricate receptor based biosensors. Patents are issued on techniques for enzyme immobilization on porous vesicles prepared from dienoate PC[100] and enzyme encapsulation in vesicles from styryl PC.[92]

Molecular Recognition

Recognition of specific species is a phenomenon well observed in nature e.g. recognition of invading cells by other cells in the body, and specificity in antigen antibody interaction to name a few. The recognition phenomenon is mimicked by using polymerizable phospholipids as model cell membranes. Monolayers at the air water interface[114] are used in the recognition and association schemes and the polymerized vesicles are used in the inhibition of enzyme activity[51] and lysis of the target cells.[56] In the former case polymerized vesicles have been prepared from **3a** and 1-acyl$_{C-16-18}$,2-octadecyl-2,4-dienoyl PE. The trypsin and soybean trypsin inhibitor (STI) were immobilized on vesicles separately. Addition of STI vesicles to trypsin bound vesicles inhibited the hydrolysis of N-tosyl-L-lysine methylester hydrochloride. In a second scheme, on vesicles made from **3** and **4a**, biotin and phospholipase A$_2$ were covalently attached, and on the vesicles made from **3**, **4a** and DMPC avidin[56] was bound. The latter vesicle population was also filled with CF. Upon mixing the two, a rapid increase in fluorescence was observed demonstrating the phenomena of recognition, association and lysis of the target cells.

Oxygen Carriers

Polymerizable lipid vesicles are studied for developing oxygen supplying medium.[46,50,52,55,63,65,66] Thus, 5,10,15,20-tetra ($\alpha,\alpha,\alpha,\alpha$-o-(2',2'-dimethyl-20'-(2"-trimethylammonioethyl) phosphonatoxyeicosanamido) phenyl) porphinatoiron(II), was embedded in the lipid bilayers of polymerizable vesicles to bind oxygen reversibly under physiological conditions. The polymerizable lipids that were used successfully in these studies are dienoyl and styryl PC. The technique of embedding lipid-heme(s) in polymerized bilayers for making artificial blood has been patented.[93] In addition to this technique, patents have also been issued on the techniques of hemoglobin encapsulation in polymerized vesicles constructed from dienate [92] and eleostearate PC.[90] The latter example utilizes the technique of redispersion of polymerized and freeze-dried vesicles in saline containing hemoglobin (Hb) to prepare Hb-containing stable vesicles. The ultimate success of these approaches for a replacement for natural oxygen carrying systems is yet to be seen.

5. DIACETYLENIC PHOSPHOLIPIDS: A Unique Class of lipids

To date, diacetylene is the only polymerizable functionality which has been used for both, stabilization and modulation of microscopic properties of lipid bilayers. Both of these properties are discussed in the following sections.

Diacetylene as Polymerizable Moiety

The incorporation of diacetylenes into lipid bilayers is prompted by its polymerization capability in crystalline state. Unlike other polymerizable moieties discussed in previous sections, polymerization in diacetylenic lipids requires chains to be ordered to facilitate diacetylene aligned in 1,4 fashion (head to tail). In phospholipid molecules hydrocarbon chains are packed in orderly fashion (i.e. the chains adopt an all *trans* conformation) once dispersed in aqueous medium and the temperature is kept below their chain melting transition temperature. This common feature of lipids which complemented the diacetylene polymerization requirement led to the synthesis of diacetylenic phospholipids for constructing polymer stabilized vesicles. The polymerization leads to a colored conjugated polymer backbone made of alternate single, double, and triple bonds.[115,116] Polymerization is achieved using UV irradiation as well as gamma radiation.[30] For the purpose of clarity in referring to a diacetylenic phosphatidylcholine, we shall use the $DC_{m,n}PC$ terminology, where $\underline{m}$ is the number of methylene units between diacetylene and carboxylate group and $\underline{n}$ refers to the number of methylenes on the terminal methyl segment. Diacetylenes as polymerizable moieties are subject matter of several reviews.[4,13]

Polymerization and Polymer Characterization. Solid state polymerization of diacetylenes initiated by absorption of ultraviolet light is well studied in non-phospholipid systems.[115,116] The reaction is topotactic, and its efficiency depends on the correct alignment of the monomers. The polymerization proceeds in solid state, monolayers, multilayers and vesicles. Because of temperature dependence in polymerization, diacetylene should in principle be a preferred polymerizable functionality for those experiments which utilize hydrocarbon chains for encapsulation. In phospholipids diacetylene stabilizes the vesicles without appreciably affecting the size of the vesicles.[9,10] After polymerization, the colorless dispersion turns red due the production of conjugated polymer backbone.[117,118] The production of a colored dispersion produces a well defined visible spectrum exhibiting absorption maxima near 485, 525 and 625 nm and the polymer characterization is based on the interpretation of their visible spectra.[9,10,44,117] The polymerization profile is determined by recording absorption spectra after exposing vesicles to UV light at different time intervals.[117,119]

Polymerization is affected by the method of vesicle preparation, temperature of the dispersions, and the time and intensity of the irradiation. It is found that highly colored polymer is formed from about 50% crosslinking. After that the irradiation bleaches the polymer resulting in a decrease in absorbance.[120] Vesicles prepared from **4b** do not polymerize well. On the other hand polymerizable mixed phospholipid made from lyso egg PC polymerize efficiently.[18] Polymerized vesicles from diacetylenic ethanolamine and phosphatidic acid[60] and phosphatidylcholines exhibit thermochromism.[8,122] Sonication produces smaller vesicles whose photoreactivity diminishes.[44,60] Probably, lipid chain order is lost as the size of the vesicles decreases. In diacetylenic lipids complete polymerization has not been achieved and no molecular weight determination has been reported. However, addition of phospholipids containing a chain length equal to the upper segment above the diacetylene in the phospholipid, by some unknown mechanism, produces an extensive polymerization.[123-124]

Technological Applications. Diacetylenic vesicles have novel properties. Vesicles from diacetylenic lipids turned slightly turbid upon polymerization but remained as stable dispersion without causing any precipitation.[60] At lower temperature polymerized vesicles from lipid, **4b**, show improved stability. At 4°C vesicle precipitation became apparent after seven days, but the later results showed an improvement in the stability to over a month duration.[18] At higher temperature (50°C) both polymerized and unpolymerized vesicles from diacetylenic PC precipitate within 1 hour.

The stability of diacetylenic vesicles is also dependent on the size of vesicles. Large vesicles ($\sim$ 10 micron diameter) are much less stable than small ones ($\sim$ 1um dia.). Thus vesicles from $DC_{8,7}PC$ were stable before and after polymerization while the vesicles prepared from $DC_{8,12}PC$ decomposed rapidly.

Entrapment and marker retaining capability of diacetylenic lipid is temperature dependent. Lipid **4a (n=12)** releases only 8% CF after 50 h. but, above T_m release occurs 1-2 order of magnitude more rapidly. However, upon cooling to 0°C before polymerization all the trapped markers are released in seconds. This is probably due to the change in morphology from vesicles to tubules.[21]

Glycerol permeability in the vesicles derived from diacetylenic lipid **4a (n=9,11)** before and after polymerization is measured by adding 2M solution of glycerol and recording the absorbance between 400-800nm.[120] Decrease in absorbance due to osmotic shrinkage is used to measure the vesicle permeability.[18] Diacetylene vesicles above their gel to liquid-crystalline phase transition temperature are impermeable to sucrose.

Diacetylenic phospholipids **4a,b** are used in the construction of vesicles filled with human adult blood (HbA) to make hemosomes.[125] The UV polymerized hemosomes remained monodispersed up to 8 hours at room temperature. For prolonged storage the hemosomes were kept at 4°C. The entrapped hemosome were capable of reversibly binding the dissolved oxygen, since the polymerized membrane is gas permeable. Hemosomes are thus reported suitable for development as surrogate erythrocytes. These may also find application as physically and chemically stable gas carriers. Diacetylenic phospholipids show reduced thrombogenicity enabling this class of lipids to become a future biomaterial.[126,127]

Polymerizable diacetylenic lipid **4a (n=12)** when mixed with non-polymerizable lipid, DLPC, forms monolayer and vesicle containing islands of monomer lipids.[128] The feasibility of using these 'islands' for incorporating proteins or using these islands (corks) to selectively open the polymerized vesicles is demonstrated by selectively hydrolyzing these islands on a monolayer by the action of phospholipase A_2 (cork screws). In DPPC and **3a** mixture two transition temperatures are observed.[129] Vesicles, made by mixing diacetylenic lipid **4** and phospholipid containing methacrylate moiety on the head group are stable and flexible due to a mosaic-like arrangement of cross-linked and monomeric domains.[24] The latter could serve as matrix for the incorporation of ligand which could be recognized by target tissues or could participate in a release mechanism triggered by pH, temperature or ions.[24,42,59,109]

Short chain phospholipids containing acyl chain lengths similar to that of the m segment of diacetylenic PC has been reported to enhance (seconds vs minutes) the diacetylene polymerization quantitatively.[129] This mixed lipid system is different from the systems reported in the literature because of the requirement of specific chain length of the non diacetylenic lipid to observe the effect. Most probably, this short chain lipid helps organize diacetylenic lipid in a manner most suitable for topotactic polymerization. DSC studies indicates that the two lipids are not phase separated. For polymerization, matching the acyl chain lengths is crucial and lipids with acyl chain shorter or longer than m segment did not enhance polymerization.[123,124] This system has various technological implications, one of which involves the insertion of the membrane protein bacteriorhodopsin.[110]

Diacetylene in Morphology Modulation

The role of the diacetylenes in phospholipids to modulate microstructure morphology is extensively studied at the Center for Bio/Molecular Science and Engineering

(CBMSE) located in the Naval Research Laboratory. These lipids are studied from two distinct points of view; the construction of stabilized vesicles and Langmuir or self-assembled monolayers, and morphology modulation. Use of diacetylenic lipids in morphology modulation began as a result of Yager and Schoen's[21] careful microscopic observation of the formation of 0.5 micron diameter hollow cylinders called "tubules" (Figure-2).

Electron Micrograph of a Tubule
Figure-2

This discovery is one of the most exciting results to emerge from the study of polymerizable lipid systems due to the intriguing theoretical aspects relating to the underlying relationship between molecular structure and the structure of the self-assembled microstructure as well as for the numerous potential applications that are now being pursued using these tubules.[11-13] Helical microstructures and hollow microcylinders have attracted the attention of experimentalists,[130-132] and theoreticians[133-135] alike. A number of synthetic,[122,136,137] mechanistic,[12,138-142] fabrication oriented,[143-146] thermotropic,[145,147,148] spectroscopic,[152-155] and application related[143-146] studies have contributed to the progress on tubules related topics. Formation characteristics and applications of tubules have been reviewed.[156]

Physical Characterization and Properties of Lipid Tubules. Diacetylenic lipids, because of their ability to transform into unique tubular microstructures, are now the focus of extensive synthetic and physicochemical studies at a large number of laboratories including the Naval Research Laboratory. These structures have close similarity to those reported from glutamate based surfactant,[130] but have the added attractions of polymerizability and easily fabrication. In an attempt to understand the molecular basis for tubule formation, a large number of diacetylenic lipids are synthesized to study their physical properties.[30,124,137,149]

Diacetylenic lipids have shown an interesting thermal phase behavior. The transition temperature (Tm) of diacetylenes in aqueous dispersions broadens and then disappears as polymerization proceeds which indicates that polymerization suppresses fluid to solid transition of $DC_{8,7}PC$. Unlike isomeric unsaturated phosphocholines, isomeric diacetylenic lipids (fixed acyl chain length) in aqueous dispersions did not exhibit influence of diacetylene position on their phase transition temperatures.[137]

Raman, FT-IR, microscopic, x-ray and molecular modelling techniques have all been used extensively to elucidate the basis for tubule formation. Helical structures are observed when lipid is dissolved in ethanol-water system by heating and cooled slowly. This observation indicated helices as a possible precursor for tubules.[130,143] Recent observation of the formation of long, thin fibers employing methanol-water system added another piece of information without any explanation.[150] Formation of right handed helices from l-$DC_{8,9}PC$[140,144] suggests that chirality is important in the formation of tubules. Formation of both left and right handed helical structures and tubules from the dl-$DC_{8,9}PC$ proposes to reconsider the mechanism of tubule formation.[138] Experimental results further provided an example of the formation of tubule-like microstructures from an achiral diacetylenic surfactant.[136] A mechanism is proposed in which tubule formation occurs by continuous transfer of lipid bilayers from vesicles by a rolling up process.[139,141,157]

FT-IR studies on the tubules and the vesicles from diacetylenic lipids have revealed a highly ordered acyl chains in the gel state.[151,152] The results are based on narrow scissoring mode at 1470 cm^{-1} and the prominent CH_2 wagging progression. Similar conclusions are also reported based on microscopic, spectroscopic and X-ray study.[140] The lipid monomers are highly ordered at the molecular level, and the trend continues to form microstructures which align and give a macroscopic order. Raman studies revealed the presence of longitudinal acoustic modes (LAM) in the diacetylenic bilayers.[140,153,154] These LAMs are not typically observed in the lipid samples. Diacetylenic PCs show the intense LAMs in the lower segment ($\underline{n}$) of the diacetylenic chain. It has been reported that $DC_{8,7}PC$, $DC_{8,9}PC$, and $DC_{8,13}PC$ show bands at 230, 194 and 144 respectively. This observation of LAMs indicate that the upper segment of the diacetylene is vibrationally decoupled from the lower segment ($\underline{n}$).[147]

Bilayers from diacetylenic PC,**8a** are extensively studied using X-Ray diffraction[155,158] and electron diffraction[159] techniques respectively. While it has not been possible to study oriented tubules, some pertinent conclusions can be drawn from these studies. It is clear that there is bilayer structure in the tubules and that the lipids readily lie at some angle (probably about 30°) with respect to the layering. Before various theories proposed so far could be verified, high resolution X-ray studies must be performed. In particular it is vital to ascertain the nature of the phase from which the tubules are formed. The present studies do not complement to all the theories proposed.[147]

Non-PC Head-Groups and Tubules. To date there is no report on the phospholipids other than diacetylenic phosphatidylcholines which undergo morphology transformations leading to tubules. Non-phospholipid amphiphiles derived from glutamic acid[130,132] or galactonamides[131] however, have produced helices and tubule like structures. Recently, diacetylenic phospholipid **4c** in which choline group is replaced with hydroxyalkanol, is reported to produce tubules.[160,161] The resultant microstructure morphologies from this lipid were both pH and ion dependent. Efficient tubule formation is observed only between pH 5.6 and 6.5. However, by altering pH, ion strength and the counterions tubules with differing diameters are produced.[162] (figure-3)

Scanning Electron Micrograph of Metal-coated Tubules (4c, n=2)
Figure-3

The ability of tubule from **4e (n=2)** in the presence of metal ions is also used in enhancing the stability of tubules. Palladium ions are used in the preparation of tubules which further acted as catalyst to electroless metal plating on the surface.[163] Metal coated tubules (nickel or cobalt) of varying diameter provide an opportunity to extend potential for tubule applications. Detailed studies on this type of system may provide enhanced understanding of the formation of microstructures. Functionalizable head groups will naturally lead the way to new applications.

Tubules and Technological Opportunities. Tubules are open ended, hollow cylindrical microstructures which have average diameter of 0.5 micron. The length of the tubules is process dependent[143-146] and may be grown as along as 600 microns. The morphology of tubules itself, control over the dimensions, and ease of fabrication of these tubules have led to a number of applications.[98]

Two approaches are used in overcoming the problem of inefficient polymerization of diacetylene in tubules and render them technologically attractive. In one approach, polymerizable moieties[30] are used to stabilize these structures. The other approach stabilizes the tubules by virtue of metallization.[84] Two polymerizable groups, olefinic and methacrylate, are placed at the terminus of acyl chains of the diacetylenic lipid. In the instance of methacrylate extensive polymerization is achieved after UV irradiation, but extensive polymerization disrupted the tubule structures. An olefinic moiety after polymerization by gamma radiation rendered more stable microstructures which are stable against organic solvents such as ethanol, methylene chloride, and chloroform.

Open ended tubules at first were not considered for encapsulation and controlled-release purposes because of their open ends. It was thought that any entrapped encapsulant would easily diffuse out the ends. Price[164] demonstrated that the encapsulant could be mixed with a thermal resin. This approach has led to exciting results with linear release rates of over 2 years been observed. A number of materials have been encapsulated in these microstructures[165] and now the use of tubules is extended to wound repair[166] and antifouling applications. Successful uniform coating of the tubules with a variety of metals[84] has already led to numerous electronic applications.

In an attempt to modulate tubule morphology the local chemical environment of diacetylenes is changed by introducing an oxygen linker between the diacetylene and alkyl segment. A linker on both side of diacetylene completely disrupted the tubule formation, while the incorporation of oxygen linker between the diacetylene and methyl terminated segment produced less crystalline, flexible tubule containing larger diameter of tubules.[148] see figure-4.

Encapsulation of antifouling agents into metal coated tubules has provided a novel and an attractive candidate for controlled-release applications. Copper coated tubules are utilized to achieve sustained release of antifouling agents in the ocean environment.[164] Tetracycline in polymer binder is incorporated into the 0.4 micron inner core of the copper coated tubules by capillary action. The tubules are then incorporated into a paint matrix and applied on surfaces. Sustained rates are observed for over 365 days in a laboratory environment and significant anti-fouling action is demonstrated in ocean conditions for the same time period. The mechanism for the release, though not well understood at this time, is thought to derive from a combination of capillarity effect, surface tension effect,

161

diffusivity of the paint matrix, and the corrosion parameters of the copper coating. This approach is suitable for applications requiring prolonged release in adverse conditions.

Optical Micrograph of Tubules From Phospholipid 4a(n=9)
Figure-4

At NRL other development and evaluation of tubule based applications are also pursued. The tubules have also been oriented in the magnetic field.[167] The ability to fabricated sub-micron size metal-cylinders in large quantities may offer significant advantages for electronic, magnetic, composite, and sustained release applications. Recent work has shown that tubule based composites can exhibit dielectric constants well over 30 at 4% weight loading in the 10 GHz regime[168] as well as the possibility of gold coated tubules for high power microwave cathode applications.[169]

6. CONCLUSIONS

Polymerizable phospholipids constitutes a unusually useful class of biomaterial. It is quite clear from the discussions in preceding sections that this class of lipids has gained the potential to become a separate field of study. The original role of stabilization on the technologically attractive vesicles has been fulfilled and has become a tool to look into chemistry and physics of bilayer structures as well as advance the applicability of polymerizable lipids to multidisciplinary areas. The ability of diacetylenic system to modulate morphologies has proven to be useful in developing new strategies to study membranes. Published trend indicates that commercial viabilities of polymerizable phospholipids are feasible in the future.

7. REFERENCES

1. Liposomes: from physical structures to therapeutic applications, 1981, C. G. Knight, ed., Elsevier/North-Holland Biomedical Press, New York.

2. A. D. Bangham, M. W. Hill, N. G. A. Miller, 1974, Preparation and use of liposome as models for biological membranes, Methods Membr. Biol., 1: 1.

3. C. Tanford, 1985, Monolayers, micelles, lipid vesicles and biomembranes in "Physics of amphiphiles: micelles, vesicles and microemulsions," V. Degiorgio and M. Corti, eds., North-Holland Physics Publishing, Amsterdam, p. 547.

4. H. Ringsdorf, B. Schlarb, and J. Venzmer, 1988, Molecular architecture and function of polymeric oriented systems: Models for the study of organization, surface recognition and dynamics of biomembranes, Angew. Chem. Int. Ed. Engl., 27: 113.

5. H. Eibl, 1984, Phospholipids as functional constituents of biomembranes, Angew. Chem. Int. Ed. Engl., 23: 257.

6. J. H. Fendler, and P. Tundo, 1984, Polymerized surfactant aggregates: characterization and utilization, Acc. Chem. Res., 17: 1.

7. S. L. Regen, B. Czech, and A. Singh, 1980, Polymerized vesicles, J. Amer. Chem. Soc., 102: 6638.

8. D. S. Johnston, S. Sanghera, M. Pons, and D. Chapman, 1980, Phospholipid polymers; synthesis and spectral characteristics, Biochim. Biophys. Acta, 602: 57.

9. H. Hub, B. Hupfer, H. Koch, and H. Ringsdorf, 1980, Polymerizable phospholipid analogues. New stable biomembrane and cell models, Angew. Chem. Int. Ed. Engl., 19: 938.

10. D. F. O'Brien, R. T. Klingbiel, and T. H. Whitesides, 1981, The polymerization of lipid diacetylenes in bimolecular-layer membranes, J. Polym. Sci., Polym. Lett. Ed., 19: 95.

11. A. Singh, and M. A. Markowitz, 1992, Phospholipids and self-assembled microstructures: accomplishments, challenges and future prospects in "Biomembrane structure and function; the state of the art", B.P. Gaber and K. R. K. Easwaran, eds., Adenine Press, New York, p. 37.

12. Biotechnological applications of lipid microstructurs, 1988, B. P. Gaber, J. M. Schnur, and D. Chapman, D., eds., Plenum Press, New York, p. 305.

13. A. Singh, and J. M. Scnur, 1993, Polymerizable phospholipids in "Phospholipid handbook, G. Cevc, ed., Marcel Dekker Inc., New York, p. 233.

14. H. Rehage, and M. Veyssie, 1990, Two dimensional model network, Angew. Chem. Int. Ed. Engl., 29:439.

15. L. Gross, H. Ringsdorf, and H. Schupp, 1981, Polymeric antitumor agents on a molecular and on a cellular level, Angew. Chem. Int. Ed. Engl., 20: 305.

16. J. H. Fendler, 1984, Polymerized surfactant vesicles: novel membrane mimetic agents, Science, 223: 890.

17. S. L. Regen, 1984, Polymerized vesicles, Polymer News, 10: 68.

18. D. S. Johnston, and D. Chapman, 1984, Polymerized liposomes and vesicles, in "Liposome Technology", G. Gregoriadis, ed., CRC Press Inc., Boca Raton, Florida, 1: 123.

19. J. H. Fuhrhop, and J. Mathieu, 1984, Routes to functional membranes without protein, Angew. Chem. Int. Ed. Engl., 23: 100.

20. A. Akimoto, K. Dorn, L. Gros, H. Ringsdorf, and H. Schupp, 1981, Polymer model membranes, Angew. Chem. Int. Ed. Engl., 20: 90.

21. P. Yager, and P. Schoen, 1984, Formation of tubules by a polymerizable surfactant, Mol. Cryst. Liq. Cryst., 106: 371.

22. A. Kusumi, M. Singh, D. A. Tirrell, G. Oehme, A. Singh, N. K. P. Samuel, J. Hyde, and S. L. Regen, 1983, Dynamic and structural properties of polymerized phosphatidylcholine vesicle membranes, J. Amer. Chem. Soc., 105: 2975.

23. R. Elbert, A. Laschewsky, and H. Ringsdorf, 1985, Hydrophilic spacer groups in polymerizable lipids:formation of biomembrane models from bulk polymerized lipids, J. Amer. Chem. Soc., 107: 4134.

24. E. Sackmann, P. Eggl, C. Fahn, H. Bader, H. Ringsdorf, and M. Schollmeier, 1985, Compound membranes of linearly polymerized and cross-linked macrolids with phospholipids: preparation, microstructure and applications, Ber. Bunsenges. phys. Chem., 89: 1198.

25. A. Laschewsky, H. Ringsdorf, G. Schmidt, and J. Schneider, 1987, Self-organization of polymeric lipids with hydrophilic spacers in side groups and main chain: Investigation in monolayers and multilayers, J. Amer. Chem. Soc., 109: 788.

26. P. Meller, R. Peters, and H. Ringsdorf, 1989, Microstructure and lateral diffusion in monolayers of polymerizable amphiphile, Colloid. Polym. Sci., 267: 97.

27. S. L. Regen, A. Singh, G. Oehme, and M. Singh, 1982, Polymerized phosphatidylcholine vesicles. synthesis and characterization, J. Amer. Chem. Soc., 104: 791.

28. S. L. Regen, A. Singh, G. Oehme, and M. Singh, 1981, Polymerized phosphatidylcholine vesicles. Stabilized and controllable time-release carriers, Biochim. Biophys. Res. Commun., 101: 131.

29. S. L. Regen, M. Singh, and N. K. P. Samuel, 1984, Functionalized polymeric liposomes. Efficient immobilization of alpha chymotrypsin, Biochem. Biophys. Res. commun., 119: 646.

30. A. Singh, R. Price, P. E. Schoen, P. Yager and J. M. Schnur, 1986, Tubule formation by heterobifunctional polymerizable lipids: synthesis and characterization, Polymer Preprints, 27(2): 393.

31. D. Bolikal, and S. L. Regen, 1984, Degree of polymerization of vesicle membrane, Macromolecules, 17: 1287

32. Y. Ohkatsu, M. Yokotu, and T. Kusano, 1988, Synthesis and polymerization of macromonomer derived from phosphatidylcholine, Makromol. Chem., 189:775.

33. J. F. Martin, W. J. Hubbell, and D. Papahadjopoulos, 1981, Immunospecific targeting of liposomes to cells: A novel and efficient method for covalent attachment of Fab fragments via disulfide bonds, Biochemistry, 20: 4229.

34. N. K. P. Samuel, M. Singh, K. Yamaguchi, and S. L. Regen, 1985, Polymerized-depolymerized vesicles. Reversible thiol-disulfide-based phosphatidylcholine membranes, J. Amer. Chem. Soc., 107: 42.

35. T. Diem, B. Czajka, B. Weber, and S. L. Regen, 1986, Spontaneous assembly of phospholipid monolayers via adsorption onto gold, J. Amer. Chem. Soc., 108: 6094.

36. S. L. Regen, 1987, Polymerized vesicles, NATO ASI Series, Ser.C., 215: 317.

37. J. Stefely, M. A. Markowitz, and S. L. Regen, 1988, Permeability characteristics of lipid bilayers from lipoic acid derived phosphatidylcholines: comparison of monomeric, cross-linked and non-cross-linked polymerized membranes, J. Amer. Chem. Soc., 110: 7463.

38. S. L. Regen, 1985, Polymerized phosphatidylcholine vesicles as drug carriers, Ann. NY Acad. Sci., 446: 296.

39. H. Bueschl, H. Ringsdorf, and U. Zimmermann, 1982, Electric field-induced fusion of large liposomes from natural and polymerized phospholipids, FEBS Letters, 150: 38.

40. V. P. Torchilin, A.L. Klibanov, N. N. Ivanov, H. Ringsdorf, and B. Schlarb, 1987, Polymerization of liposome-encapsulated hydrophilic monomers, Makromol. Chem., Rapid Commun., 8: 457.

41. M. Takane, K. Shigehara, and E. Tsuchida, 1986, Polymerized phospholipids and their polymeric liposomes, Makromol. Chem., 187: 853.

42. Y. Okahata, K. Ariga, and T. Seki, 1988, Polymerized lipid-corked capsules membranes. Polymerization at different positions of corking lipid bilayers on the capsule and effect of polymerization on permeation behavior, J. Amer. Chem. Soc., 110: 2496.

43. K. Dorn, R. T. Klingbiel, D. P. Specht, P. N. Tyminski, H. Ringsdorf, and D. F. O'Brien, 1984, Permeability characteristics of polymeric bilayer membranes from methacryloyl and butadiene lipids, J. Amer. Chem. Soc., 106: 1627.

44. D. F. O'Brien, R. T. Klingbiel, D. P. Specht, and P. N. Tyminski, 1985, Preparation and characterization of polymerized liposomes, Ann. NY Acad. Sci., 446: 282.

45. P. N. Tyminski, I. S. Ponticello, and D. F. O'Brien, 1987, Polymerizable dienoyl lipids as spectroscopic bilayer membrane probes, J. Amer. Chem. Soc., 109: 6451.

46. H. Nishide, M. Yuasa, Y. Hashimoto, and E. Tsuchida, 1987, Amphiphilic and polymerizable porphyrins and their copolymerization with phospholipid: oriented fixation of porphyrins in a bilayer membrane, Macromolecules, 20: 459.

47. H. Ohno, S. Takeoka, and E. Esuchida, 1987, Unequivalent chemical environment of diene groups in 1- and 2-acyl chains of polymerizable lipids analyzed by radical polymerization, Bull. Chem. Soc. Jpn.,60: 2945.

48. H. Ohno, Y. Ogata, and E. Tsuchida, 1987, Polymerization of liposomes composed of diene containing lipids by uv and radical initiators: evidence for the different chemical environment of diene groups on 1- and 2-acyl chains, Macromolecules, 20: 929.

49. E. Hasegawa, K. Eshima, Y. Matsushita, H. Ohno, and E. Tsuchida, 1987, Characterization of polymerized vesicles derived from polymerizable 1,3-diglycero-2-phoshocholines: Formation of large unilamellar vesicles by ultrasonication, Polymer Bull.(Berlin), 18: 65.

50. M. Yuasa, H. Nishide, E. Tsuchida, and A. Yamagishi, 1988, Oriented fixation of synthetic heme complexes in phospholipid bilayer membranes: Electro optic measurement, J. Phys. Chem., 92: 2987.

51. H. Kitano, N. Kato, and N.Ise, 1988, Mutual recognition between polymerized liposomes:Enzyme and enzyme inhibitor system, Biochim. Biophys. Acta, 942: 131.

52. E. Tsuchida, H. Nishide, M. Yuasa, T. Babe, and M. Fukuzumi, 1989, Synthesis of polymerizable amphiphiles (porphinato) irons and their copolymers with polymerizable phospholipid, Macromolecules, 22: 66.

53. E. Wang, E. Shouji, H. Ohno, and E. Tsuchida, 1989, Fluorescence behavior of porphinato zinc derivative in the molecular assembly of polymerized lipid, Polym. Bull.(Berlin), 21: 195.

54. H. Ohno, S. Takeoka, H. Iwai, and E. Tsuchida, 1989, Effect of phase transition on photosensitized radical polymerization of diene-containing lipids as liposomes, Macromolecules, 22: 61.

55. E. Tsuchida, H. Nishide, M. Yuasa, E. Hasegawa, K. Eshima, and Y. Matsushita, 1989, Polymerizable liposome/lipid-heme as an oxygen transporter under physiological conditions, Macromolecules, 22: 2103.

56. H. Kitano, N. Kato, and N. Ise, 1989, Mutual recognition between polymerized liposomes: Macrophage model system by polymerized liposomes, J. Amer. Chem. Soc., 111: 6809.

57. P. N. Tyminski, L. H. Latimer, and D. F. O'Brien, 1988, Reconstitution of rhodopsin and cGMP cascade in polymerized bilayer membranes, Biochemistry, 27: 2696.

58. E. Hasegawa, Y. Matsushito, K. Eshima, H. Ohno, and E. Tsuchida, 1986, Polymerizable glycerophosphocholines containing terminal 2,4-hexadienoyloxy groups and their polymerized vesicles, Polym. Bull.(Berlin), 15: 397.

59. D. A. Frankel, H. Lamparski, U. Liman, and D. F. O'Brien, 1989, Photoinduced destabilization of bilayer vesicles, J. Amer. chem. Soc., 111: 9262.

60. B. Hupfer, H. Ringsdorf, and H. Schupp 1983, Liposome from polymerizable phospholipids, Chem. Phys. Lipids, 33: 355.

61. E. Hasegawa, K. Eshima, Y. Matsushita, H. Nishide, and E. Tsuchida, 1985, Synthesis of polymerizable glycerophosphocholines and their polymerizable vesicles, Polym. Bull., 14: 31.

62. E. Hasegawa, K. Matsushita, and K. Eshima, 1984, Synthesis of novel styrene groups containing glycerophosphocholines and their polymerization as liposomes, Makromol. Chem., Rapid Commun., 5: 779.

63. E. Tsuchida, E. Hasegawa, Y. Matsushita, K. Eshima, M. Yuasa, and H. Nishide, 1985, Polymerized liposomes as the carrier of heme. A physically stable oxygen carrier under physiological conditions, Chemistry Lett., 969.

64. H. Ohno, S. Takeoda, and E. Tsuchida, 1985, Skeletonized hybrid liposomes, Polym. Bull., 14:487.

65. M. Yuasa, E. Hasegawa, H. Nishide, and E. Tsuchida, 1987, Oxygen-exchange reaction between artificial lung device: The heme embedded in polymerized lipo liposome as an artificial oxygen carrier, J. Macromol. Sci.-Chem., A24: 661.

66. Y. Matsushita, E. Hasegawa, and K. Eshima, 1987, Synthesis and properties of polymerizable phospholipids. 5. Molecular weight of polymeric liposomes, Makromol. Chem., Rapid Commun., 8: 1.

67. K. Yamaguchi, S. Watanabe, and S. Nakahama, 1989, Emulsion polymerization of styrene using phospholipids as emulsifier. Immobilization of phospholipids on the latex surface, Makromol. Chem., 190: 1195.

68. M. C. Cleij, M. F. M. Roks, and R. J. M. Nolte, 1987, Effect of polymerization on properties of vesicles derived from isocyano surfactant, Polymer Prepr., 28(2): 432.

69. M. F. M. Roks, R. S. Dezentje, V. E. M. Kaats-Richters, W. Drenth, J. Verkleij, R. J. M. Nolte, 1987, Synthesis and characterization of vesicles stabilized by polymerization of isocyano functions, Macromolecules, 20: 920.

70. C. M. Gupta, C. C. Costello, and H. G. Khorana, 1979, Site of intermolecular crosslinking of fatty acyl chains in phospholipids carrying a photoactivable carbene precursor, Proc. Natl. Acad. Sci. USA, 76: 3139.

71. M. L. Tsirenina, T. N. Simonova, N. A. Koltovaya, E. F. Golubeva, and A. N. Ushakov, 1981, A study of lipid-lipid and lipid-protein interactions in membranes using phospholipids containing photoreactive groupings. Synthesis of new photoreactive phosphatidylcholines, Soviet J. Bioorg. Chem., 7: 671.

72. C. M. Gupta, R. Radhakrishnan, G. E. Gerber, W. L. Olsen, S. L. Quay, and H. G. Khorana, 1979, Intermolecular crosslinking of fatty acyl chains in phospholipids:Use of photocleavable carbene precursors, Proc. Natl. Acad. sci., U.S.A., 76: 2595.

73. C. M. Gupta, R. Radhakrishnan, and H. G. Khorana, 1977, Glycerophospholipid synthesis: Improved general method and new analogues containing photoactivable groups, Proc. Natl. Acad. Sci., U.S.A., 74: 4315.

74. Y. G. Molotkovskii, D. A. Dergousov, and L. D. Bergel'son, 1988, New type of polymerizable phosphatidylcholines: Synthesis and properties, Soviet J. Biorg. Chem., 14: 849.

75. K. A. Aliev, H. Ringsdorf, B. Schlarb, and K. H. Leister, 1984, Liposome in net: spontaneous polymerization of 4-vinylpyridine on acidic liposomal surfaces, Makromol. Chem., Rapid Commun., 5: 345.

76. N. Higashi, T. Adachi, and M. Niwa, 1988, Molecular weight control of photopolymerization at an oriented bilayer surface using phase separation of fluorocarbon- and hydrocarbon-amphiphiles, J. Chem. Soc., Chem. Commun., 1573.

77. H. Ringsdorf, and B. Schlarb, 1986, Liposomes in a net from lipids with ionically or covalently bound polymerizable headgroups, Polymer Prepr., 27(2): 195.

78. S. L. Regen, J. S. Shin, J. F. Hainfeld, and J. S. Wall, 1984, Ghost vesicles, J. Amer. Chem. Soc., 106: 5756.

79. H. Fukuda, T. Diem, J. Stefely, F. J. Kezdey, and S. L. Regen, 1986, Polymer-encased vesicles derived from dioctadecyldimethylammonium methacrylate, 108: 2321.

80. R. Mehta, M. J. Hsu, R. L. Juliano, H. J. Krause, and S. L. Regen, 1986, Polymerized phospholipid vesicles containing amphotericin B: Evaluation of toxic and antifungal activities in vitro, J. Pharm. Sci., 75: 579.

81. H. J. Krause, R. L. Juliano, and S. Regen, 1987, In vivo behavior of polymerized lipid vesicles, J. Pharm. Sci., 76: 1.

82. F. Bonte, M. J. Hsu, A. Papp, K. Wu, S. L. Regen, and R. L. Juliano, 1987, Interaction of polymerizable phosphatidylcholine with blood components: Relevance to biocompatibility, Biochim. Biophys. Acta, 900: 1.

83. F. S. Ligler, T. L. Fare, K. D. Seib, J. W. Smuda, A. Singh, M. E. Ayers, A. W. Dalziel, and P. Yager, 1988, Fabrication of key components of receptor based biosensor, Med. Inst., 22: 247.

84. J.M. Schnur, P.Yager, R. Price, J. M. Calvert, P.E. Schoen, and J. H. Georger, Metal clad lipid microstructures, U.S.Patent #4,911,981, March 27, 1990.

85. Y. Nagata, A. Akimoto, Y. Muneda, A. Miyamoto, and F. Shichino, 1987, Preparation of mixed acid polymerizable phospholipid derivatives, Jpn. Kokai Tokkyo Koho JP 62 081,394 14 April, 10 pp.

86. Y. Noguchi, O. Nakachi, Polymerizable glycerophospholipids, Jpn. Kokai Tokkyo Koho JP 61129190, June 17, 1986, 6 pp.

87. H. Eibl, and A. Nicksh, Polymerizable phospholipids, Ger. Offen. DE 3010185, September 24, 1981, 35 pp.

88. N. Hasegawa, K. Ejima, Y. Matsushita, and H. Tsuchida, Polymerizable phosphatidylcholines. Jpn. Kokai Tokkyo Koho JP 61000091, January 6, 1986, 10 pp.

89. T. Nakaya, M. Yasuzawa, and M. Imoto, Polymerizable phospholipids, Jpn. Kokai Tokkyo Koho JP 61205291, September 11, 1986 Showa, 7 pp.

90. K. Suzuki, and H. Yoshioka, Polymerizable liposome-forming lipid and its use. Eur. Pat. Appl. EP 186211, July 2, 1986, 22 pp.

91. H. Tsuchida, K. Ejima, Y. Matsushita, and H. Tsuchida, Polymerizable phospholipids, Jpn. Kokai Tokkyo Koho JP 61129192, June 17, 1986, 8 pp.

92. H. Ono, T. Takahashi, and H. Tsuchida, Manufacture of liposomes containing polymerized phospholipids, Jpn. Kokai Tokkyo Koho JP 63232841, September 28, 1988, 6 pp.

93. K. Ejima, E. Hasegawa, Y. Matsushita, H., Nishide, and E. Tsuchida, Preparation of polymerizable -omega -imidazole-1-ylalkadienoates and their use for preparation of polymerized liposome-embedded heme-imidazole complexes. Jpn. Kokai Tokkyo Koho JP 62292762, December 19, 1987, 7 pp.

94. D. Chapman, 1981, Polymerizable phospholipids and polymers, their use in coating substrates and forming liposomes and the resulting coated substrates and liposome compositions. Eur. Pat. Appl. EP 32622, July 29, 38 pp.

95. S. L. Regen, 1985, Assembling multilayers of polymerizable surfactant on a surface of a solid material, Eur. Pat. Appl. EP 153133, 28 Aug, 27 pp.

96. S. L. Regen, 1985, Assembling multilayers of polymerizable surfactant on a surface of a solid material, US Patent 4,560,599, December 24, 16 pp.

97. T. Shigehara, M. Takane, and H. Tsuchida, Polymerizable phosphatidylcholines, Jpn. Kokai Tokkyo Koho JP 60214794, October 28, 1985, 8 pp.

98. P. E. Schoen, P. Yager, and J. M. Schnur, Lipid tubules, U. S. Pat. Appl. US 852596, October 24, 1986, 24 pp.

99. K. D. Schmitt, Polymerizable surfactants for permeability control in water flooding. U.S. Patent 4582137, April 15, 1986, 6 pp.

100. H. Ono, K. Ukaji, and H. Tsuchida, Porous liposomes for enzyme immobilization, Jpn. Kokai Tokkyo Koho JP 62,104,844, May 15, 1987, 6 pp.

101. R. L. Juliano, M. Hsu, D. Peterson, S. L. Regen, and A. Singh, 1983, Interaction of conventional or photopolymerized liposomes with platelets in vitro, Exp. Cell. Res., 146: 422.

102. R. L. Juliano, S. L. Regen, M. Singh, M. J. Hsu, and A. Singh, 1983, Stability properties of photopolymerized liposomes, Biotechnology, 1: 882.

103. A. Sadownik, J. Stefely, and S. L. Regen, 1986, Polymerized liposomes formed under extremely mild conditions, J. Amer. Chem. Soc., 108: 7789.

104. K. Dorn, E. V. Patton, R. T. Klingbiel, D. F. O'Brien, and H. Ringsdorf, 1983, Molecular weight of polymers from methacryloyl lipids in bilayer membranes, Makromol. Chem. Rapid Commun., 4:513.

105. E. Hasegawa, K. Ejima, Y. Matsushita, and H. Tsuchida, Polymerizable phospholipids, Kokai Tokkyo Koho JP 61178996, August 11, 1986, 8 pp.

106. H. Yoshioka, and K. Suzuki, Radiation - sensitive polymerizable coating material containing eleostearate residues for testing surfaces to make them biocompatible, Eur. Pat. Appl. EP 245,799, November 19, 1987, 35 pp.

107. Toyo Soda Mfg. Co., Ltd., Manufacture of polymerizable phospholipids, Jpn. Kokai Tokkyo Koho JP 60067489, April 17, 1985, 5 pp.

108. K. Hirotake, S. Kobayashi, H. Matsumura, H. Yokoyama, M. Aizawa, and Y. Katayama, Preparation of stabilized liposomes or micelles for use in agglutination tests, Jpn. Kokai Tokkyo Koho JP 63,274,870, Nov. 11, 1988, 8 pp.

109. S. Takeoda, H. Ohno, N. Hayashi, and E. Tsuchida, 1989, Control of release of encapsulated molecules from polymerized mixed liposomes induced by physical or chemical stimuli, J. Controlled Release, 9: 177.

110. P. L. Ahl, A. Singh, R. Price, J. Schmuda, and B. P. Gaber, 1990, Insertion of bacteriorhodopsin into polymerized diacetylenic phosphatidylcholine bilayers, Biochim. Biophys. Acta, 1028: 141.

111. A. W. Dalziel, J. Georger, R. R. Price, A. Singh, and P. Yager, 1986, Progress report on the fabrication of an acetylcholine receptor-based biosensor, in "Membrane Proteins", S. C. Goheen, ed., Bio-Rad Laboratories Pub., p. 643.

112. T. L. Fare, A. Singh, K. D. Seib, J. W. Smuda, P. L. Ahl, F. S. Ligler, and J. M. Schnur, 1989, Incorporation of ion channels in polymerized membranes and fabrication of a biosensor, in "Molecular Electronics", F. T. Hong, ed., Plenum Publishing Corporation, New York, 305.

113. R. Pabst, H. Ringsdorf, H. Koch, and K. Dose, 1983, Light-driven proton transport of bacteriorhodopsin incorporated into a long-term stable liposome of a polymerizable sulfolipid, FEBS Lett., 154: 5.

114. D. W. Grainger, J. Sunamoto, K. Akiyoshi, M. Goto, and K. Knutson, 1992, Langmuir, 8: 2479.

115. B. Tieke, G. Lieser, and G. Wegner, 1979, Polymerization of diacetylene in multilayers, J. Polym. Sci., Polym. Chem. Ed., 17:1631.

116. R. H. Baughman, 1972, Solid state polymerization of diacetylenes, J. Appl. Phys., 43:4362.

117. M. Pons, D. S. Johnston, and D. Chapman, 1982, A study of the spectra of diacetylenic phospholipid polymers in solvents and dispersions, J. Polym. Sci.: Polym. Chem. Ed., 20: 513.

118. M. Pons, C. Villaverde, and D. Chapman, 1983, A ^{13}C-NMR study of 10,12-tricosadiynoic acid and the corresponding phospholipid and phospholipid polymer, Biochim. Biophys. Acta, 730: 306.

119. E. Lopez, D. F. O'Brien, and T. H. Whitesides, 1982, Structural effects on the photopolymerization of bilayer membranes, J. Amer. Chem. Soc., 104: 305.

120. J. Leaver, A. Alonso, A. A. Durrani, and D. Chapman, 1983, The physical properties and photopolymerization of diacetylene-containing phospholipid liposomes, Biochim. Biophys. Acta, 732: 210.

121. E. L. Chang, 1985, Magnetically localizable polymerized lipid vesicles containing pharmaceuticals, and a method of releasing them, U. S. Patent Appl., US 714711, August 16, 17 pp.

122. A. Singh, R. B. Thompson and J. M. Schnur, 1986, Reversible thermochromism in photopolymerized phosphatidylcholine vesicles, J. Amer. Chem. Soc., 108:2785.

123. A. Singh, and B. P. Gaber, 1988, Influence of short chain lipid spacers on the properties of diacetylenic phosphatidylcholine Bilayers in "Applied bioactive polymeric materials", C.G. Gebelein, C.E. Carraher, Jr. and V.R. Foster, eds., Plenum Press, New York, 239.

124. D. G. Rhodes, and A. Singh, 1991, Syructure ofpolymerizable lipid bilayers IV. mixtures of long chain diacetylenic and short chain saturated phosphatidylcholines and analogous asymmetric isomers, Chem. Phys. Lipids, 58:215.

125. J. A. Hayward, M. L. Daniel, N. Lawrence, R. S. Sanford, D. S. Johnston, and D. Chapman, 1985, Polymerized liposomes as stable oxygen-carriers, FEBS Letters, 187: 261.

126. J. A. Hayward, D. S. Johnston, and D. Chapman, 1985, Polymeric phospholipids as new biomaterials, Ann. NY Acad. Sci., 446: 267.

127. B. Hall, R. le R. Bird, and D. Chapman, 1989, Phospholipid polymers and new haemocompatible materials, Die Angew. Makromol. Chem., 166/167: 169.

128. R. Bueschl, B. Hupfer, and H. Ringsdorf, 1982, Mixed monolayers and liposomes from natural and polymerizable lipids, Makromol. Chem., Rapid Commun., 3: 588.

129. N. Seki, E. Tsuchida, K. Ukaji, T. Sekiya, and Y. Nozawa, 1985, Phase separation of polymerized lipids in hybrid liposomes, Polymer Bull., 13: 489.

130. N. Nakashima, S. Asakuma, and T. Kunitake, 1985, Optical microscopic study of helical superstructures of chiral bilayer membranes, J. Amer. Chem. Soc., 107: 509.

131. J.-H. Fuhrhop, P. Scneider, E. Boekema, and W. Helfrich, 1988, Lipid bilayer fibers from diastereomeric and enantiomeric N-octylaldonamides, J. Amer. Chem.Soc., 110: 2861.

132. K. Yamada, H. Ihara, T. Ide, T. Fukumoto, and C. Hirayama, 1984, Formation of helical super structure from single-walled bilayers by amphiphiles with oligo-L-glutamic acid head group, Chem. Lett., 1713.

133. W. Helfrich, 1986, Helical bilayer structures due to spontaneous torsion of the edges, J. Chem. Phys.,85: 1085.

134. P. G. De Gennes, 1987, Surface and interphase physics. Electrostatic buckling of chiral lipid bilayers, C. R. Acad. Sc. Paris, 304: 259.

135. J. S. Chappell, and P. Yager, 1991, A model for crystalline order within helical and tubular structures of chiral bilayers, Chem. Phys. lipids, 58:253.

136. A. Singh, P. E. Schoen, and J. M. Schnur, 1988, Self-assembled microstructures from a polymerizable ammonium surfactant: di (hexacosa-12,14-diynyl) dimethylammonium bromide, J. Chem. Soc., Chem. Commun., (18): 1222.

137. A. Singh, B. P. Singh, B. P. Gaber, B. Herendeen, R. Price, T. G. Burke, P. E. Schoen, J. M. Schnur, and P. Yager, 1989, Synthesis and characterization of positional isomers of 1,2 bis heptacosadiynoyl phosphatidylcholine, Surfactants in Solution (K. L. Mittal, Ed.), Plenum press, New York, 8: 467.

138. A. Singh, T. G. Burke, J. M. Calvert, J. H. Georger, B. Herendeen, R. R. Price, P. E. Schoen, and P. Yager, 1988, Lateral phase separation based on chirality in a polymerizable lipid and its influence on formation of tubular microstructures, Chem. Phys. Lipids, 47: 135.

139. P. Yager, R. R. Price, J. M. Schnur, P. E. Schoen, A. Singh, and D. G. Rhodes, 1988, The mechanism of formation of lipid tubules from liposomes, Chem. Phys. Lipids, 46: 171.

140. P. Schoen, P. Yager, J. P. Sheridan, R. R. Price, J. M. Schnur, A. Singh, D. G. Rhodes, and S. L. Blechner, 1987, Order in diacetylenic microstructures, Mol. Cryst. Liq. Cryst., 153: 357.

141. P. Yager, P. E. Schoen, J. H., Georger, R. R. Price, and A. Singh, 1986, Two mechanisms for forming novel tubular microstructures from polymerizable lipids, Biophys. J., 49: 320a.

142. A. L. Plant, D. M. Benson, and G. L. Trusty, 1990, Probing the structure of diacetylenic phospholipid tubules with fluorescent lipophiles, Biophys. J., 57: 925.

143. P. Yager, P. E. Schoen, C. Davies, R. Price, and A. Singh, 1985, Structure of lipid tubules formed from a polymerizable lecithin, Biophys. J., 48: 899.

144. J. H. Georger, A. Singh, R. R. Price, J. M. Schnur, P. Yager, and P. E. Schoen, 1987, Helical and tubular microstructures formed by polymerizable phosphatidylcholines, J. Amer. Chem. Soc., 109: 6169.

145. A. S. Rudolph, J. M. Calvert, M. E. Ayers, and J. M. Schnur, 1989, Water-free self-assembly of phospholipid tubules, J. Amer. Chem. Soc., 111: 8516.

146. J. M. Schnur, R. R. Price, P. Yager, P. Schoen, J. H. Georger, and A. Singh, 1989, Process for fabrication of lipid microstructures, U.S. Patent # 4,877,501, Oct. 31.

147. T. G. Burke, A. S. Rudolph, R. R. Price, J. P. Sheridan, A. W. Dalziel, A. Singh, and P. E. Schoen, 1988, Differential scanning calorimetric study of the thermotropic phase behavior of a polymerizable, tubule-forming lipid, Chem. Phys. Lipids, 48: 215.

148. A. Singh, Li-I Tsao, and M.A. Markowitz, 1993, Modulation of bilayer structures derived from diacetylenic phosphocholines containing oxygen linker beta to diacetylene, Chem. Phys. lipids, 63:0000.

149. A. Singh, and J. M. Schnur, 1985, Polymerized diacetylenic phosphatidylcholine vesicles: synthesis and characterization, <u>Polymer Preprints</u>, <u>26(2)</u>:184.

150. A. S. Rudolph, B. R. Ratna, and B. Kahn, 1991, Self-assembling phospholipid filaments, <u>Nature</u>, <u>352</u>:52.

151. A. S. Rudolph, B. P. Singh, A. Singh, and T. G. Burke, 1988, Phase characteristics of positional isomers of 1,2 bis heptacosadiynoyl-sn-glycero-3-phosphocholines, <u>Biochim. Biophys. Acta</u>, <u>943</u>: 454.

152. A. S. Rudolph, and T. G. Burke, 1987, A Fourier-transform infrared spectroscopic study of the polymorphic phase behavior of 1,2-bis (tricosa-10,12-diynoyl) -sn-glycero-3-phosphocholine, a polymerizable lipid which forms novel microstructures, <u>Biochim. Biophys. Acta</u>, <u>902</u>: 349.

153. A. S. Rudolph, P. E., Schoen, M. Nagumo, F. Behroozi, T. G. Burke, M.E. Ayers, A. Singh, and R. Treanor, 1989, Spectroscopic studies of tubule-forming polymerizable lecithins, <u>SPIE</u>, <u>1057</u>: 57.

154. J. P. Sheridan, 1988, Conformational order in lipid tubules formed from a diacetylenic Lecithin: A Raman spectroscopic study, <u>NRL Memorandum Report 5975</u>, Published by the Naval Research Laboratory, Washington, D.C.

155. D. G. Rhodes, S. L. Blechner, P. Yager, P. E. Schoen, 1988, Structure of polymerizable lipid bilayers.I. 1,2-Bis (10,12-tricosadiynoyl) -sn- glycero -3- phosphocholine, a tubule-forming phosphatidylcholine, <u>Chem. Phys. Lipids</u>, <u>49</u>: 39.

156. R Shashidhar, and Joel Schnur, 1993, Self-assembling tubules from phospholipids in "ACS symposium series", A.J. Bard ed., American Chemical Society, Washington D.C. in press.

157. M. H. Lu, C. Rosenblatt, and P. Yager, 1993, Influence of pH on the precursors of phospholipid tubules in methanolic solution, <u>Chem. phys. Lipids,</u> in press.

158. S. L. Blechner, W. Morris, P. E. Schoen, P. Yager, A. Singh, and D. G. Rhodes, 1991, Structure of polymerizable lipid bilayers III: two heptcosadiynoyl phosphocholine isomers, <u>Chem. Phys. lipids</u>, <u>58</u>:41.

159. J. B. Lando, and R. V. Sudiwala, 1990, Structural investigations of Langmuir-Blodgette films and tubules of 1,2 bis (10,12-tricosadiynoyl)-*sn*-glycero-3-phosphocholine ($DC_{8,9}PC$) using electron diffraction techniques, <u>Chem. matr.</u>, <u>2</u>:594.

160. A. Singh, and S. Marchywka, 1989, Synthesis and characterization of headgroup modified 1,3 diacetylenic phospholipids, Polym. Mat. Sci. Eng., 61: 675.

161. M. Markowitz, and A. Singh, 1991, Self-assembling properties of 1,2-diacyl-sn-glycero-3-phosphohydroxyethanol: A headgroup modified diacetylenic phospholipid, Langmuir, 7:16.

162. M. A. Markowitz, J. M. Schnur, and A. Singh, 1992, The influence of polar headgroups of acidic diacetyelnic phospholipids on tubule formation, microstructure morphology and Langmuir film behavior, Chem. Phys. Lipids, 62:193.

163. M. A. Markowitz, S. Baral, S. Brandow, and A. Singh, 1993, Palladium ion assisted formation and metallization of lipid tubules, Thin Solid Films, 0000.

164. R. R. Price, and m. Patchan, 1991, Controlled release from cylindrical microstructures, J. Microencapsulation, 8:301.

165. T. G. Burke, A. Singh, and P. Yager, 1987, Entrapment of 6-carboxyfluorescein within cylindrical phospholipid microstructures, Ann. N. Y. Acad. Sci., 507: 330.

166. B. J. Spargo, G. E. Stillwell, R. O. Cliff, R. L. Monroy, F. M. Rollwagen, and A. S. Rudolph, 1992, Tissue inducing bio aterials, Proc. Mat. Res. Soc.,000

167. C. Rosenblatt, P. Yager, and P. E. Schoen, 1987, Orientation of lipid tubules by a magnetic field, Biophys. J., 52: 295.

168. F. Behroozi, M. Orman, W. Stockton, J. Calvert, F. Rochford, P. Schoen, 1990, Interaction of metallized tubules with electromagnetic radiations, J. Appl. Phys., 58:3688.

169. D. A. Kirkpatrick, R. Price, and J. M. Schnur, 1989, Measurement of vacuum field emission from biomolecular and semiconductor metal eutectic composite microstructures, IEEE Trans. Plasma Science,19:749.

CELLULAR ADHESION TO SOLID SURFACE:
EFFECT OF THE PRESENCE OF CATIONIC
ELECTROLYTES IN THE SUSPENSION MEDIUM

You-Im Chang[1] and Jyh-Ping Hsu[2]

[1]Dept. of Chemical Engineering
Tunghai Univeristy, Taichung, Taiwan 40704
[2]Dept. of Chemical Engineering
National Taiwan University, Taipei, Taiwan 10764

INTRODUCTION

The relationship between defective leucocytes and the incidence of inflammation has been increasingly explored since Louis Pasteur and Robert Koch. It is generally accepted that during the adhesion of leucocytes to the inflammatory sites of vascular tissue, in order to achieve an intimate contact, leucocytes must overcome a substantial energy barrier established by the interaction potential between the surfaces of leucocyte and tissue. This interaction potential comprises the van der Waals attraction and electrical double-layer interactions [1]. By assuming that the Brownian motion of cells over the energy barrier of the interaction potential is the rate-determining step, Ruckenstein and Prieve[2] show that the rate of adhesion is expotentially sensitive to the height of the barrier. According to the classical DLVO theory [3], the height of this barrier depends on the Hamaker constant, ionic strength of the suspension medium, and the surface potentials of the two interacting surfaces. In general, if the Hamaker constant remains unchanged, the height of the energy barrier and the electrostatic repulsion force can be lowered either by increasing the ionic strength of the suspension medium or by decreasing the surface potential of cells through a specific cationic adsorption. In these cases, the rate of adhesion of cells is increased.

As pointed out by Ninham and Parsegian [4], since the DLVO theory assumes that the interacting surfaces are fixed at either constant surface charge or constant surface potential, it is inappropriate for the description of cell adhesion to surfaces. This is beacuse the condition of cell surface depends upon the degrees of dissociation of the ionizable groups on the surface, which in turn, is affected by the property of the bulk fluid. Since the pioneer work of Ninham and Parsegian[4], several attempts have been made to investigate the adhesion of charge-regulated cells to surfaces [5,6]. In a recent study, Chang [7] examined the effect of the presence of multi-valent cations on the electrostatic double-layer force between a cell bearing with ionogenic groups and a solid surface with either constant potential or constant charge. It is found that the presence of cations in the suspension medium (e.g., Ca^{+2} or Fe^{+3} in plasma) reduces the replusive force between cells and solid surface only if the separation distance between them is greater than some critical value. If the separation distance is smaller than this critical value, the repulsive force is greater than that if the cations are absent. It is suggested that the existence of the critical separation distance may be due to the increase of osmotic pressure caused by the overloaded cations in the interaction region when a cell is near the surface[7,8].

In this study, the effect of the presence of cationic electrolytes in the suspension medium on cell adhesion is examined. Here, the kinetic model proposed by Ruckenstein and Prieve [2] is extended to the case that cell surface bears with ionogenic groups. In other word, the charge of the surface of a cell is continuously regulated as it approaching the solid surface. The effects of Hamaker constant and the ionic strength of the suspension medium on adhesion are discussed.

THEORETICAL FORMULISM

Adhesion Time Constant

Ruckenstein and Prieve [2] assume that the adhesion of cells to a solid surface occurring in a stagnant fluid consists of two processes in series: (i) cells move down to a region near the surface by the gravitational force, and (ii) diffusion of these cells over a potential barrier to the surface. If the second process is the rate-determining one, the rate of adhesion of cells, characterized by the adhesion time constant τ, is calculated by

$$\tau = \int_0^\delta \frac{\exp\left(\frac{\phi}{kT}\right)}{D(x)} dx \int_0^L \exp\left(-\frac{\phi}{kT}\right) dx \tag{1}$$

The adhesion time constant is defined as the time required to reduce the number of cells in the suspending medium to 1/e of the original amount. The greater its value, the slower the adhesion. In eq. (1), x, L, and δ are, respectively, the minimum distance between cell surface and the solid surface, the height of liquid above the solid surface, and the smallest value of x for which cells can be reached by the first process, k denotes the Boltzmann constant, T is the absolute temperature, ϕ represents the interaction potential between cell and solid surface, and D(x) is the local diffusion coefficient of cells. The value of D(x) is evaluated by

$$
\begin{aligned}
D(x) &= m(x)\, kT \\
&= \frac{f_1\left(\dfrac{x}{a_p}\right)}{6\,\pi\,\mu\,a_p} kT
\end{aligned}
\tag{2}
$$

where m(x) is the cellular mobility and μ and a_p are respectively, the viscosity of liquid, and the radius of cell. The hydrodynamic retardation factor $f_1(x/a_p)$ is estimated by the procedure suggested by Brenner[9]. If $x/a_p \ll 1$, the value of $f_1(x/a_p)$ tends to x/a_p; on the other hand, if $x/a_p \gg 1$, it approaches unity. In the case ϕ has a deep secondary minimum, eq.(1) reduces to

$$
\tau \cong \frac{1}{D(x_{max})} \cdot \frac{2\pi kT}{(\gamma_{max}\cdot\gamma_{min})^{\frac{1}{2}}} \exp([(\phi_{max}-\phi_{min})]
\tag{3}
$$

where

$$
\phi_{min} = \phi(x_{min})\,,\quad \gamma_{min} = \left.\frac{d^2\phi}{dx^2}\right|_{x=x_{min}}
$$

$$
\phi_{max} = \phi(x_{max})\,,\quad \gamma_{max} = -\left.\frac{d^2\phi}{dx^2}\right|_{x=x_{max}}
$$

On the other hand, if ϕ has a shallow secondary minimum, only the contribution due to gravity is significant for $x > \delta$. In this case,

$$
\phi(x) \cong \frac{4}{3}\pi a^3{}_p(\rho_p - \rho_f)g \cdot x
\tag{4}
$$

Equation (1) becomes

$$\tau \cong \frac{3kT \cdot \left(\frac{2\pi kT}{\gamma_{max}}\right)^{\frac{1}{2}}}{4a_p^3 \,(\rho_p - \rho_f)\, g\, D(x_{max})} \, \exp\left(\frac{\phi_{max}}{kT}\right) \tag{5}$$

Equations (3) and (5) reveal that the adhesion time constant is exponentially sensitive to the height of ϕ_{max} and the depth of ϕ_{min}.

Interaction Energy Barrier

According to the DLVO theory, $\phi(x)$ in eq.(1) is the sum of the unretarded van der Waals potential $\phi_{vdw}(x)$ and the electrostatic repulsive potential between cell and surface $\phi_{DL}(x)$. That is

$$\phi(x) = \phi_{vdw}(x) + \phi_{DL}(x) \tag{6}$$

The variation of $\phi_{vdw}(x)$ as a function of x is[10]

$$\phi_{vdw}(x) = \frac{Ad}{3}\left[\frac{1}{2}\ln\left(\frac{x+a_p}{x}\right) - \frac{a_p(x+a_p)}{x(x+2a_p)}\right] \tag{7}$$

where $Ad = A/kT$, and A is the Hamaker constant. Derjaguin [11] suggests that $\phi_{DL}(x)$ can be evaluated by

$$\phi_{DL}(x) = 2\pi a_p \int_x^\infty \int_{x'}^\infty F(x')\, dx'dx \tag{8}$$

where $F(x')$ is the repulsive force per unit area exerted between two interacting surfaces separated by a distance x'. Suppose that the approaching of cells to the solid surface is sufficiently slow so the electrochemical equilibrium is always maintained during the course of adhesion. Then the value of F can be calculated by [4]

$$\frac{F}{nkT} = -\left[\frac{\partial\left(\frac{\psi}{kT}\right)}{\partial\left(\frac{x}{a_p}\right)}\right]^2 \frac{1}{(\kappa a_p)^2} + \left[\left(\exp\left(\frac{\psi}{kT}\right) - 1\right)\right] + (1-\eta)\left[\left(\exp\left(\frac{\psi}{kT}\right) - 1\right) + \frac{\eta}{q}\left(\exp\left(\frac{-q\psi}{kT}\right) - 1\right)\right] \tag{9}$$

where ψ is the charge regulated potential on the cellular surface, κ is the reciprocal of Debye-Huckel length, n is the ionic strength of the suspension medium, η is the fraction of cationic electrolytes present in the suspension medium, and q is the valence for cations. The first term on the right-hand side of eq.(9) represents the contribution due to the Maxwell stress, and the second term denotes the contribution due to the osmotic pressure caused by the presence of cations in the suspension medium[12]. The variation of ψ as a function of x is described by the following Poisson-Boltzmann equation[4,5]

$$\frac{d^2\left(\frac{\psi}{kT}\right)}{d\left(\frac{x}{a_p}\right)^2} = \frac{(\kappa a_p)^2}{2}\left[\exp\left(\frac{\psi}{kT}\right) - (1-\eta)\cdot\exp\left(-\frac{\psi}{kT}\right) - \eta\cdot\exp\left(-\frac{q\psi}{kT}\right)\right] \qquad (10)$$

The boundary conditions associated with this equation are

$$\frac{d\left(\frac{\psi}{kT}\right)}{d\left(\frac{x}{a_p}\right)} = \frac{\kappa^2 a_p}{2n}\left[\frac{[H^+]_s}{K_b+[H^+]_s}\cdot S_b^{-1} - \frac{K_a}{K_a+[H^+]_s}\cdot S_a^{-1}\right] \qquad (11)$$

with

$$[H^+]_s = [H^+]_r\cdot\exp\left(-\frac{\psi}{kT}\right)$$

on the cellular side, and

$$\left.\frac{\psi}{kT}\right|_{x=0} = -1 \qquad (12)$$

on the solid surface. In these expressions $[H^+]_s, [H^+]_r, K_a, K_b, S_a$ and S_b are the concentration of hydrogen ions on the cell surface, the concentration of hydrogen ions in the suspension medium, the dissociation equilibrium constant for acid groups on cell surface, the dissociation equilibrium constant for base groups on cell surface, the reciprocal of the density of acidic groups on cell surface and the reciprocal of the density of base groups on cell surface respectively.

NUMERICAL SIMULATION

The behavior of cell adhesion is investigated numerically on the basis of the above discussion. The value of $\psi(x)$ is obtained by solving eq.(10) numerically, subject to the boundary conditions shown in eqs.(11) and (12). $\phi(x)$ is then calculated by substituting eqs. (7), (8) into eq.(6). The value of τ is estimated either by eq.(3) or by eq.(5), depending on the nature of $\phi(x)$. If $\phi(\delta)-\phi(x_{min})>>kT$ and $-\phi(x_{min})>>kT$, eq.(3) is used; if $-\phi_{min}<<kT$ and $\phi_{max}>>kT$, eq.(5) is adopted. For illustration, the adhesion of sheep leucocytes to a solid surface of constant potential is estimated. The relevant surface properties of sheep leucocytes can be found in Prieve and Ruckenstein[6], and a brief summary is given in Table 1.

Table 1. Surface properties of sheep leucocytes[6]

n(M)	pK_a	pK_b	S_a^{-1} (10^{13} groups cm^{-2})	S_b^{-1} (10^{13} groups cm^{-2})
0.145	2.7	≥ 12	0.70	0.00
0.00145	3.4	≥ 12	0.61	0.40

It is worth noting that the nature of the surface of leucocytes is dependent on the ionic strength of the suspension medium. It is suggested that, due to hemolysis, there may be some structural changes in the surface of leucocytes at low ionic strength (n=0.00145M). In the numerical simulation, the value of q is assumed as 2 (divalent cations) and the pH of the suspension medium is assumed as 7.

RESULTS AND DISCUSSION

Two levels of ionic strength of the suspension medium are examined: n=0.00145 M and n=0.145 M.

Low Ionic Strength (n=0.00145 M)

The variation of the dimensionless electrostatic interaction force between cell and solid surface, F/nkT, as a function of the dimensionless separation distance, x/a_p, at n = 0.00145 M is shown in Fig.1. In the case the cations are absent (q=1 and η =0.00), the electrostatic repulsion force increases with the decrease of the separation distance, passes through a maximum, and then decreases rapidly with the further decrease of the separation distance. If the separation distance is small enough, the contribution to the interaction force between cell and surface due to the Maxwell stress is less significant than that due to the osmotic pressure. Equation (9) indicates that the electrostatic force is negative. This means that cell will experience an attractive force. In the case there is a small amount of cations present in the suspension medium (q=2 and η=0.05 in Fig. 1), the presence of cations in the suspension medium has the effect of decreasing the repulsion force if the separation distance is greater than a critical value (x/a_p>3.4$\times$10^{-3}). However, the separation distance is less than this critical value (if x/a_p<3.4$\times$10^{-3}), the presence of cations increases the replusion force between cell and solid surface. This may be due to that as the separation distance reduces, the osmotic pressure required to screen out those overloaded cations in the region

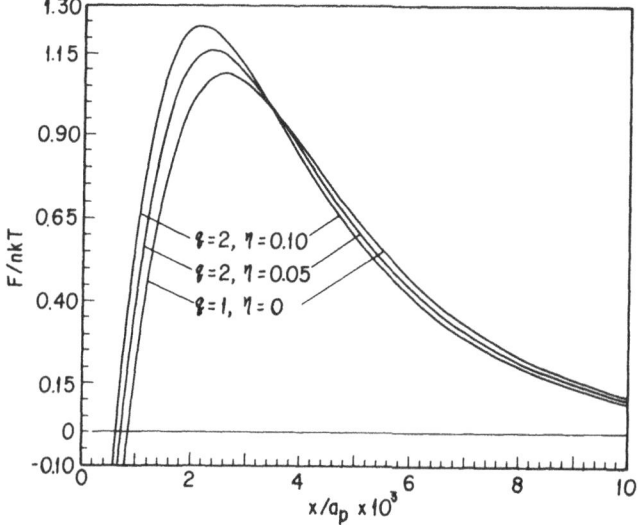

Fig. 1. Variation of the dimensionless electrostatic interaction force between cell and solid surface as a function of dimensionless separation distance at a low ionic strength n=0.00145 M.

between cell and surface when cations are present in the suspension medium is higher than that if they are absent[7,8]. The result shown in Fig.1 reveals that the higher the concentration of cations, the less the repulsive force if the separation distance is greater than the critical distance, and the greater the repulsive force if the separation distance is small than the critical distance.

Fig. 2 illustrates the dependence of adhesion time constant on Hamaker constant. As suggested by the result shown in this figure, the adhesion time constant desceases with the increase of Hamaker constant. It is worth noting that the variation of the rate of adhesion is rather sensitive to the value of Hamaker constant. This is consistent with the experimental observations[13,14]. The presence of cations in the suspension medium has the effect of increasing the adhesion time constant. The greater the concentration of cations, the greater the adhesion time constant. In other words, the presence of cations in the suspension medium has a negative effect on cell adhesion.

This is contrary to that predicated by the DLVO theory, which assumes either constant potential or constant surface charge on both surfaces.

High Ionic Strength (n=0.145 M)

The variation of the dimensionless electrostatic interaction force between cell and solid surface as a function of the demensionless separation distance between them at n=0.145 M is shown in Fig. 3. The increase of the ionic strength of the suspension

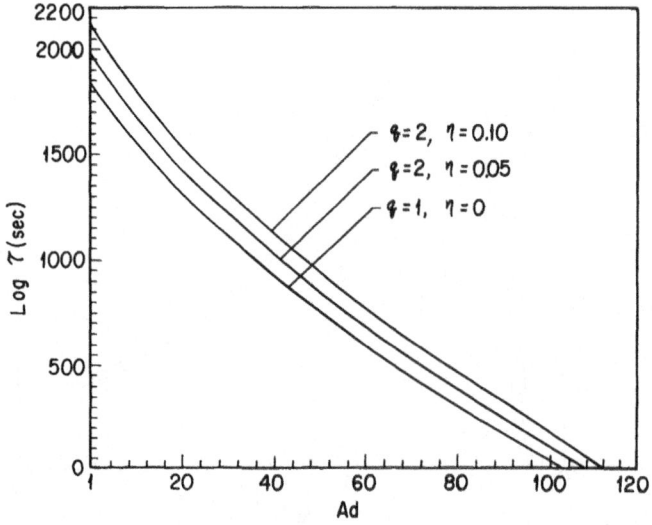

Fig. 2. Variation of the adhesion time constant as a function of Hamaker constant at a low ionic strength n = 0.00145 M.

medium has the effect of compressing the thickness of the double layer near the surfaces. As a result, the peak of the curve representing the variation of the electrostatic force shifts more closely to the solid surface than that for the case n=0.00145 M. The same conclusions as those obtained for the case of Fig. 1 can be drawn. The critical distance, however, reduces to $x/a_p = 3.0 \times 10^{-4}$.

The variation of the adhesion time constant as a function of Hamaker constant is illustrated in Fig. 4. The same conclusions as those obtained from Fig. 2 can be drawn. Note that the concentration of cations has a less significant effect on the adhesion time constant. A comparison of Figs. 2 and 4 indicates that the adhesion time constant decreases with the increase of the ionic strength. The result shown in Fig. 4 provides useful information for the prediction of the behavior of the adhesion of leucocytes to the inflammatory sites of vascular tissue. Typically, the ionic strength of physiologic saline solution is on the order of 0.145 M. Thus, an increase in the concentration of Ca^{+2} in plasma from $\eta = 0.05$ (normal condition) to $\eta = 0.10$ results in a decrease of the rate of adhesion of leucocytes to the surface of injured tissue. This is unfavorable to the subsequent phagocytosis reaction.

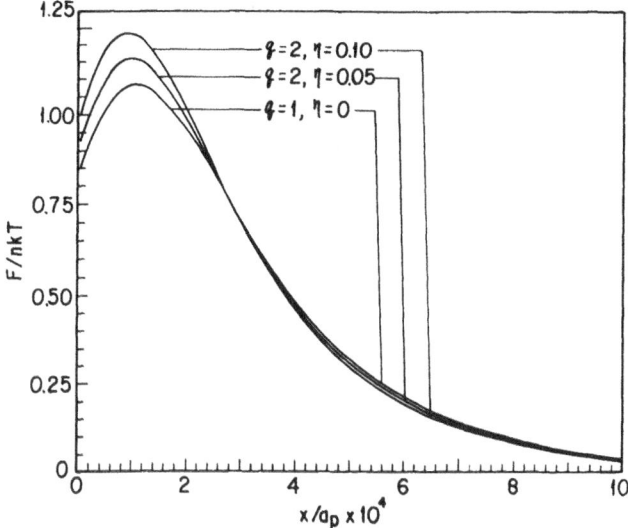

Fig. 3. Variation of the dimensionless electrostatic interaction force between cell and solid surface as a function of dimensionless separation distance at a high ionic strength n = 0.145 M.

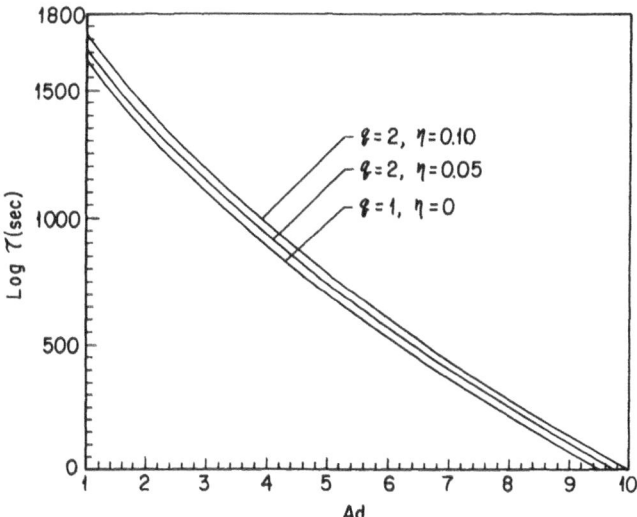

Fig. 4. Variation of the adhesion time constant as a function of Hamaker constant at a high ionic strength n = 0.145 M.

CONCLUSION

The cell adhesion model presented in this study is an extension of the theories developed by Ninham & Parsegian[4] and Ruckenstein & Prieve[2] on the interaction between two identical colloidal particles bearing with the ionogenic groups on their surfaces. Here, the surface of a cell is regulated by either dissociation or association of the ionizable groups on its surface, but the electrical potential on the substratum surface remains constant. It is found that the presence of cations in the suspension medium has a negative effect on the rate of cell adhesion. This is contrary to that predicted by the classical DLVO theory. The result of numerical simulation reveals that: (i) The effect of the presence of cations in the suspension medium on the interaction force between cell and surface depends upon the separation distance between them. If the separation distance is greater than a critical value, the repulsive force is reduces. On the other hand, if the separation distance is smaller than the critical value, the repulsive force is enhanced. For a fixed ionic strength, the rate of adhesion increases with the decreases of the concentration of cations present in the suspension medium. (ii) The effect of the presence of cations in the suspension medium on cell adhesion is pronounced at a low ionic strength. (iii) For a fixed concentration of cations, the rate of adhesion increases with the increase of ionic strength. (iv) The rate of adhesion is sensitive to the value of Hamaker constant.

ACKOWLEDGEMENT

This work is financially supported by the National Science Council of the Republic of China under project number NSC-79-0421-E002-44Z.

REFERENCE

1. G. Bitton, and K.G. Marshall, "Adsorption of micro-organisms to surface,"John Wiley Sons Publishing Corp. , New York (1980).
2. E. Ruckenstein,and D.C. Prieve, Dynamics of cell deposition on surface, J.Theor.Biol. 51:429 (1975).
3. E.J.W.Verwey, and J.Th.G. Overbeek,"Theory of the stability of lyophobic colloids," Elsevier Publishing Co., Amsterdam , Holland (1948).
4. B.W. Ninham and V.A. Parsegian, Electrostatic potential between surfaces bearing ionizable groups in ionic equilibrium with physiologic saline solution, J .Theor.Biol. 31:405(1971).
5. D. Chan, J. W. Perram, L.R. White and T.W. Healy, Regulation of surface potential at amphoteric surfaces during particle-particle interaction, J. Chem.Soc.Farad.Trans. I. 71: 1046 (1975).
6. D.C. Prieve and E. Ruckenstein, The surface potential and double-layer interaction

force between surfaces characterized by multiple ionizable groups, *J. Theor.Biol.* 56:205 (1976).

7. Y.I. Chang, The effect of cationic electrolytes on the electrostatic behavior of cellular surface with ionizable groups, *J . Theor.Biol.* 139:561 (1989).

8. D.C. Prieve and E. Ruckenstein, The double-layer interaction between dissimilar ionizable surfaces and its effect on the rate of deposition, *J. Colloid and Interface Sci.* 63:317 (1978).

9. H. Brenner, The slow motion of a sphere through a viscous fluid towards a plane surface, *Chem.Eng. Sci.* 16:242 (1961).

10. H.C. Hamaker, The London-van der Waals attraction between spherical particles *Physica.* 4:1058 (1937).

11. B.V. Derjaguin, Untersuchungen uber die reibung und adhesion, *Kolloid Z.* 69:155 (1934).

12. D.C. Prieve and E. Ruckenstein, Role of surface chemistry in particle deposition, *J. Colloid and Interface Sci.* 60:337 (1977).

13. M. Marmur, Kinetics of particle deposition from a stagnant suspension:Application to blood platelets, *J. Colloid and Interface Sci.* 106:360 (1985).

14. E. Ruckenstein and J.H. Chen, Kinetically caused saturation in the deposition of cell: Effect of saturation at the secondary minimum and of excluded area, *J. Colloid and Interface Sci.* 128:592 (1989).

APPLICATIONS OF BACTERIORHODOPSIN IN MEMBRANE MIMETIC CHEMISTRY

Mow S. Lin and Eugene T. Premuzic

Department of Applied Science
Biosystems and Process Sciences Division
Brookhaven National Laboratory
Upton, NY 11973

INTRODUCTION

The chemistry of the photoreceptors of higher animals is a topic that reaches back more than forty years. By 1970, the photoreceptors of more than four hundred species of animals had been studied. All these species have in their visual pigments a common chromophore, retinal, connected to the protein opsins through a Schiff base bond. This protein complex, called rhodopsin, with its associated lipids makes up the photosensitive membrane in photoreceptor cells.[1,2]

About the same time, in 1968, a purple-colored membrane was isolated from a halophilic bacteria Halobacterium halobium.[3] This membrane was found to be photosensitive and had a retinyl chromophore connected to a protein in much the same manner as rhodopsin in the animal world.[4] The photosensitive membrane was named bacteriorhodopsin because of its resemblance to the photoreceptive pigments of the higher animals. It also can be classified as a retinal protein, because later more retinyl-containing proteins were found in nature.[5] Follow-up studies have shown that the principal role of bacteriorhodopsin in the cell is that of a light-driven proton pump which generates a proton gradient across the membrane for ATP regeneration. Such comparative studies of bacteriorhodopsin and rhodopsin have been extensively reviewed.[5,6,7]

In recent years bacteriorhodopsin became one of the most studied membrane systems.[5,6,7,8,9,10] In this paper, we discuss recent structure and photochemical studies in terms of membrane mimetic chemistry. Recent applications will be emphasized.

Structure of Bacteriorhodopsin

Bacteriorhodopsin (BR) is usually isolated from the purple membrane of H. halobium grown in a medium with a high salt concentration (4M NaCl) and a low level of dissolved oxygen. The lipid bilayer membrane consists of a matrix of protein BR and lipids in a 3:1 ratio. The lipid portion is composed of a diether analogue of phosphatidyl glycerophosphate (50%), sulfoglycolipid (30%), carotenoids (6%), and various apolar lipids.[5]   Bacteriorhodopsin itself is composed of a retinylidene chromophore and a protein of 26,000 dalton (Figure 1).   The protein's primary structure was determined by amino-acid sequencing in 1979,[10,11] and confirmed two years later by DNA sequencing.[12]

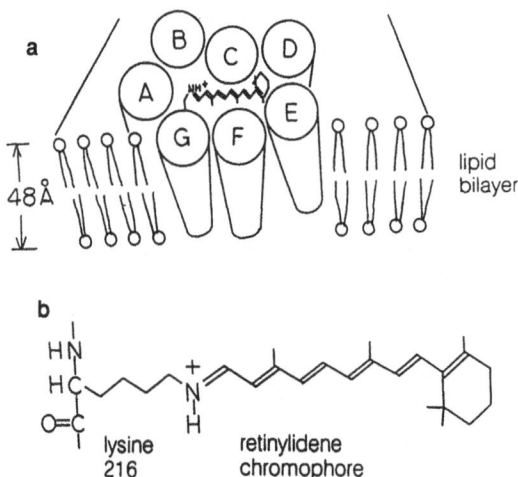

Figure 1. (a) Purple membrane of BR embedded in lipid bilayer (b) chromophore of protonated N-retinylidene-n-lysine 216 residue.

Compared to most cellular membranes, the purple membrane has a highly ordered structure and can be fused to form large two-dimensional crystals for structural studies.[13] Electron microscopy revealed the protein's backbone to be made up of seven connected helical sections spanning across the membranes, with retinyl section in the middle[13,14,15,16] (Figure 2). The seven α-helices represented by tilted rods A.B.C.D.E.F.G. are nearly perpendicular to the top cytoplasmic side and the bottom extracellular side. The chromophore retinyl section lies in the middle and nearly parallel to the membrane's two sides. It is connected to lysin (residue 216) through a Schiff base bond with a 1:1 ratio of retinyl to protein.[13]

Photochemical Cycle and Proton-Pumping Mechanism of BR

Photon absorption associated with events in ATP generation has been extensively reviewed in terms of the kinetic and spectroscopic analysis of intermediates in the photochemical cycles of BR.[8,17,18,19,20,21] Figure 3 depicts a simplified scheme based on the cycle proposed by Lozier et al.[22]

The retinyl chromophore is a polyene which can exist in several isomeric forms. At room temperature and in a light-adapted ground state, it is a protonated Schiff base in an all-trans configuration, with an absorption maximum at 570 nm. On absorption of a photon, it isomerizes within picoseconds to a K610 state which is the 13-cis isomer with an absorption maximum at 610 nm. At temperatures of 77°K or lower, the reaction is reversible; the K610 intermediate may absorb a photon and isomerize back to BR 570. At higher temperatures, thermal reactions occur so that K610 rapidly changes to the L550 state in two microseconds with an absorption maximum at 550 nm. After about forty microseconds, L550 changes to the important M412 state, accompanied by the release of a proton to the extracellular side. Then, the M412 intermediate with an absorption maximum at 412 nm picks up a proton from the cytoplasmic side and returns to its initial state within 10 milliseconds.

Figure 2. BR of retinyl chromophore setting in seven α-helices connected by β-structures. CT is carbon terminal and NT is nitrogen terminal (taken from [ref. 13] with permission).

Figure 4 shows a recently proposed scheme for the overall proton-pumping mechanism.[13] In this scheme, the proton on the Schiff base nitrogen is transferred to the carboxylate of residue 85 in the L550 to M412 reaction. During the M412 to N transition, the carboxylic proton of aspartic residue 96 is transferred to the Schiff base nitrogen, accompanied by the ejection of the carboxylic proton on residue 85 to the extracellular side. When returning to the BR 570 state, a proton is picked up from the cytoplasmic interior to re-protonate the carboxylate of aspartic residue 96. It was proposed that the proton gating from the cytoplasmic to extracellular side was mediated by hydrogen bonds within the channel.[13,17]

PREPARATIONS OF BACTERIORHODOPSIN MEMBRANE MIMETIC SYSTEMS

Highly organized biomembranes provide microenvironments for fast and stereoselective reactions. Various membrane mimetic systems have been investigated; however, the performance of the most simple systems falls short of real biomembranes. The problems common to all simple systems are a lack of stability, poor reaction rates, and selectivity.

To illustrate these shortcomings, the properties of retinal proteins and their mimetic models can be compared. The photon absorption in the visible range is of practical and theoretical interest. The common structural feature of retinal proteins, e.g., bacteriorhodopsin (λ max 570 nm) and rhodopsin (λ max 498 nm) is the protonated retinyl-lysin Schiff base chromophore. Chromophore models of retinyl Schiff base absorb maximally near 370 nm, which is shorter than the λ max of both membranes. This spectroscopic discrepancy has been an interesting topic of discussion. The observed red shift in λ max from the Schiff base models to the natural photosensitive pigments can be partially explained by a protonation hypothesis. Because protonation to the Schiff base bond can shift the λ max to 450 nm, it was proposed that counter ions from the environment around the chromophore site could affect the absorption maxima and shift it further into the red towards that of the natural pigments.[23,24,25] Indeed, protonated all-trans retinyl Schiff bases embedded in dextrin as ω-amino ethyl amino β-cyclodextrin absorb at the same λ max (497 nm) as rhodopsin.[26] These results indicate that the spectroscopic absorption maxima of protonated Schiff bases can be affected by the chemistry of the surrounding environment and the spectroscopic absorption maximum could be mimicked by providing a suitable environment for a simple model.

Other than spectroscopic properties, the main difference between retinal proteins are the diversified cellular functions which include proton pumping for bacteriorhodopsin (BR), generating neuro-signals for rhodopsin, and chloride-ion pumping for halorhodopsin (HR). To mimic these different functions requires a more complex microenvironment. For practical purposes, readily organized cellular parts would be good building blocks with which to assemble a membrane mimetic system with a reasonable lifespan. The crystalline nature of BR makes it a good candidate for a membrane mimetic medium, because it is stable against thermal, chemical, and photochemical degradation. For example, BR extracted from purple membrane can be stored for years,[27] and photo-switched between absorption states several hundred thousand times without degradation.[28]

The detailed chemistry for the preparation and application of various membrane mimetic systems can be found in several texts.[29] For the purpose of this discussion, the preparation of BR membrane mimetic systems will be briefly described next.

Fusion of Purple Membrane (FPM)

FPM was developed for electron microscopic study where large single crystals are needed.[13,14] Native purple membranes are round or oval patches of 0.5 μm in diameter. To make a larger membrane sheet, the native purple membrane was first suspended in 0.1 M potassium phosphate buffer at pH 5.2 at a concentration of 3 mg/ml. Then, the suspension was mixed with octyl glucoside (OG) and dodecyl trimethylammonium chloride (DTAC) to a final concentration of 6 mM OG and 200 μM DTAC. The mixture was incubated at 20°C for several weeks. The small patches were then fused into a large membrane sheet, 10 μm in diameter. Occasionally, sheets up to 20 μm in diameter were obtained. In the process, not all the fused sheets were annealed into a single crystal. Half of the sheets were annealed with one half of the area upside down in relation to the other half. In either case, an entwined area large enough for high-resolution electron microscopic study (5 to 7 μm in diameter) could be selected.[13,14]

Bacteriorhodopsin Solubilization by Miceller Formation and Reconstitution

In isolating membrane proteins, it is a common practice to use a surfactant to extract membrane-bound proteins. In the early studies of BR isolation and purification, the purple membrane (PR) was extracted with 2% Triton x-100 by gentle shaking[30] or sonication.[31] The BR extracted in this manner is stable for several days at room temperature. The structure of BR extracts was found to be consistent with that of Triton-lipid-BR micelles.[30]

Using prolonged dialysis at 4°C for several days[42], or gel filtration[50], most of the Triton X-100 could be removed[33] and the BR micelles could then be reconstituted by incorporating other surfactants such as octylglucoside (OG).[34,42] The Stoke radius of the reconstituted BP-lipid -OG micelles was determined to be 28$\pm$5 Å by gel-filtration.[34] Similarly, other surfactants such as dipalmitoyl- and dimyristoyl-phosphatidyl choline (DPPC and DMPC), egg phosphatidylcholine (EPC), and asolectin have been used to reconstitute BR and build various membrane mimetic systems.[35,42]

Incorporation of BR into Mono- and Multi-Layer Membranes

A monolayer of purple membrane was formed by compressing the floating layer at the air-water interface using a procedure described by Hwang et al.[36] The formed monolayer can be subsequently sandwiched between two electrodes,[37,72] or transferred onto millipore filters,[48] silicon wafers,[33,61] or teflon films.[39] Mono-layer or multi-layer thin films can also be made by suspending small patches of prepared purple membrane in dimethylformamide to a concentration of 25%. The suspension is spread over the surface of a $CaCl_2$ solution as a subphase, and then compressed, which results in a monolayer of purple membrane (PM). This PM monolayer can be transferred horizontally onto an indium tin oxide or aluminum (ITO or AL) electrode.[61,70] Multilayers can be formed by successive application onto the electrodes;[40] higher photo-voltaic potentials, up to ten volts, were obtained from a multi-layer BR film.

Owing to its intrinsic dipole moment, BR can be incorporated into the membranes in an orderly way. For example, the reconstituted bacteriorhodopsin vesicle BR-DMPC-DHPC was mixed with a bilayer forming and optically inactive dye p-[(ω- (trimethylammonio) pentyl) oxy] - p'- (dodecyloxy) - azobenzene bromide (A) on a solid lattice at 25°C. The ordering of optically inactive azobenzene bromide and BR mixture was investigated by circular dichroism (CD). When the temperature of the mixture was lowered from 25°C to 5°C, the appearance of CD bands at 330 and 380 nm indicated the accompanying ordered formation of azobenzene bromide on the BR lattice.[41]

Incorporation of BR into Vesicles and Liposomes

Bacteriorhodopsin liposomes have been obtained by sonicating BR directly with soybean phospholipids in 0.15 M KCl.[31] Bacteriorhodopsin also can be delipidated of its endogenous lipids and then reconstituted with exogenous lipids to form vesicles of long-lasting stability. Typically, regularly prepared purple membrane patches were dissolved in 5% (v/v) Triton x-100/ 0.1 M sodium acetate at pH 5.0.[50] On standing in the dark for two days with occasional agitation, the solution was centrifuged and then applied to a column (Bio - Gel A) pre-equilibrated with 0.01M Tris. HCl/0.15

M NaCl/0.25% deoxycholic acid /0.025% NaN$_3$ at pH 8.0. The purified BR fraction is 99% free of endogenous lipid and suitable for reconstitution with exogenous lipids.

The delipidated BR was then mixed with required amount of phospholipids and clarified by sonication in 2% sodium cholate in 0.15 M KCl of pH 8.0 at 4°C. The clear solution was dialyzed against 0.01 M Tris.HCl pH 8.0/0.15 M NaCl/0.025% NaN$_3$ at room temperature for 2-3 days, then dialyzed against 0.15 M NaCl/0.025% NaN$_3$. The reconstituted BR vesicles can be concentrated by ultrafiltration. The proton pumping capacity of reconstituted BR vesicles was several times higher than vesicles prepared directly from the purple membrane sheet.[50]

Triton x-100-solubilized purple membrane can be reconstituted with either DMPC or DPPC by mixing them and then removing the Triton by dialysis against 0.1 M sodium acetate, 3 mM sodium azide at pH 5.0. Consequently, the dialysis method incorporates BR into a large unilamellar lipid vesicle[42] suitable for enhancing signal-to-noise ratio.

Incorporation of BR into Solid Film

Having a large dipole moment, BR can self-assemble in water because solubilized BR becomes orderly oriented during the drying process. It can be used to obtain photovoltaic BR films by directly drying it on glass.[32,62] To further improve its mechanical strength and structural integrity, BR can be also incorporated into polymeric films. The polymeric matrices used included polyvinyl alcohol,[32] gelatine (17-2) poly(hydroxy)-methyl methacrylate,[43] propyleneoxide based prepolymer[43] acrylamide-bisacrylamide,[58] and polyethyleneglycol-derived prepolymers.[72] Orientation of BR in the film can be enhanced by applying an electric field[43,72] or a magnetic field[57] during the incorporation process for maximum mimetic effect.

APPLICATIONS

Because of their light-driven proton pumping mechanisms, photoelectric and photochromic properties, many possible applications of BR membrane or membrane mimetic systems have been studied and reported in recent years;[45] continuation of such studies may eventually lead to the development of BR-based products for molecular electronics and material sciences. Typical examples are summarized in Table 1.

Membrane Mimetic Systems for Light-Driven Proton or Ion Pumps

Since the early discovery of the proton pumping function of BR in H. halobium, several membrane mimetic systems have been used to demonstrate the proton-pumping phenomenon. These systems include small patches of BR-containing purple membrane incorporated in micelles,[46] inverted micelles,[50] liposomes,[48,50] fused purple membranes,[5,13] black lipid membranes,[47,49] and thin solid film.[50]

The light-driven proton pumping BR which generated a proton gradient across the membrane was proposed for use in the conversion of an alkali ion gradient by adding a carrier for transferring chloride ions through membrane. The application of such a membrane can be found in the desalination of seawater.[51] For this particular application, the second prominent retinal protein in H. halobium, namely halorhodopsin (HR), which has a 62% homolog sequence with BR has been reported[60,68] to be responsible for the transfer of sodium and chloride ions across the membrane. An understanding of the difference in structure and ion-pumping mechanisms of BR and HR will help in the design of specific ion-pumping membranes and specific ion sensors.

Table 1

Possible Applications of BR Membrane

Applications	References
Light-driven ion pump.	51
Proton gradient for ATP synthesis.	85
Desalinalination of sea water.	51
Biosensor.	53,87
Conversion of solar to electrical energy.	5,38,49,82
Photochemical generation of hydrogen.	52,59
Photoelectric switch.	83,84
Random access memory.	81
Holographic recording and polarizable hologram.	28,45,76,77,78
Optical phase conjugation.	45
Real-time interferometry.	79
Second and third harmonic generation.	73,74
Piezoelectric devices.	75,78
Four-wave mixing.	80

BR-Based Biosensor

A conceptual biosensor based on BR has been proposed that would incorporate a membrane-receptor protein into a PM reconstituted on an electrode.[53,87] When specific molecular binding occurs on the receptor, the protein may transmit a signal for its detection.

To test this concept, a BR membrane biosensor was constructed by first covalently binding hydrophobic alkylsilanes of either n-octadecyltrichlorosilane (OTS) or n-dimethyloctadecylcholrosilane (DMOCS) on a p-type (Si/SiO2) electrode. The alkylsilanated surface served both as a conductive support and as one layer of the bilayer membrane. The other complementary layer was made by placing the alkylsilanated surface in one compartment of a dialysis chamber with a solution of 0.5 mg/ml BR, 30 mM octylglucoside in a buffer of 100mM KCl and 20 mM HEPES, then dialyzing it against the KCl/HEPES buffer for eight hours. A BR membrane-mimetic structure was then surface-bound to the alkylsilanated electrode, and thus, a specific molecule-binding biosensor was formed. The results showed successful bilayer-surface binding on the p-type electrode but failure on the n-type electrode.

BR Membrane System for Converting Solar to Electrical Energy

Due to the thermostability and availability of BR, it has been incorporated into various membrane mimetic systems, such as fused sheets,[13] black lipid films,[49] and vesicles.[55,56] These membrane systems are very thin, fragile, and small and therefore, their uses are limited. However, the mechanical stability of such systems can be enhanced by incorporating them into films of colloidin,[38] loaded on millipore filters,[48] on SnO_2[54] or an indium tin oxide (ITO) electrode[5,61] or fixed in acrylic-acrylamide polymers.[58] Their use can be extended by forming mono or multilayer membranes, which then can be sandwiched between two electrodes.

Other functional improvements have also been tried, such as for example fixing BR in the same direction during processing. Since the purple membrane isolated from H. halobium has a large, permanent, electric dipole moment, the molecules can be lined up in solution under an external electric field with their sheet surface perpendicular to the applied field. Such oriented membranes can be immobilized in a gel[57] or transferred on an ITO[61] transparent electrode and then dried.[58] In the first preparation, two platinum electrodes could be attached to two sides of the gel slab. In the second preparation, metals could be evaporated on the other side of the dried film. Electrodes prepared in this way have been reported to be stable and functional for many years.[27]

During electric-field orientation of PM in an electric conducting polymer, difficulties have been encountered in trying to orient the PM in polyacrylamide hosts. The problem was solved by supplying an intensive magnetic field for orientation through interaction with paramagnetic purple membrane sheets. A vectorial orientation greater than 90% was achieved under a magnetic field of eleven Tesla.[57]

Several photoelectroactive BR membrane systems are now known.[82] Comparative studies of BR based systems and commercial photovoltaic devices used in solar energy conversion have indicated that in terms of cost-efficiency, there is a need for further research to develop a practical BR-based photoelectric device.

BR for Photoelectrochemical Hydrogenation or the Production of Hydrogen

Solar energy can be converted directly into electric energy by BR membrane systems or into electrochemical energy stored in the form of gaseous hydrogen. In the latter method, the photo-driven proton-pumping of BR is used to pump a proton across the cell wall of H. halobium. Then, an electron donor hydrogenase is used to reduce the activated proton into molecular hydrogen. Since H. halobium lacks the enzyme hydrogenase, a hydrogenase-containing Escherichia coli was reported to couple with H. halobium in the continuous production of hydrogen under illumination.[52] Cultured at 40°C under 0.3 mW/cm^2 light intensity, a maximum hydrogen production rate of 40.105 μ mole/mg cell min-1 for 30 minutes was achieved from 0.05 mg (wet weight) of the best strains of H. halobium and E. coli tested. Thus, it is conceivable that an existing BR-based membrane system can be developed for harvesting solar energy in the form of hydrogen. The success of such a process will depend on its cost-efficiency.

The produced hydrogen can be used in chemical reactions. Thus, the photolysis of water also can be coupled with the photochemical hydrogenation reaction for reducing cyclohexanone to cyclohexane,[59] as shown below:

$$\text{Cyclohexanone } (C_6H_{10}O) \xrightarrow{\text{H}_2} \text{Cyclohexanol } (C_6H_{12}O) \xrightarrow{\text{H}_2} \text{Cyclohexane } (C_6H_{12})$$

BR for Ultrafast Switch and Random Access Memory

When BR is irradiated, it is converted into a series of photo intermediates with life-times ranging from sub picosecond to millisecond, as shown in Figure 3. Each photo intermediate has its own characteristic absorption spectrum that results in the photochromic properties. The photocycle can be simplified in terms of a cycle of three intermediates, as shown in Figure 4. This arrangement is important to optical signal processing, whose absorption spectrum is shown in Figure 5.

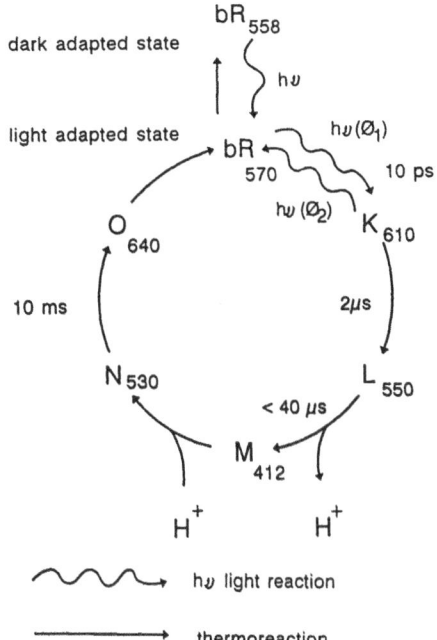

Figure 3. Simplified photocycle of BR.

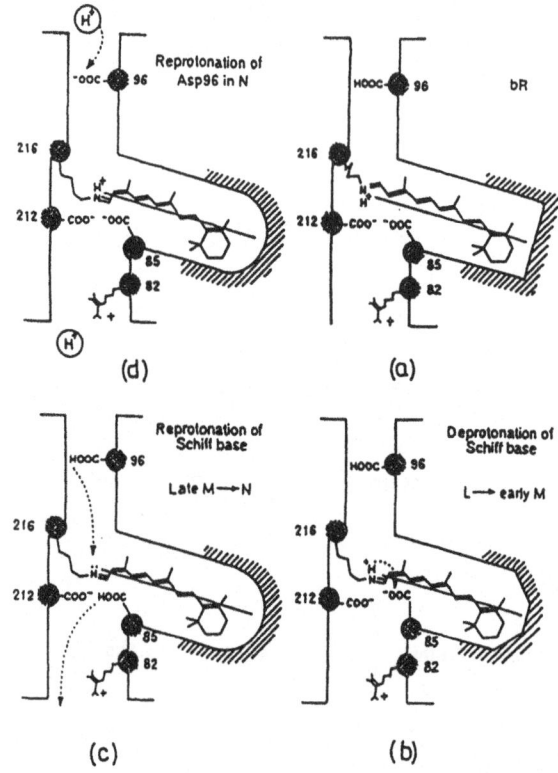

Figure 4. Proton pumping mechanism of BR (taken from [ref. 13] with permission).

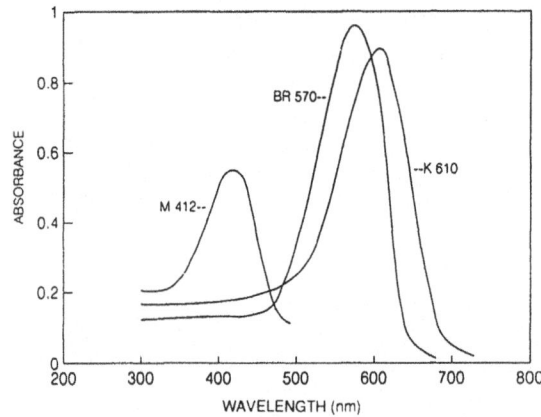

Figure 5. Absorption spectrum of BR 570, K610, and M412.

If BR 570 is illuminated by a short pulse of laser light, it is converted into a K(610) intermediate in a picosecond. At 77°K, the K(610) intermediate is indefinitely stable and it can be photoconverted back to BR 570 within a picosecond by irradiation at 610 nm. This reversible photoconversion occurs with a high quantum conversion efficiency of 0.7 and 0.3 for forward and reverse directions, respectively. At 77°K, the cyclic conversion can be repeated for 10×10^6 cycles without degradation.[45] This photochromic effect serves as a basis for the design as an ultrafast switch and random-access memory. This was accomplished by assigning 0 bit to the BR 570 state, and 1 bit to the K(610) state. The information can be stored for years[45] at 77°K.

At temperatures above 200°K, the photolytic intermediate truncates to a M(412) intermediate in 40 microseconds so the change can be used for a slower switch or for single-bit data storage. The advantage of using the M(412) state is its long life and the small amount of light needed to convert all the BR(570) into the M(412) state without producing a mixture of photointermediates. The memory storage of the M(412) state depends on its life time, which can be hours at 200°K. The life time of the M(412) state can be extended by using BR isolated from a mutant strain which differs from the regular BR by having the amino-acid residue aspartic 96 replaced by asparagine.[45,71] The life time of the M(412) state can be further extended by using the dehydrated BR incorporated into a polymeric film.[28,63]

These phenomenon serve as the basis for the development of biochips, a possibility still in its infancy. Several conceptual designs of and methods of fabrication of such biochips have been recently described.[64,65,66,67,86]

BR Film for Hologram Recording

Due to its extreme thermal and photochemical stability[78] and high storage density (10^8 bit/cm²),[45,77] several applications of BR in hologram and polarizable hologram tests have been reported.[28,45,69,77,78] For example, the recording mechanism is shown in which two lasers are used (Figure 6) in which the writing laser 1 of wavelength 1 is split by beam splitter (BS) into a sample and a reference beam.[45] The sample beam

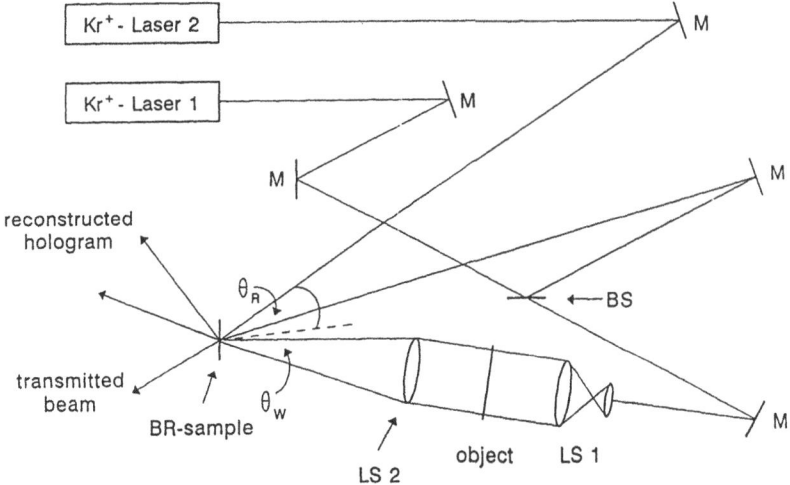

Figure 6. Setup of BR-film for transmission hologram recording (taken from [ref. 45] with permission). (M: Mirror, BS: Beam splitter, LS: Lens system).

Figure 7. USAF - test pattern recorded with light of wavelength of 412 nm and reconstructed with green light of wavelength 530 nm (taken from [ref. 45] with permission).

is expanded by lens (LS 1), then transmitted through a transparent object, and becomes spatially modulated. Upon refocusing by lens (LS 2), it is then incorporated with the reference beam to form an interference impressed into the BR film. The hologram can be reconstructed by a reading beam from laser 2 of wavelength 2 irradiating on the BR film at the Bragg angle.

If the recording wavelength of 1 is in the absorption band of BR 570 for B570--M412 conversion, then the reading beam of wavelength of 2 near 1 is limited in intensity so that it does not diminish interference grating.

In another example, a strong reading beam is used to pump BR 570 into the M(412) state, and the blue light from M412--BR 570 is used for recording. The result is an intense reconstructed hologram, as shown in Figure 7, using a USAF test pattern.[45]

BR for Optical Phase Conjugation

A conventional mirror reflects a beam of light in accordance with the reflection law. However, a phase-conjugating mirror reflects by reversing the light beam back to its own path. Thus, a medium distorted parallel light can be restored by using a phase-conjugating BR system to reverse it back to its original parallel path. The photochromic property of BR has been demonstrated for optical phase conjugation, which has many possible applications in photolithography, image processing, and telecommunication systems.

A representative set up taken from Hampp and Brauchle[45] is shown in Figure 8, in which BE is the beam expander, BS 1,2,3 are beam splitters, Di is the distorting object, and M 1,2,3 are mirrors. Pc is the phase-conjugated beam, and Sc is the screen. The opposing beams reflected from M2 and M3 hit the BR sample, and at the same time, a probe beam reflected from BS3 is distored by Di. The distored beam can be restored by the phase-conjugating property of BR, shown as the image on the screen SC. Photographs of distored and restored images are shown in Figure 9.

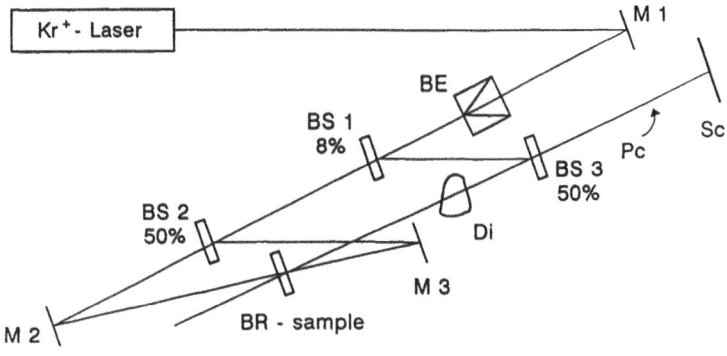

Kr⁺- Laser
M 1
BE
BS 1
8%
Sc
Pc
BS 3
50%
BS 2
50%
Di
M 3
M 2
BR - sample
BE: beam expander
BS: beam splitter
Di: distorting object
M: mirror
Pc: phase conjugated beam
Sc: screen

Figure 8. Setup for the demonstration of the phase conjugating properties of BR-films (taken from [ref. 45] with permission).

Figure 9. Photographs of distorted (a) and phase conjugated (b) beam (reproduced from [ref. 45] with permission).

ACKNOWLEDGMENT

This work is performed under the auspices of the U.S. DOE Contract No. DE-AC02-76CH00016. The authors wish to thank Mitzi McKenna and Corinne Messana for preparing the manuscript, and Avril Woodhead for editing.

REFERENCES

1. H.J.A. Dartnall (ed.). "Photochemistry of Vision, Handbook of Sensory Physiology VII/1," Springer, Berlin (1972).
2. H. Langer (ed.) "Biochemistry and Physiology of Visual Pigments," Springer, Berlin (1973).
3. W. Stoeckenius and W.H. Kunau, Further characterization of particulate fractions from lysed cell envelopes of *Halobacterium halobium* and isolation of gas vacuole membranes, J. Cell Biol., 38:337-357 (1968).
4. D. Osterhelt and W. Stoeckenius, Rhodopsin-like protein from the purple membrane of *Halobacterium halobium*, Nature New Biology, 233:149-151 (1971).
5. F. Siebert, Retinal proteins, *in*: "Photochromism Molecules and Systems," H. Durrand and H. Bouas-Laurent, Eds., Elsevier, New York (1990).
6. R.R. Birge, Nature of the primary photochemical events in rhodopsin and bacteriorhodopsin, Biochim. Biophys. Acta, 1016:293-327 (1990).
7. T.G. Ebrey, H. Frauenfelder, B. Honig, and K. Nakanishi, "Biophysical Studies of Retinal Proteins," University of Illinois Press, Urbana (1987).
8. W. Stoeckenius, R.H. Lozier, and R.A. Bogomolni, Bacteriorhodopsin and the purple membrane of Halobacteria, Biochim. Biophy. Acta, 505:215-278 (1979).
9. W. Stoeckenius and R.A. Bogomolni, Bacteriorhodopsin and related pigments of halobacteria, Ann. Rev. Biochem., 52:587-616 (1982).
10. Yu A. Ovchinnikov, N.G. Abdulaev, M. Yu Feigina, A.V. Kiselev, and N.A. Lobanov, The structural basis of the functioning of bacteriorhodopsin: An overview, FEBS Lett, 100:219-224 (1979).
11. H.G. Khorana, G.E. Gerber, W.C. Herlihy, C.P. Gray, R.J. Anderegg, K. Nihei, and K. Biemann, Amino acid sequence of bacteriorhodopsin, Proc. Natl. Acad. Sci., U.S.A. 76:5046-5050 (1979).
12. R.J. Dunn, N.R. Hackett, K.S. Huang, S. Jones, H.G. Khorana, D.S. Lee, M.J. Liao, K.M. Lo, J.J. McCoy, S. Noguchi, R. Radhakrishnan, and U.L. RajBhandry, Studies on the light-transducing pigment bacteriorhodopsin, Cold Spring Harbor Symp. Quant. Biol. 48:853-862 (1983).
13. R. Henderson, J.M. Baldwin, T.A. Ceska, F. Zemlin, E. Beckmann, and K.H. Downing, Model for the structure of bacteriorhodopsin based on high-resolution electron cryo-microscopy, J. Mol. Biol. 213:899-929 (1990).
14. J.M. Baldwin and R. Henderson. Measurement and evaluation of electron diffraction patterns from two-dimensional crystals, Ultramicroscopy 14:319-336 (1984).
15. J.M. Baldwin, R. Henderson, E. Beckman, and F. Zemlin, Images of purple membrane at 2.8 Å resolution obtained by cyro-electron microscopy. J. Mol. Biol. 202:585-591 (1988).
16. R. Henderson, J.M. Baldwin, K.H. Downing, J. Lepault, F. Zemlin, Structure of purple membrane from *Halobacterium halobium*: Recording, measurement and evaluation of electron micrographs at 3.5 Å resolution, Ultramicroscopy 19:147-178 (1986).

17. G. Souvignier and K. Gerwert, Proton uptake mechanism of bacteriorhodopsin as determined by time-resolved stroboscopic-FTIR-spectroscopy. Biophys. J. 63:1393-1405 (1992).

18. S.J. Doig, P.J. Reid, and R.A. Mathies, Picosecond time-resolved resonance Raman spectroscopy of Bacteriorhodopsin: Structure and kinetics of the J.K. and KL intermediates, SPIE Vol. 1432, Biomolecular Spectroscopy II, 184-196 (1990).

19. G. Varo and J.K. Lanyi, Pathways of the rise and decay of the M photointermediate(s) of bacteriorhodopsin, Biochemistry 29:2241-2250 (1990).

20. T. Kobayashi, M. Terauchi, T. Kouyama, M. Yoshizawa, and M. Taiji, Femto second spectroscopy of acidified and neutral bacteriorhodopsin, SPIE Vol. 1403 Laser Application in Life Sciences, 407-416 (1990).

21. M. Lin and S. Seltzer, The consequences of a deuterium exchange test on proposed mechanism for the purple membrane proton pump, FEBS. Lett. 106:135-139 (1979).

22. R.H. Lozier, R.A. Bogomolni, and W. Stoeckenius, Bacteriorhodopsin: A light-driven proton pump in Halobacterium halobium, Biophys. J. 15:955-962 (1975).

23. M. Akhtar, P.T. Blosse, and P.B. Dewhurst, Studies on Vision, Biochem. J. 110:693-702 (1968).

24. P.E. Blatz, J.H. Mohlerand and H.V. Navanguo, Anion-Induced Wavelength Regulation of Absorption Maxima of Schiff Bases of Retinal, Biochemistry 11:848-855 (1972).

25. P.E. Blatz and J.H. Mohler, Effect of Selected Anions and Solvents on the Electronic Absorption, Nuclear Magnetic Resonance and Infrared Spectra of the N-Retinylidene-n-butylammonium cation, Biochemistry 14:2304-2309 (1975).

26. I. Tabushi, Y. Kuroda, and K. Shimokawa, Cyclodexin having an amino group as a rhodopsin model, J. Am. Chem. Soc. 101:-4759-4760 (1979).

27. G. Varo and L. Keszthely, Photoelectric signals from dried oriented purple membranes of *Halobacterium halobium*, Biophys. J. 43:47 (1983).

28. N.N. Vsevolodov, G.R. Ivanitskii, M.S. Soskin, and V.B. Taranenko, Biochrome films: Reversible media for optical recording, Optoelectron, Instrumm. Data Process 2:39-46 (1986).

29. J.H. Fendler, "Membrane Mimetic Chemistry," John Wiley & Sons, New York (1982).

30. J.A. Reynolds and W. Stoeckenius, Molecular Weight of bacteriorhodopsin solubilized in Triton X-100, Proc. Natl. Acad. Sci. 74:2803-2804 (1977).

31. E. Racker, A New Procedure for the Reconstitution of Biologically Active Phospholipid Vesicles, Biochem. Biophys. Res. Commun. 55:224-230 (1973).

32. S. Kunugi, S. Nakaizumi, K. Ikeda, N. Itoh, and A. Nomura, Effect of ionic amphiphiles and poly (vinyl alcohol) on the sidedness of purple membrane in dried films, Langmuir, V.7 N.8:1576-1578 (1991).

33. T. Furuno and H. Sasabe, Denaturation of purple membranes at the air/water interface studied by SEM, J. of Colloid and Interface Science, 147:225-232 (1991).

34. N.A. Dencher and M.P. Heyn, Formation and properties of Bacteriorhodopsin monomers in the non-ionic detergents octyl-β-D-glucoside and Triton X-100, FEBS Letters 96:322-326 (1978).

35. N.A. Dencher and M.P. Heyn, Bacteriorhodopsin Monomers Pump Protons, FEBS. Letters 108:307-310 (1979).

36. S.B. Hwang, J.I. Korenbrot, and W. Stoeckenius, Structural and spectroscopic characteristics of bacteriorhodopsin in air-water interface films, J. Membr. Biol. 36:115-135 (1977).

37. A.R. McIntosh and F. Boucher, On the action spectrum of the photoelectric transients of bacteriorhodopsin in solid-state films, Biochim. Biophys. Acta 1056:149-158 (1991).

38. L.A. Drachev, A.D. Kaulen, and V.P. Skulachev, Time resolution of the intermediate steps in the bacteriorhodopsin-linked electrogenesis, FEBS, Letters 87:161-167 (1978).

39. S. Michaile and F.T. Hong, Signal modulation via interfacial processes in molecular optoelectronic devices, IEEE Engineering in Medicine & Biology Society 11th Annual International Conference, 1333-1335 (1989).

40. H. Sasabe, T. Furuno, and K. Takimoto, Photovoltaics of photoactive protein/polypeptide LB films, Synthetic Metals 28: C787-C792 (1989).

41. T. Katsura, Ordering of cationic amphiphiles on two-dimensional lattice of bacteriorhodopsin, IEEE Engineering in Medicine & Biology Society 11th Annual International Conference, 1328 (1989).

42. R.J. Cherry, U. Muller, R. Henderson, and M.P. Heyn, Temperature-dependent aggregation of Bacteriorhodopsin in dipalmitoyl- and dimyristoylphosphatidylcholine vesicles, J. Mol. Biol. 121:283-298 (1978).

43. S. Kunugi, K. Tatsukawa, T. Nakajima, A. Nomura, and A. Tanaka, Orientation of bacteriorhodopsin in non-aqueous polymer membrane, Polymer Bulletin 21:59-62 (1989).

44. S. Kungi, S. Nakaizumi, K. Ikeda, N. Itoh, and A. Nomura, Effect of ionic amphiphiles and poly (vinyl alcohol) on the sidedness of purple membrane in dried film, Langmuir 7:1576-1578 (1991).

45. N. Hampp and C. Brauchle, Bacteriorhodopsin and its functional variants: Potential applications in modern optics, in: "Photochromism, Molecules and Systems," H. Durr and H. Bouas-Laurent (ed.), Elsevier, New York, 954-975 (1990).

46. B. Lorber, L.J. DeLucas, and J.B. Bishop, Changes in the physico-chemical properties of the detergent octyl glucoside during membrane protein crystallization using a salt as the precipitant, J. of Crystal Growth 110:103-113 (1991).

47. Z. Danzshazy and B. Karvaly, Incorporation of bacteriorhodopsin into a bilayer lipid membrane: A photoelectric-spectroscopic study. FEBS Lett. 72:136-138 (1976).

48. K.J. Hellingwerf, J.C. Arents, B.J. Scholte, and H.V. Westerhoff, Bacteriorhodopsin in liposomes, Biochim. Biophys. Acta 547:561-582 (1979).

49. K. Seifert, K. Fendler, and E. Bamberg, Charge transport by ion translocating membrane proteins on solid supported membranes. Biophys. J. 64:384-391 (1993).

50. K.S. Huang, H. Bayley, and H.G. Khorana, Delipidation of bacteriorhodopsin and reconstitution with exogenous phospholipid, Proc. Natl. Acad. Sci. USA, 77:323-327 (1980).

51. D. Oesterhelt, Bacteriorhodopsin as a light driven ion exchanger, FEBS Lett. 64:20 (1976).

52. M.M. Taqui Khan and J.P. Bhatt, Light dependent hydrogen production by *Halobacterium halobium* coupled to *Escherichia coli*, Int. J. Hydrogen Energy 14:643-645 (1989).

53. A.S. Lader, C. Tamanaha, J. Li, N.W. Downer, H.G. Smith, Y. Mendelson, and R.A. Peura, An investigative study of membrane-based biosensors, IEEE CH2997-5/91, 253-254 (1991).

54. T. Miyasaka and K. Koyama, Rectified photocurrents from purple membrane Langmuir-Blodgett films at the electrode-electrolyte interface, Thin Solid Films, 210/211:146-149 (1992).

55. K.J. Hellingwerf, B.J. Scholte, and K. Van Dam, Bacteriorhodopsin vesicles an outline of the requirements for light-dependent H^+ pumping. Biochim. et Biophys. Acta 513:66-77 (1978).

56. N.A. Dencher, Spontaneous transmembrane insertion of membrane protein into lipid resides facilitated by short-chain lecithins. Biochemistry 25:1195-1200 (1986).

57. G.W. Rayfield, Nonlinear absorbance effects in bacteriorhodopsin, SPIE Vol. 1436 photochemistry and photoelectrochemistry of organic and inorganic molecular thin films, 150-159 (1991).

58. S. Kunugi, T. Kusano, H. Yamada, and Y. Nakamura, Orientation and immobilization of bacteriorhodopsin in polyacrylamide gel membranes, Polymer Bulletin 19:417-421 (1988).

59. M.M. Taqui Khan and J.P. Bhatt, Efficient reduction of cyclohexanone to cyclohexanol and cyclohexane by Halobacterium halobium MMT22, J. of Mol. Catalysis, 63:L15-L19 (1990).

60. S. Michaile, A. Duschl, J.K. Lanyi, and F.T. Hong, Chloride ion modulation of the fast photoelectric signal in halorhodopsin thin films, Annual International Conference of the IEEE Engineering in Medicine and Biology Society 12:1721-1723 (1990).

61. T. Furuno, K. Takimoto, T. Kouyama, A. Ikegami, and H. Sasabe, Photovoltaic properties of purple membrane Langmuir-Blodgett films, Thin Solid Film 160:145-151 (1988).

62. N. Hampp, C. Brauchle, and D. Oesterhelt, Bacteriorhodopsin wildtype and variant aspartate-96 → asparagine as reversible holographic media. Biophys. J. 58:83-93 (1990).

63. N.N. Vsevolodov, A.B. Druzhko, and T.V. Djukova, Actual possibilities of Bacteriorhodopsin in: "Molecular Electronics, Biosensors, and Biocomputers," F.T. Hong (Ed.), Plenum Press, New York (1989) 381-384.

64. F.T. Hong (Ed.), "Molecular Electronics, Biosensors and Biocomputers," Plenum Press, New York (1989).

65. R.M. Metzger, P. Day, G.C. Papavassiliou, "Lower-dimensional System and Molecular Electronics," Plenum Press, New York (1991).

66. P.I. Lazarev (Ed.), "Molecular Electronics, Materials and Methods," Kluwer Academic Publishers, Boston (1991).

67. A. Aviram (Ed.), "Molecular Electronics-Science and Technology," American Institute of Physics, New York (1992).

68. E. Bamberg, P. Hegemann, and D. Oesterhelt, The chromoprotein of halorhodopsin is the light-driven electrogenic chloride pump in Halobacterium halobium. Biochemistry 23:6216-6221 (1984).

69. R.B. Gross, K.C. Izgi, and R.R. Birge, Holographic thin films, spatial light modulators, and optical associative memories based on bacteriorhodopsin. Proceedings of SPIE - The International Society for Optical Engineering 1662:186-196 (1992).

70. H. Sasabe, T. Furuno, A. Sato, and K.M. Ulmer, 2-Dimensional protein crystals for bioelectronics, IEEE Engineering in Medicine and Biology Society 10th Annual International Conference CH2566-8/88 P. 1003 (1988).

71. A.K. Mitra, L.J.W. Miercke, M.C. Betlach, R.F. Shand, and R.M. Stroud, High resolution electron diffraction study in projection on bacteriorhodopsin mutants: Ground state structure in D96N is unaltered. Abstracts 37th Annual Meeting of Biophysical Society, Washington, DC, Feb. 14-18, 1993, A246 (1993).

72. S. Kunugi, H. Yamada, Y. Nakamura, F. Tokunaga, and A. Tanaka, Oriented immobilization of bacteriorhodopsin in synthetic polymer membranes by use of electric static field, Polymer Bulletin 18:87-90 (1987).

73. A.V. Sharkov and T. Gillbro, Second harmonic generation in oriented purple membrane films under picosecond light excitation, Thin Solid Films, 292:L9-L14 (1991).

74. O.A. Aktsypetrov, N.N. Akhmediev, N.N. Vsevolodov, D.A. Esikov, and D.A. Shutov, Photochromism in nonlinear optics: Photocontrolling generation of the second harmonics by bacteriorhodopsin molecules, Dokl. Akad. Nauk. USSR 293:594 (1987).

75. B.P. Kethis, Piezoelectric mechanism of charge active transport in purple membrane from *Halobacterium halobium*, Biolog. membrane 1:1307 (1984).

76. N.N. Vsevolodov, G.R. Ivanitsky, M.S. Soskin, and V.B. Taranenko, "Biochrom" films-reversible medium for optical recording, Avtometriya 2:41-48 (1986).

77. N.M. Kozhevnikov, Y.O. Barmenkov, and M.Y. Lipovskaya, Holographic recording in photorefractive media containing bacteriorhodopsin (PMBR), SPIE Vol. 1507 Holographic Optics III: Principles and Applications 517-524 (1991).

78. N. Hampp, C. Brauchle, and D. Oesterhelt, Bacteriorhodopsin as a reversible holigraphic medium in optical processing, Annual International Conference of the IEEE Engineering in Medicine and Biology Society 12:1719-1720 (1990).

79. Y.O. Barmenkov, V.V. Zosimov, N.M. Kozhevnikov, O.I. Kotov, L.M. Lyamshev, and V.M. Nikolaev, Detection of a phase-modulation signal from a fiber-optic interferometer by means of a dynamic hologram in bacteriorhodopsin. Sov. Phys. Acoust. 33:334-335 (1987).

80. E.Y. Korchemskaya, M.S. Soskin, and V.B. Taranenko, Spatial polarization wavefront reversed under conditions of four-wave mixing in biochrome films. Sov. J. Quantum Electron. 17:450-454 (1987).

81. R.R. Birge, Optical random access memory based on bacteriorhodopsin. Bull. Am. Phys. Soc. 34:483 (1989).

82. K. Singh and S.R. Caplan, The purple membrane and solar energy conversion. TIBS 5:62-64 (1980).

83. F.T. Hong, The bacteriorhodopsin model membrane system as a prototype molecular computing element. Biosystems 19:223-236 (1986).

84. C. Mobarry and A. Lewis, Implementation of neural networks using photoactivated conducting biological materials SPIE J. 700:304-308 (1986).

85. H.S. Van Walraven, R.L. Van Der Bend, M.J.M. Hagendoorn, N.P. Haak, A. Oskam, A. Oostdam, K. Krab, and R. Kraayenhof, Comparison of ATP Synthesis Efficiencies in ATPase proteoliposones of different complexities. Bioelectrochemistry and Bioenergetics 16:167-180 (1986).

86. Computer, Nov. 1992, Vol 2, No. 11, published by IEEE Computer Society, Los Alamitos, CA, 90720-1264.

87. P. Yager, Functional reconstitution of a membrane protein in a diacetylenic polymerizable lecithin. Biosensors 2:363-373 (1986).

SOLID STATE pH AND pNa GLASS ELECTRODES

Naila Ashraf, Kahkashan Hamdini, and K.L. Cheng

Department of Chemistry, University of Missouri-Kansas City
Kansas City, Missouri 64110

ABSTRACT

Simple solid state pH and pNa glass electrodes have been prepared and studied. The conventionally used internal solution is replaced by a conducting wire or graphite fiber string which is firmly connected to the inner surface of the glass bulb or flate glass membrane. Both the conducting glue and internal solution have a contact junction potential at the glass surface and the conducting wire or internal solution serves as a conducting medium. The solid state pH and pNa electrodes studied showed a slope of 58 mV/pH and 55 mV/pNa, respectively and were stable in a period of more than one-month. The sensitivity of pNa electrodes is better than the commercial one (linear to 1×10^{-5} M Na^+). The slopes of pNa electrodes were affected by pH. They behaved best in a buffer of pH 8. The pNa electrode mechanism is attributed to the charge adsorption of both Na^+ and OH^- ions on the zwitterionic glass surface. The measured potential is the result of the net charge.

INTRODUCTION

Thompson[1] in 1932 pioneered a glass electrode with a comparatively thick wall and a direct metal connection to the glass. Two types of electrodes were reported, a dipping type and a containing type. Either tinfoil or silver coating was applied to the inner side of the glass (dipping type) or outer side of pH sensitive glass (containing type). In the dipping type, the electrode is dipped in the solution whose pH is to be determined, while in the containing type the solution is placed inside the electrode and the reference electrode is then dipped in this solution. Thompson observed that electrodes of the dipping type were less accurate than those of the containing type. Bates[2] has pointed out that electrodes of these types have not been found as satisfactory as those with inner reference electrodes, and have never been widely adopted. Early patents were also issued on the use of metals and alloys for electrical contact with the glass[3,4]. A relatively complicated solid state pH glass electrode that exhibited near so-called Nernstian response (>50 mV/pH) has recently been reported by Parr et al[5]. Here a modernized version of the Thompson electrode is presented[6].

Glass electrodes are the electrodes of choice for H^+, Na^+, Ag^+, and Li^+, for which their selectivity is sufficiently high. The cation response of pH glasses can be enhanced[7] by suitable modifications of the composition, to the point that practical glass electrodes for the measurement of alkali cations can be fabricated. Eisenman, Rudin, and Casby investigated[8,9] the cation response of sodium aluminum silicate glasses with a wide range of compositions. The preferred composition was 11 mole per cent Na_2O, 18 mole per cent Al_2O_3, 71 mole per cent SiO_2.

In our studies we prepared solid state Na^+ ion selective electrodes with two different internal references. The responses of these electrodes were then compared with the responses of the commercial sodium electrode (with internal solution). We also studied the response of these solid state electrodes and commercial sodium electrode at different pH's.

EXPERIMENTAL
pH ELECTRODE

The pH sensitive glass bulbs with glass stems were obtained from Phoenix Electrode Company. Four kinds of electrodes were prepared having different internal references using these bulbs.

1) A copper wire cleaned with sandpaper was attached to the inner side of the pH sensitive glass bulb using silver epoxy paste. This was then left for overnight to dry. The electrode was then soaked in a buffer solution of pH 5 for an hour before using.

2) Graphite fibers were used to make the connection between the inside of the electrode bulb and the potentiometer. These graphite fibers were attached to the bulb with the help of liquid solder (a minimal amount). bundles of graphite fibers were rolled to form a thick thread, which was then used for the connection instead of using a copper wire (Fig. 1a). After the electrode dried overnight, it was soaked in the pH 5 buffer for an hour before using.

3) The inner surface of pH sensitive bulb was covered with graphite fibers using a minimum amount of liquid solder. Copper wire was then used to make a connection between the inner side of bulb and potentiometer (Fig. 1a). After air drying for twenty four hours, the electrode was soaked in pH 5 buffer for an hour before using.

4) The copper wire was attached to the inside of pH sensitive bulb with Duco cement and air dried overnight. It was soaked in buffer of pH 5 for an hour before using.

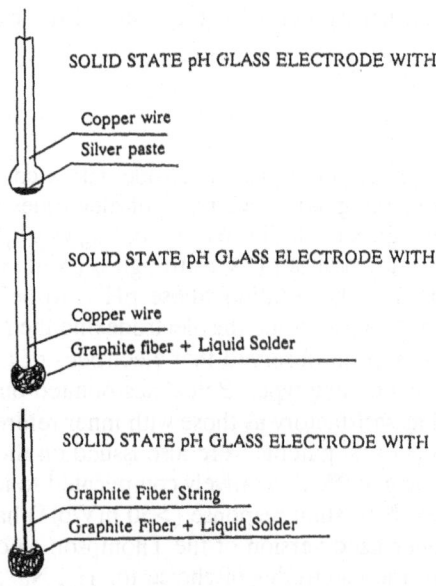

Fig 1a. Solid state pH glass electrodes.

The silver past used above was a conductive adhesive resin from Johnson Matthey, Inc. Graphite fibers were obtained from AESAR (Johnson Matthey Inc.) and were of 8 micron in diameter and 25.4 mm long. Duro liquid solder was used in all these studies.

The potentials of buffer solutions were measured with all these electrodes versus SCE and then, for sake of comparison, the potentials of the same solutions were recorded by a Fisher pH glass electrode (catalog #13-639-3) against SCE. the same reference electrode (SCE from Fisher (catalog #13-639-52)) was used throughout all the experiments.

To study the salt effect, the potentials of 10 mL of various pH buffer solutions were measured with the solid state pH glass electrode (with solver paste and copper wire). To these solutions was then added 0.2 g of sodium chloride. Potentials of these solutions were then measured again with the solid state pH glass electrode.

To study the effect of internal surface area coverage by silver paste, four electrodes were prepared. The internal surface area of the pH sensitive glass bulb covered by silver paste was different in each electrode. Copper wire was used for connection of inner surface of pH sensitive glass bulb to the potentiometer. The potentials of buffer solutions of various pHs were measured with these electrodes and with the Fisher pH glass electrode.

For electrode life-time and stability studies, electrodes were used repeatedly with different intervals over a time span of 30 days to measure the potentials of various buffers versus SCE. The potentials of these buffers were also measured with the Fisher pH glass electrode. These solid state pH glass electrodes were made of copper wire and silver paste and were kept dry for a week. After the measurements, the electrodes were rinsed and again kept dried for two weeks and then used and so on. Each time before using they were soaked in a buffer of pH 6 for twenty minutes. Three electrodes were used for these studies.

A solid state pH glass electrode with copper wire and Duco cement was also checked for stability and reproducibility. The electrode was kept soaked in pH 5.8 buffer for all the time during these studies. The potentials of pH buffers were measured for four days with this electrode and the Fisher pH glass electrode. The same experiment was repeated with a solid state electrode with graphite fibers, Duco cement and graphite fibers for connection.

Two pH glass electrodes from China with internal solution were used to measure the potential of buffers of various pHs. One of these was then cut from the top and the internal reference electrode along with the internal reference solution was removed. After washing and drying the inside of the pH sensitive glass bulb, it was converted to solid state pH glass electrode with silver paste and copper wire. The potentials of the same buffer solutions were measured with this electrode after overnight drying.

pNa ELECTRODE

Sodium ion sensitive glass bulbs with stems were obtained from Phoenix Electrode company (Houston) and were used to prepare two solid state Na^+ ion selective electrodes with different internal references.

1) Copper wire was used as an internal reference (Fig. 1b) and was attached to the inner surface of the bulb using silver paste (Johnson Matthey conductive adhesive resin, A 500 H). A pH sensitive glass electrode of a similar kind has already been reported[10] by Cheng and Ashraf.

2) Graphite fibers mixed with liquid solder were used to glue the string of graphite fiber (internal reference) to the inner surface of the bulb. (Fig. 1b).

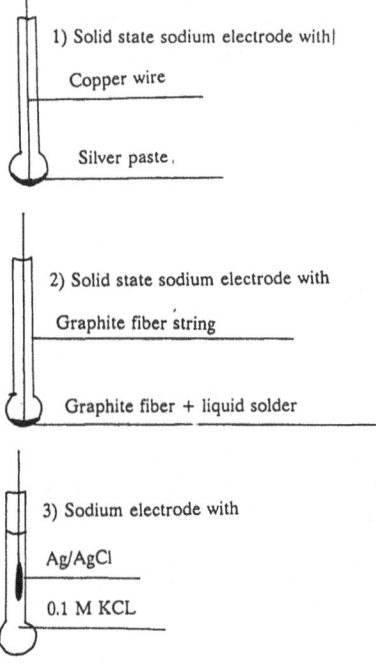

Fig. 1b Structure of three sodium electrodes with different internal references.

Another electrode was prepared using these stemmed sodium sensitive glass bulbs, in which KCl solution (0.1 M) with Ag/AgCl was used as an internal reference.

Sodium nitrate stock solutions (0.1 M) of different pH's (8, 9, 10, 11 and 12) were prepared in ethylenediamine buffer[11]. Sodium nitrate solutions (of different pH) of concentrations ranging from 10^{-2} to 10^{-6} M were prepared by the successive dilution method.

All the potential measurements were made with a Fisher Accument pH meter (Model 825 MP). A saturated calomel electrode (Fisher Scientific company, catalog.#. 13-639-52) was used as a reference electrode. Potential measurements of the entire sodium concentration range were done at different pH's using the solid state sodium electrode, the sodium electrode with internal solution, and commercial sodium electrode (Markson Science Incorporation. catalog.#.1001).

RESULTS AND DISCUSSION

Measurements of potentials with different solid state pH glass electrodes of various buffer solutions are shown in Figures 2. a-e. For comparison, results obtained by a commercially available Fisher glass electrode are also shown. As is clear from these graphs, both electrodes showed exactly the same response as far as slopes are concerned (Table I).

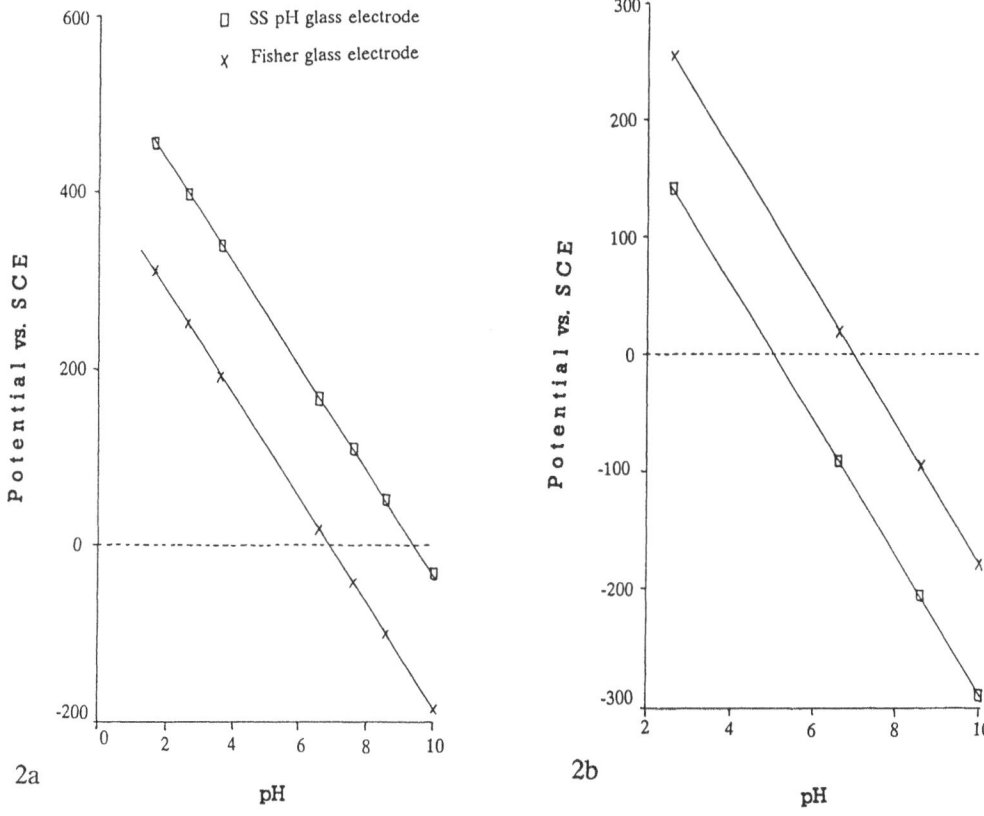

Fig. 2a   Potential vs. pH response of solid state (SS) pH glass electrode (with Cu wire and silver paste) and fisher glass electrode.

Fig. 2b   Potential vs. pH response of solid state pH glass electrode (with graphite fiber + liquid solder + graphite fiber for connection) and fisher electrode.

However, the absolute potential values are different, which is not surprising as absolute potential values depend on type of glass used and the internal reference. In glass electrodes with internal solution there is a constant potential on the inner side of an electrode. In the case of solid state pH glass electrode there is also a potential on the inner surface of electrode which is due to the junction of metal with the glass. Therefore, as the internal reference changes the absolute potential value also changes. Addition of salt showed no pronounced effect on the solid state pH glass electrode (Fig. 3), although in the neutral region there seems to be some tendency towards an increase in the potential value. This could be due to the fact that in this region concentration of H^+ and OH^- is very low. There is a possibility that when salt (NaCl) is added, the Na^+ also gets adsorbed at the glass surface, resulting in increased potential value.

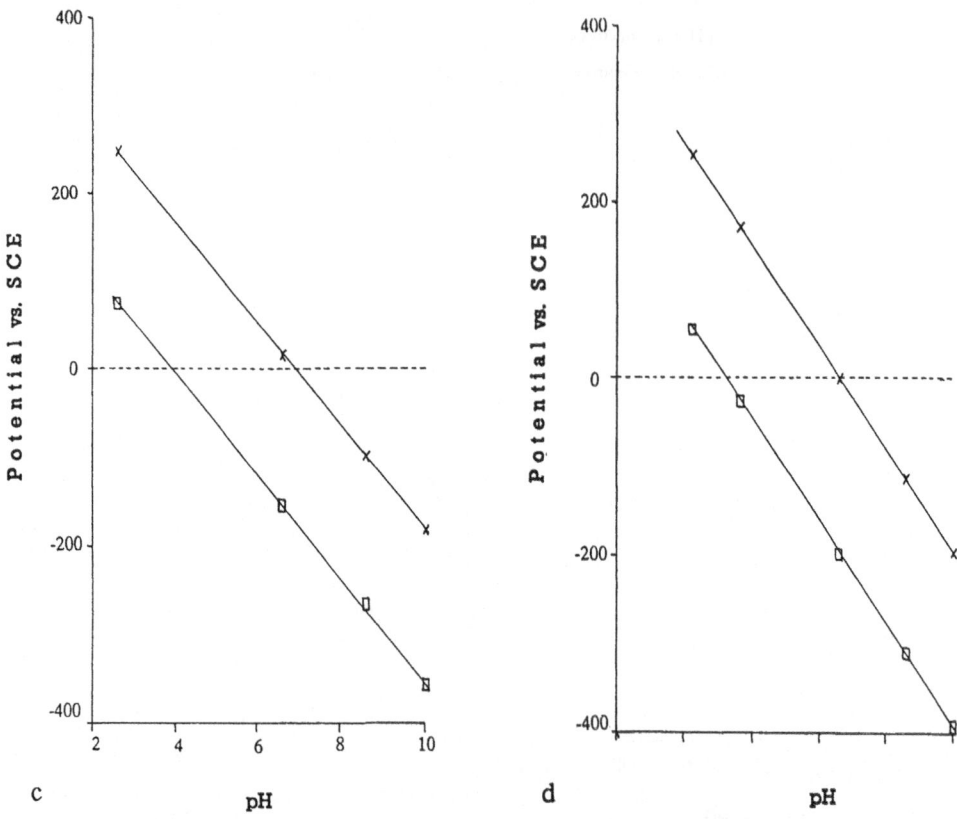

Fig. 2c Potential vs.pH response of solid state (SS) pH glass electrode (with graphite fiber + liquid solder + Cu wire) and fisher glass electrode.

Fig. 2d Potential vs. pH response of solid state (SS) pH glass electrode (with Cu wire and Duco cement) and fisher glass electrode.

The internal surface area coverage by silver paste showed no effect on electrode response (Fig. 4). All the electrodes showed similar behavior (Table II). This result proves that only a proper connection between the internal surface of glass bulb and the copper wire is needed and the whole internal surface need not be covered entirely.

The junction potential between the copper wire and the internal glass surface needs to be constant in order for the electrode to be used for pH measurements. This potential also should not change with time. The results of such time studies for different solid state

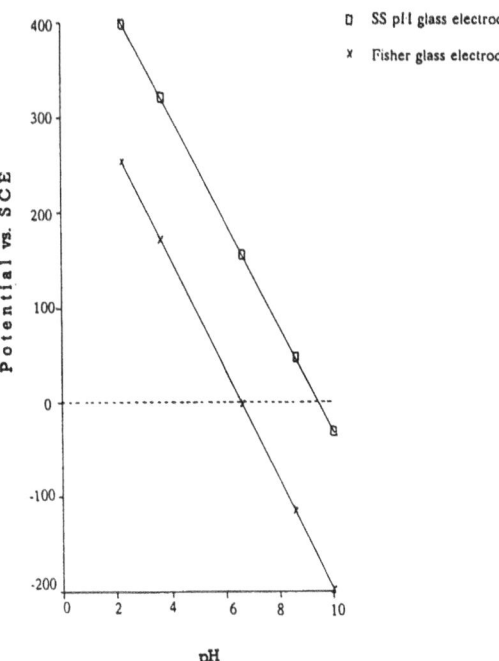

Fig. 2e   Potential vs.pH response of solid state (SS) pH
glass electrode (with graphite fibers + Duco
cement and graphite fibers) and fisher electrode.

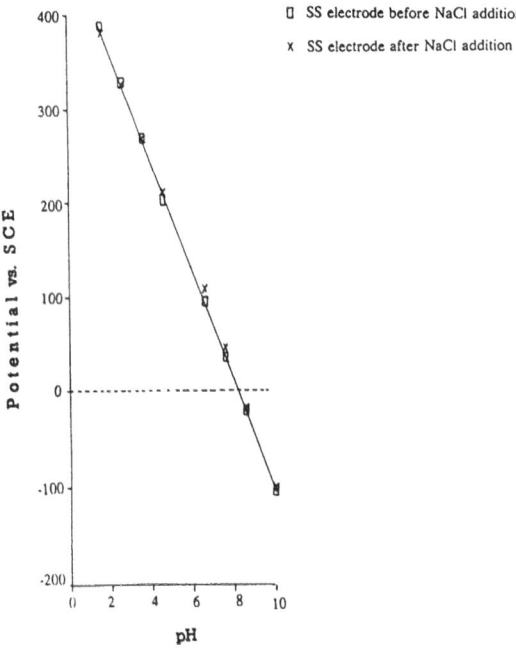

Fig. 3   Effect of salt on response of solid state pH glass
electrode.

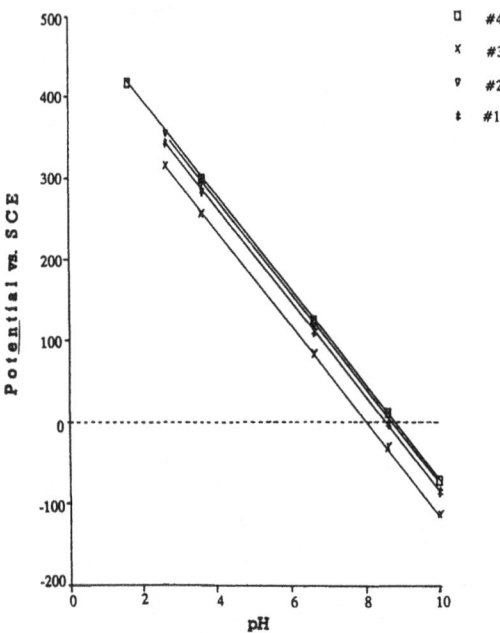

Fig. 4 Effect of covered internal surface area with
silver paste on solid state pH glass electrode
(decreasing order of covered area #4> #3> #2> #1>).

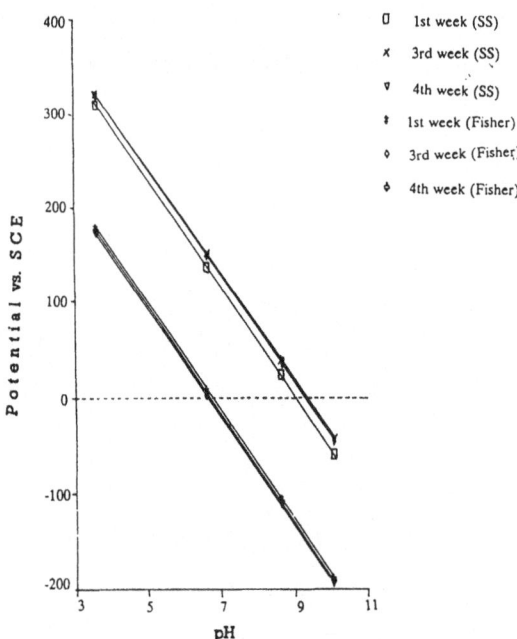

Fig. 5 Reproducibility of solid state (SS) pH glass
electrod (with Cu wire and silver paste) after
different time intervals. (1/4 of the internal
surface area covered by silver paste).

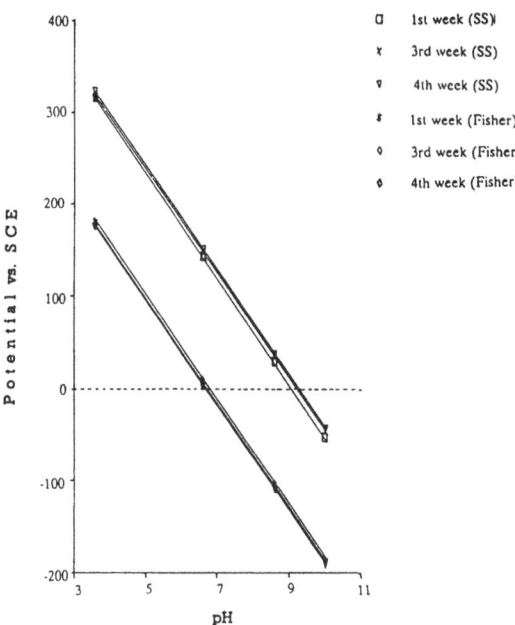

Fig. 6   Reproducibility of solid state (SS) pH glass electrod (with Cu wire and silver paste) after different time intervals. (1/2 of the internal surface area covered by silver paste).

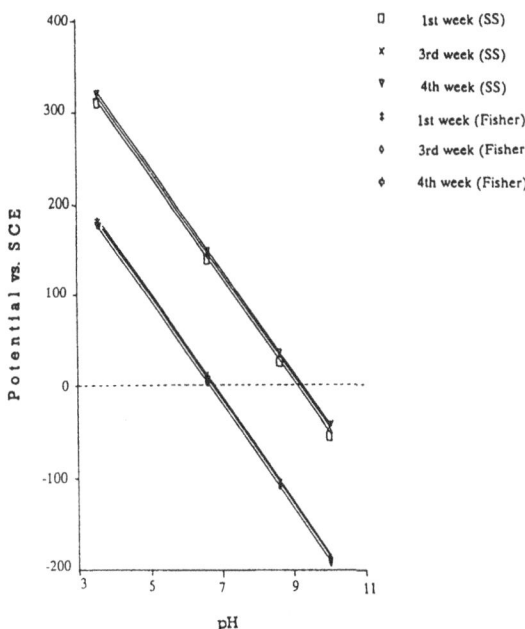

Fig. 7 . Reproducibility of solid state (SS) pH electrode (with Cu wire and silver paste) after different time intervals. (with maximum internal surface area covered by silver paste).

Table I. Response of Fisher and various Solid State (SS) pH glass electrode.

Type of Electrode	Slope mV/pH
Fisher Glass Electrode	58.0
(SS) Cu wire + Ag paste	58.0
(SS) Graphite fiber + Liquid solder + Graphite fiber	58.0
(SS) Cu wire + Duco cement	57.1
(SS) Graphite fiber + Duco cement + Graphite fiber	55.0

Table 2. Effect of covered internal surface area on Solid State pH glass electrode response.

Fraction of Internal Surface Area covered	Slope mV/pH
1/4	57.7
1/2	58.3
3/4	58.3
1	58.8

pH glass electrodes are shown in Figures 5-7. The solid state pH glass electrode with silver paste and copper wire having different internal surface areas covered showed no effect with time (Table III). However, the solid state pH glass electrode made with Duco cement and copper wire and the one with Duco cement and graphite fibers showed changes in absolute potential values (Figs. 8-9) although the value of the slope (mV/pH) did not change (Table IV). These results showed that there is a slow change in the internal junction potential. A change in color of the copper wire was noticed as it changed from rust to greenish blue at the junction of the glass surface and the copper wire, which could be the reason of changing internal potential; means of preventing such air oxidation of copper should be provided.

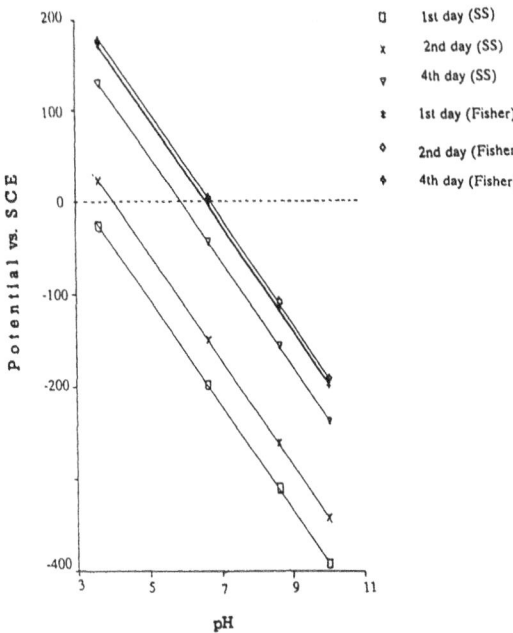

Fig. 8   Reproducibility of solid state (SS) pH glass
electrode (with Cu wire and Duco cement) after
different time intervals.

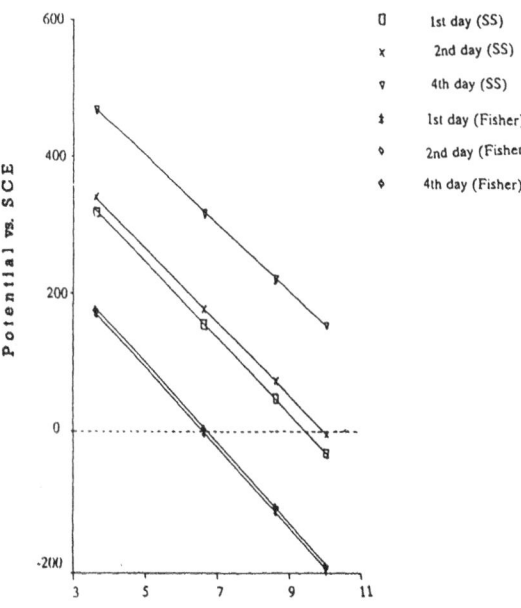

Fig. 9 . Reproducibility of solid state (SS) pH glass
electrode (with graphite fibers and Duco cement)
after different time intervals.

Table 3. Stability of Solid State pH glass electrode (with Cu wire and Duco cement) response after different time intervals.

Time Interval	Slope mV/pH Fraction of Internal Surface Area Covered		
	1/4	1/2	1
After One Week	57.7	57.5	57.7
After Two Week	57.5	58.0	57.7
After Fours Weeks	57.7	58.0	58.0

Table 4. Stability of Solid State pH glass electrode response after different time intervals.

Time Interval	Slope mV/pH		
	Duco Cement + Cu wire	Duco Cement +Graphite Fiber	Fisher
After One Day	57.5	55.8	58.0
After Two Day	58.0	55.5	58.0
After Fours Days	57.3	55.3	58.0

A Chinese pH glass electrode with silver/silver chloride as an internal reference electrode showed linear behavior (Fig. 10). When it is changed to a solid state pH glass electrode (with silver paste and copper wire), it showed promising results, and a linear response was observed (Fig. 10). The absolute potential values were changed, which was due to the change in the internal reference.

The responses of these new sodium electrodes were found to be similar to the commercial sodium electrode (Table V) under all conditions, i.e.; with and without internal solution.

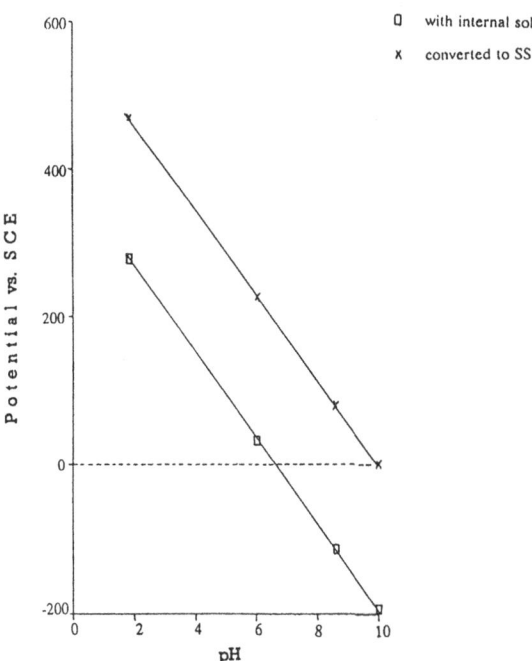

Fig. 10 Potential vs. pH response of chinese glass electrode (with internal solution and after conversion to solid state with Cu wire and silver paste).

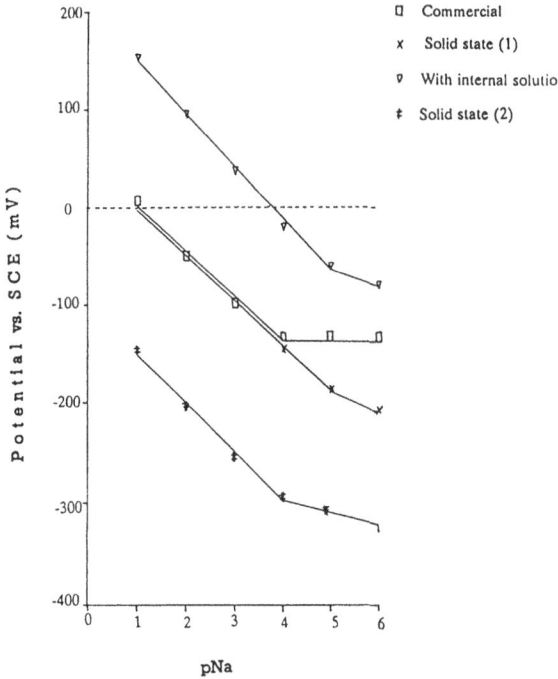

Fig. 11 Potential vs. pH respond of various sodium ion selective electrodes at pH 8.

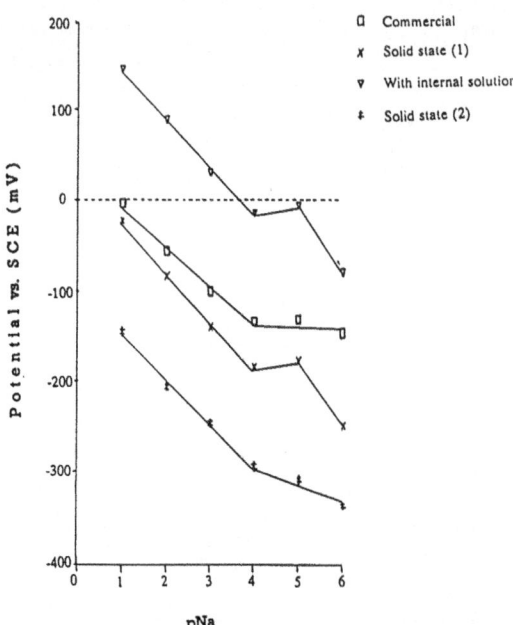

Fig. 12   Behaviour of various sodium ion selective
electrodes at pH 9.

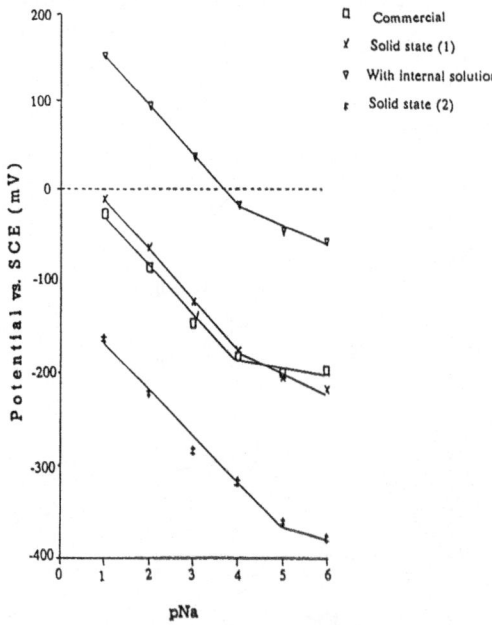

Fig. 13   Behaviour of different sodium ion selective
electrodes at pH 10.

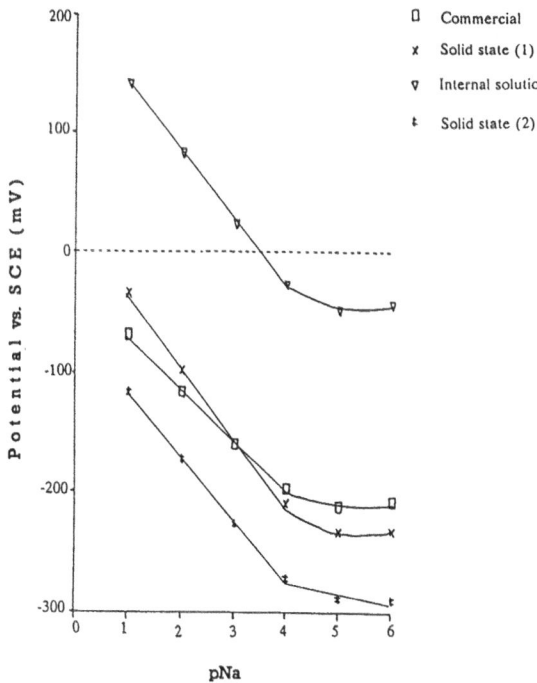

Fig. 14   Behaviour of different sodium ion selective
          electrodes at pH 11.

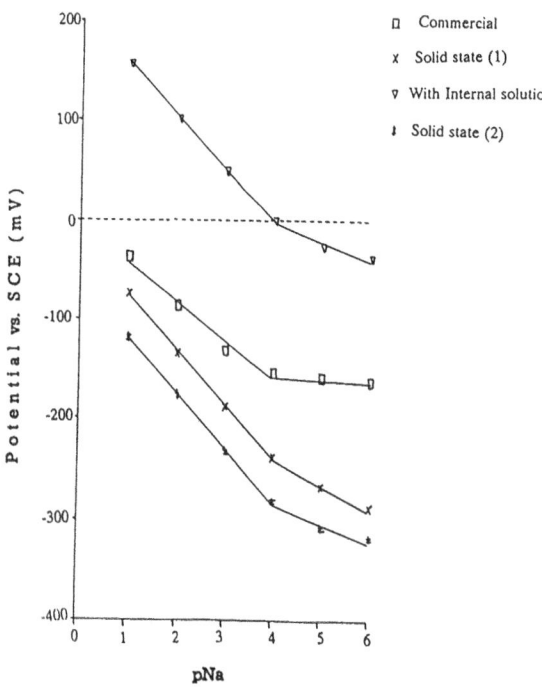

Fig. 15   Behaviour of various sodium ion selective
          electrodes at pH 12.

Table V Response of various sodium ion selective electrode at different pH's.

Electrode	Slope mV/pNa				
	pH = 8	pH = 9	pH= 10	pH = 11	pH = 12
Commercial	46	45	50	54	43
Solid State[1]	51	54	55	54	58
Solid State[2]	50	49	49	48	52
0.1 M KCL+ (Ag/AgCl)	54	54	55	51	56

(1) Copper wire attached to the inner surface of the bulb with silver paste was used as internal reference.

(2) Graphite fiber string glued to the inner surface of the electrode bulb with graphite fiber mixed in liquid solder was used as an internal reference.

The best response obtained from the sodium solid state (1) electrode (with copper wire + silver paste) was at pH 8 (Fig. 7.). At this pH, a slope of 51 mV/pNa was obtained for the pNa range of 1-5. The behavior of the other electrodes at pH 8 was also found to be similar.

The response of all the electrodes did not change much with a change in the pH (Figs. 11-15).

The sodium glass electrode is also a pH and pOH electrode with a zwitterionic surface which can adsorb H^+, OH^- and Na^+ ions[12], and that is why it is essential to make all the sodium ion measurements with the sodium electrode in a basic buffer solution.

CONCLUSION

There have been no laboratory supports for that the pH glass electrode mechanism is attributed to the passing the proton through a thin glass membrane, the concentration cell, the ion exchange of sodium ion with proton, the Donnan equilibrium of osmosis, or the ion hoping. All the above proposed mechanisms have been based on an assumption that an internal solution is necessary and plays an important role in the electrode potential development. Now, the evidence of solid state pH electrodes without an internal solution has again disproved all the myths which have confused us for decades.

The results from the solid state pNa electrodes point out that they are not a cell, but a chemical capacitor. Its outer glass surface is adsorbed by both Na^+ and OH^- ions. Its charge phenomenon has been demonstrated. In a buffer solution, the number of OH^- ions adsorbed is kept constant, we measure the net charge due to various amounts of Na^+ adsorbed. We avoid the buffer containing alkali metals which can be adsorbed on the glass surface to different degrees, the organic alkyl amine cation does not adsorb on the glass surface. Our pNa electrodes showed better sensitivity (linear to 1×10^{-5} M) than the commercial pNa electrode, assuming that the conductivity of glass may affect the electrode sensitivity.

REFERENCES

1. Thompson, M. R. J. Res. Bur. Stds. 1932, 9, 833.
2. Bates, R. G. Determination of pH; Theory and Practice; John Wiley and Sons: New York, 1973; p 377.
3. Kryubov, P. A.; Kryubov, A. A. Russian Patent, 51, 509, July 31, 1937.
4. Bender, H.; Pye, D. J. U. S. Patent, 2, 117, 596, May 17, 1938.
5. Parr, R. A.; Wilson, J. C.; Kelly, R. G. Anal. Proc. 1986, 23, 291.
6. Cheng, K. L.; Ashraf, N. Talanta 1990, 37(6), 659.
7. Cheng, K. L.; In Electrochemistry. Past and Present; Stock, J. T.; Orna, M. V. EDS; ACS Symposium Series, 390, 1989, 289.
8. Eisenmen, G.; Rudin, D.O.; Casby, J.U. Science. 1957, 126, 831.
9. Isard, J.O. Natuure. 1959, 184, 1616.
10. Ashraf, N.; Dissertion: New Developments in Understanding the Mechanism of pH Glass Electrode; University of Missouri-Kansas City, 1991.
11. Katsuaki, O.; Goto, F.; Abe, K.; Yoslida, T. J. Electrchem. Soc. 1986, 136, 1989.
12. Durst, R.A. Ion Selective Electrodes; National Bureau of Standards Special Publication. 314, Washington D.C., 1969.6.

EFFECT OF STIRRING ON PH MEASUREMENTS

Ching-I Huang, Hsuan Jung Huang, and K. L. Cheng*

Department of Chemistry
National Sun Yat-sen University
Kaohsiung, Taiwan 80424 R.O.C.

ABSTRACT

Potential variations in pH measurements using a pH glass electrode and SCE during agitation by a magnetic bar, compressed gas, a glass rod, and ultrasound in the presence and absence of colloidal particles have been studied. The so-called streaming potential is partly related to the disturbance of the charge density of glass electrode double layer as a result of stirring. The observed potential under stirring (observed disturbed potential) is the potential difference between the undisturbed potential and the stirring potential which may be more positive or negative than the undisturbed potential depending on the testing conditions. The stirring effect derived from both glass electrode and SCE have been studied individually. The junction potential of SCE is affected by stirring or ultrasound for different reasons, chiefly for changes in charge density of the electrode interface. Electrolyte solution flowing in a vortex mode at the interface carries part of double layer charges and produces an electric field. The stirring of an electrolyte solution may also be viewed as a reversed electro-osmosis. That the stirring may result in the change of the double layer capacitance is suggested. An accurate pH measurement using a glass electrode should be done in a quiet solution containing no colloidal particles. For the potentiometric titration using a pH glass electrode as an indicator electrode should have no adverse effect, because only the relative potential jump signifies the end point.

INTRODUCTION

In potentiometric measurements using a membrane electrode, stirring a test solution has been a common practice for assuring a uniform solution and fast response. The stirring speed varies with each individual. Though the effect of stirring on the electrode potential has been known for a long time and discussed in literature, we have found conflicting recommendations on stirring in the pH measurement (5,6). It seems that most chemists

---

*Was a visiting professor at NSYU 1990-2.
Author for correspondence at
Department of Chemistry
University of Missouri-Kansas City
Kansas City, Missouri   64110
U.S.A.

Advances in the Applications of Membrane-Mimetic Chemistry
Edited by T.F. Yen *et al.*, Plenum Press, New York, 1994

Stirring for half an hour (stirring speed : 5)

pH=1 solution

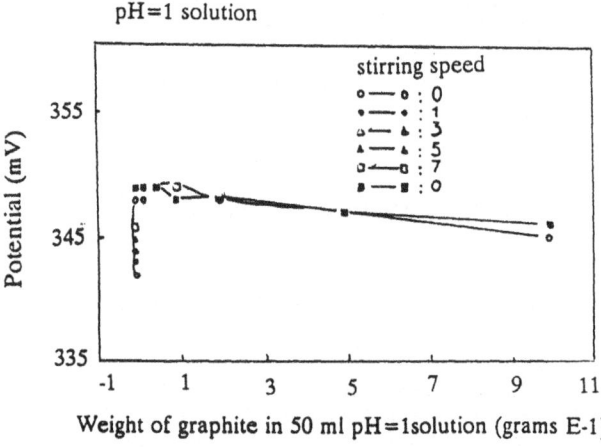

Weight of graphite in 50 ml pH=1solution (grams E-1)

Stirring for half an hour (stirring speed : 5)

pH=2 solution

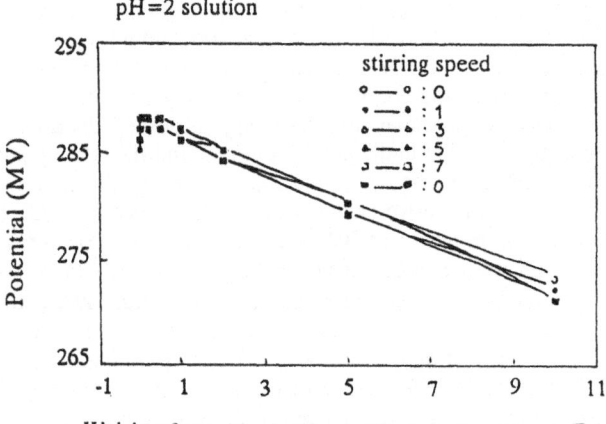

Weight of graphite in 50 ml pH=2solution (grams E-1)

pH=3 solution

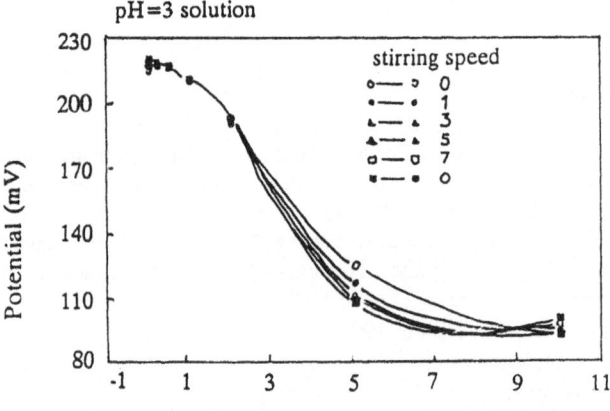

Weight of graphite in 50 ml pH=3 solution (grams E-1)

Figure 1

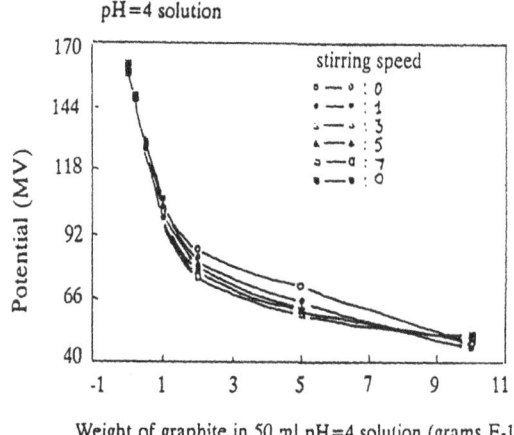

pH=4 solution

Weight of graphite in 50 ml pH=4 solution (grams E-1)

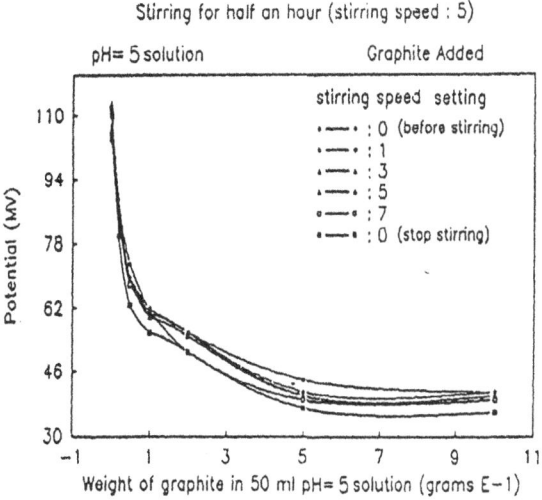

Stirring for half an hour (stirring speed : 5)

Weight of graphite in 50 ml pH= 5 solution (grams E−1)

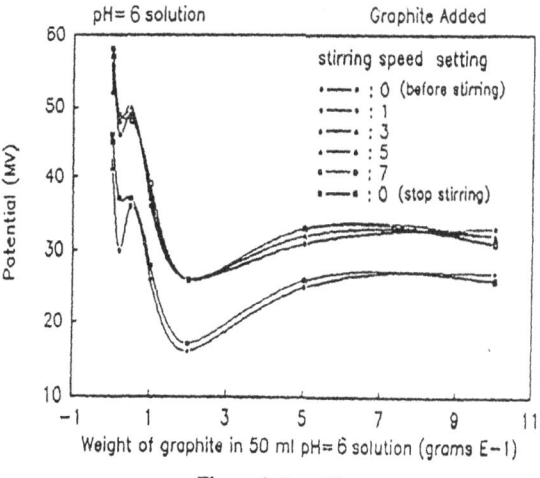

Weight of graphite in 50 ml pH= 6 solution (grams E−1)

Figure 1. (cont'd)

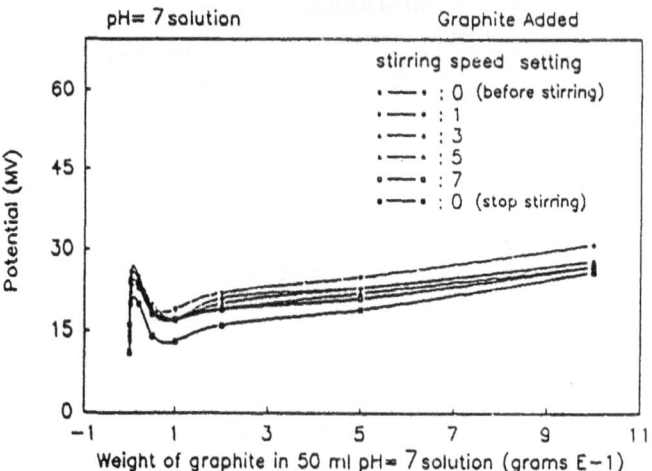

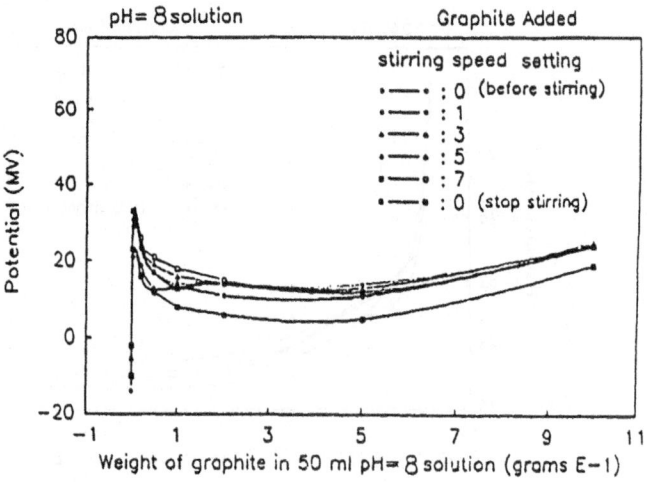

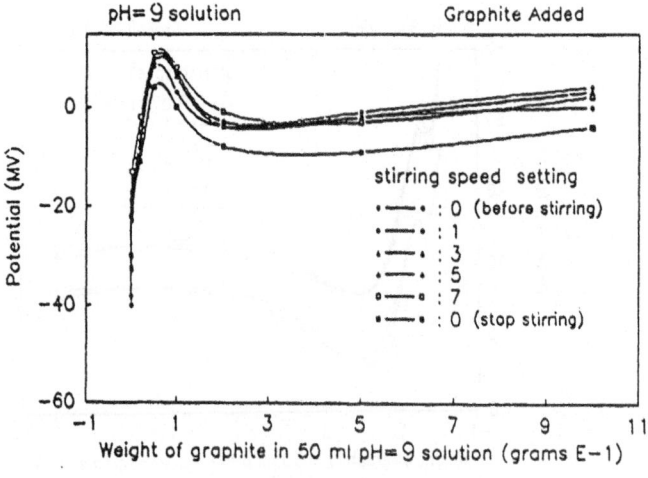

Figure 1. (cont'd)

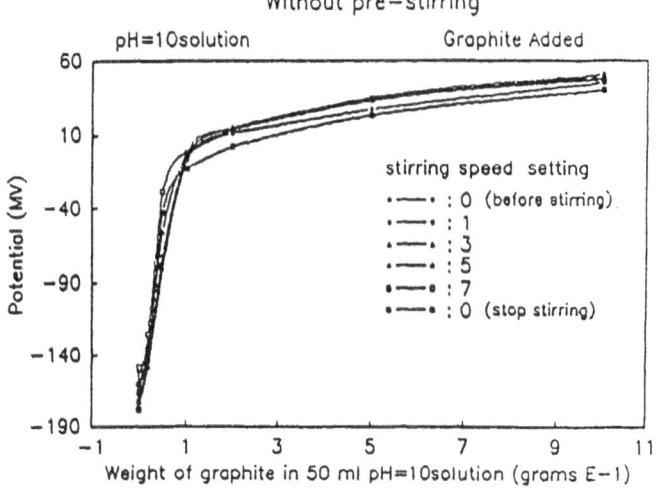

Without pre-stirring

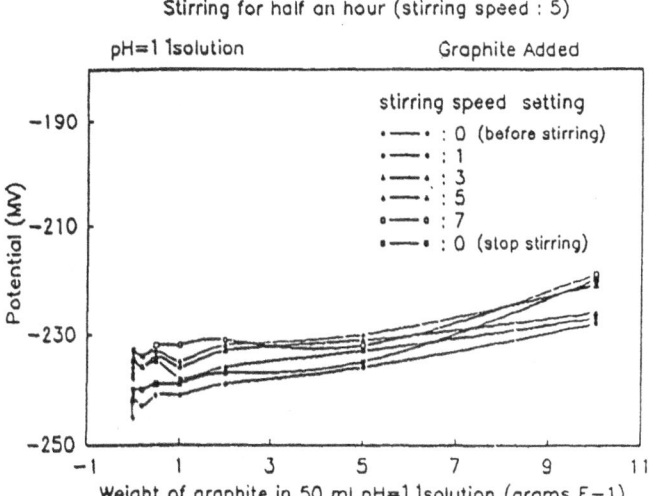

Stirring for half an hour (stirring speed : 5)

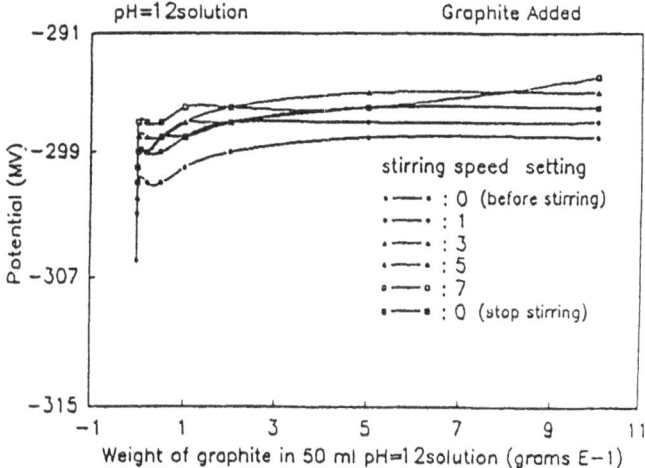

Figure 1. (cont'd)

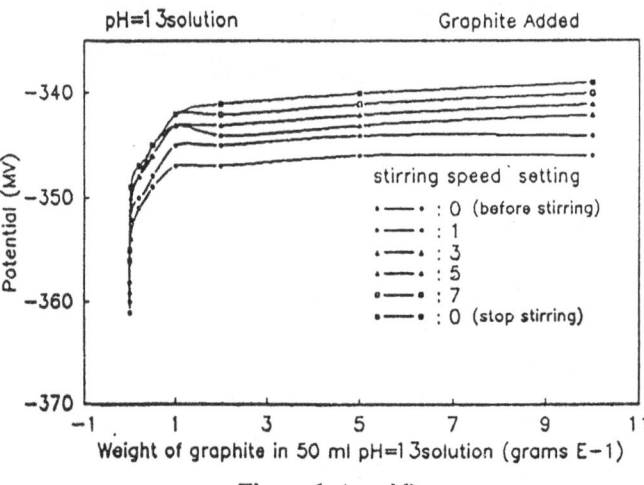

Figure 1. (cont'd)

have never paid much their attention to any stirring effect on the pH measurement.

The present studies focus on the factors affecting the electrode potential caused by various stirring techniques and the mechanism of stirring potential relating to the pH glass electrode (1-4).

Stirring a suspension containing colloidal particles is another serious matter as they are charged and cause a complicated disturbance on the double layer and the interface of electrode which acts a capacitor (3.7). Stirring causes the change in the electrode capacitance because the flowing electrolyte solution moving in a vortex mode carries part of double layer resulting in an electric field and a stirring current. The observed disturbed potential is the difference between the undisturbed potential and the stirring potential.

Most analytical books have avoided to discuss the stirring potential and its mechanism. The present results may help get an insight understanding of the stirring potential, particularly, in the pH measurement. Chemists have often a misconception that the pH measurement is specific for hydrogen ion concentration, rarely subject to interferences from other conditions. This may not be true. The present results will demonstrate some physical conditions which affect the electrode potential.

During the ultrasonic agitation of an electrolyte solution, solvent molecules are forced flowing into an oscillatory motion due to the periodic explosive cavity formation at the electrode surface. Ions in the double layer move with solvent molecules but may lag somewhat behind. When the lag in a cation (or anion) motion differs from that in anion (or cation) motion, because of a different frictional coefficients, a small ultrasonic potential difference may develop at the frequency of the ultrasonic wave (2,4). In the case of SCE, the ultrasonic explosive cavity may drastically change the KCl concentration at the interface.

EXPERIMENTAL

APPARATUS

A Corning pH/ion meter 150 was used for pH and potential measurements. A magnetic stirrer was used for stirring. A Branson ultrasonic cleaner B-12, 50/60 Hz was used for agitation. A KIPP & ZONEN recorder was used for recording potential results.

pH = 4 buffer solution:	Potassium biphthalate buffer, 0.05 M. Mold inhibited with 0.05% formalin.
pH = 7 buffer solution:	Potassium phosphate monobasic-sodium hydroxide buffer, 0.05 M.
pH = 10 buffer solution:	Mold inhibitor added, potassium carbonate-potassium borate-potassium hydroxide buffer, 0.05 M.

PROCEDURE

A. Stirring Effect on Graphite Suspension Solution

Prepared different pH solutions (from pH 1 to 13). Added different weight of graphite (0, 0.002, 0.02, 0.05, 0.1, 0.2, 0.5, 1 g) into each 50 mL of solution, after pre-stirring for 30 minutes with stirring speed 5, used different magnetic stirring speed (0, 1, 3, 5, 7) to stir the solution, and measured the potential. The results are shown in Fig. 1.

B-1. pH Measurement in Different Stirring Methods

Stirring the pH 4 solution (unbuffered) with different methods. 1. magnetic bar stirring, 2. ultrasonic stirring, 3. manual stirring, 4. stirring with compressed liquid nitrogen gas, 5. magnetic field (thermolyne stir-mate magnetic motor without magnetic bar).

B-2.

Added 0.2 g graphite to a pH 4 (unbuffered) solution, and carried out the same stirring procedure like B-1 to measure the potential. The results are shown in Fig. 3.

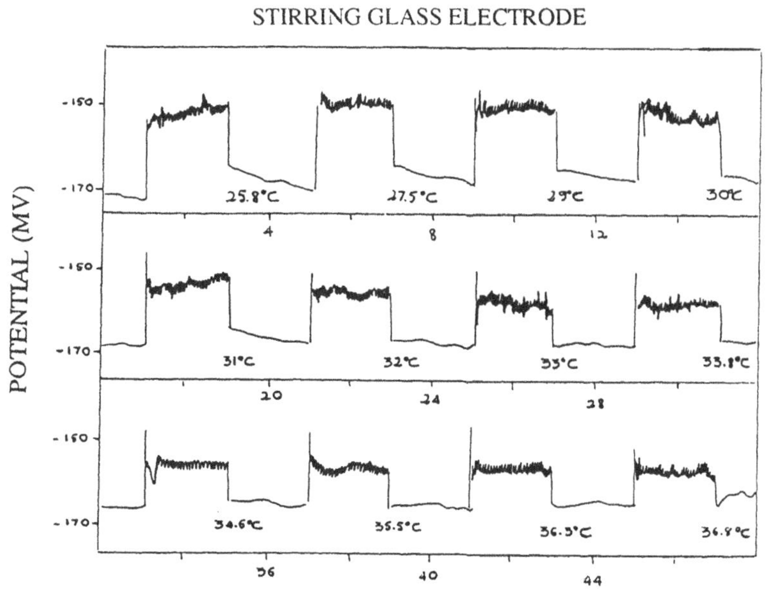

STIRRING GLASS ELECTRODE

Figure 2a. Effect of temperature on potential of pH 10.0 solution

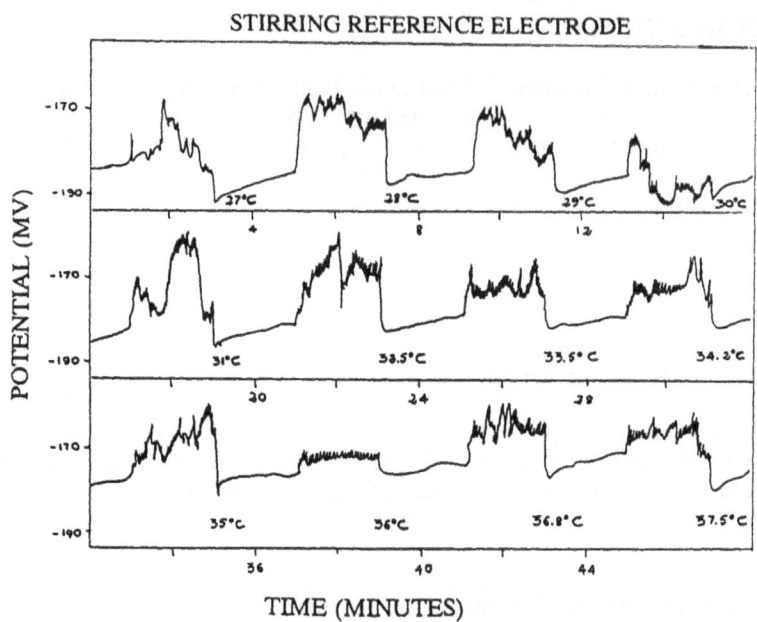

Figure 2b. Effect of temperature on potential of pH 10.0 solution

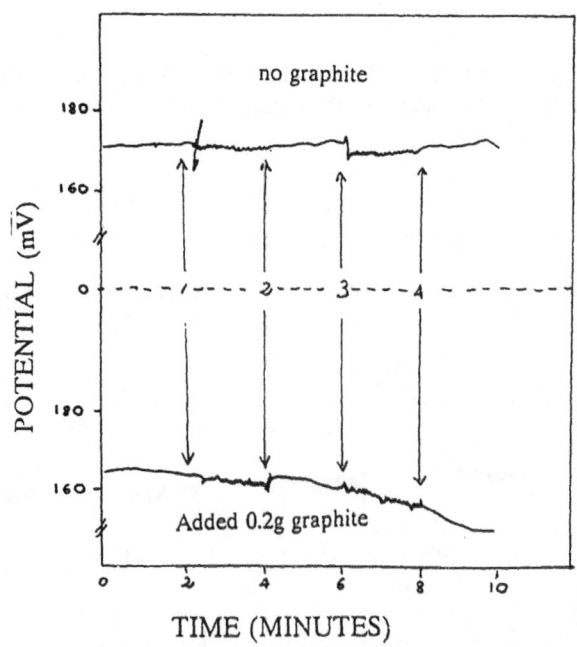

Figure 3. Effect of stirring with compressed liquid nitrogen gas on potential of pH 4.0 solution (unbuffered)

1: stirring applied with liquid N_2 gas
2: stirring stopped
3: stirring applied with liquid N_2 gas again
4: stirring stopped

C-1. Stirring Effect On Glass And Reference Electrodes

Putting each electrode (glass electrode and reference electrode) in two separate beakers containing the same 50 mL solution. For studying the stirring effect with each individual electrode, one beaker was stirred while the other remained unstirred. The two beakers were connected with an insulated copper wire.

RESULTS AND DISCUSSION

If an electrolyte flows through a tube, a potential difference develops between the inflowing and outflowing solution. The application of a pressure in an electrolyte solution should give rise to a potential difference and a corresponding electric field. This is the phenomenon of streaming potential (1-4) which has been applied to the stirring effect (2,4). At 25 °C, the streaming potential for an aqueous solution is represented by (4)

$$\Phi = P\zeta/k(10^{-6}) \tag{1}$$

where P is measured in mm Hg pressure and Φ (streaming potential) and ζ (zeta potential) are in mV. The streaming potential has been described in the literature with a capillary mode. It may not fit into non-capillary situation and the disturbance without liquid flow. The reasons will be explained in the latter part of this paper.

Effect of stirring speed. According to Eq.[1] the streaming potential should be increasing with increasing solution flow rate, or pressure P, however, our results in Fig. 1 indicate that increasing stirring speed does not alter the potential significantly. Some small difference in potential due to the stirring speed were noted for very basic solutions. This is one of important reasons that the streaming potential equation as described in Eq.[1] does not explain the stirring effect in the present studies.

Effect of pH. Stirring gives rise to different effects on potential for different pH solutions. The electrode potential tends to shift to less positive in acid solution; on the other hand, it tends to shift to more positive in basic solutions. Consider that the adsorption of H^+ or OH^- ions on the electrode double layer interface produces a positive or negative potential, i.e., surface charges determine the potential shift directions. When liquid flows pulling away some charges or ions from the electrode interface, or changing the capacitance. In a neutral region where there are few H^+ or OH^- ions, so the number of ions being disturbed at the double layer is much less than that in acid or basic region, resulting in a less degree of vibration (see Fig. 1, pH 7). For pH 6, an up and down shift takes place, it may be due to the fact that the glass has its isoelectric point of around pH 6, the shift of charge direction from the neutral point may cause the instability of potential. this is to be further studied.

Effect of stirring with a magnetic bar. The results in Figs. 1 and 2 were obtained with a magnetic bar stirring. By using manual stirring or stirring with compressed nitrogen gas similar results as those with a magnetic bar were obtained (Fig. 3). This shows that the stirring effect is purely mechanical without influence from the magnetic motor placed underneath the beaker. It was also found that an unstirred solution placed inside a 200 MHz NMR sample well no stirring effect on potential was noticed.

Effect of temperature on stirring effect. Experiments were carried out separately for SCE and glass electrodes between 25.8 and 36.8 °C for pH 10 buffer solutions. Results in Fig. 2a and 2b show that the stirring potential shifts higher with decreasing temperature. It could be that at higher temperature, the adsorbed ions are moved away from the double layer glass interface faster than those are returned to the electrode surface with increasing temperature. the SCE shows much noisy shift curves at pH 10 and high temperatures than the glass electrode.

Effect of colloidal particles. It has been known that the pH measurement is affected in suspensions(7). Different suspensions such as graphite, silica, silver sulfide, and soil have been studied.

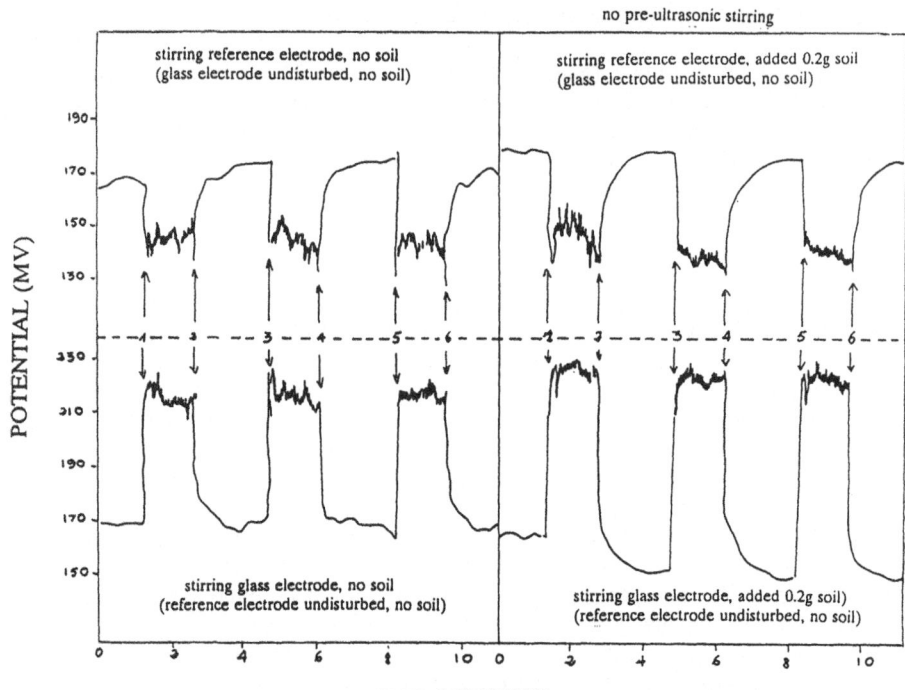

Figure 4. Effect of ultrasonic stirring on potential of pH 4.0 solution

1: ultrasonic stirring on
2,4,6: stirring off
3,5: ultrasonic stirring on again

The stirring potential effect may be considered as a reversed electro-osmosis(4). during the electrophoresis, we have the ionic movement in an electric field upon the applied potential. Reversely, without an applied potential, charged particles are moved by other mechanical means such as stirring or flowing solution, a potential is then generated. This is called a stirring potential.

An aqueous very fine powder suspension of graphite gives a pH 9.2 with the pH glass electrode, its clear filtrate fails to any color test with phenolphthalein. This misleading pH glass measurement is attributed to adsorption of charged graphite on the glass surface providing an extra negative potential reading. Qualitatively, the moving suspension affects the electrode potential through removing part of ions from the double layer and creating an electric field. When stirring stops, the removed ions immediately return to proper positions in the double layer at the interface. The undisturbed potential is reappeared. Because of the presence of charged particles, the stirring effect is much enhanced(7). Such an additional effect will be leveled off after a maximum amount of particles has been added. The start and stop stirring effect are demonstrated in Figs. 2-5.

Stirring effect by ultrasound. During the ultrasonic agitation of an electrolyte solution or suspension, solvent molecules are forced into an oscillatory motion due to the periodic explosive cavity formation at the electrode surface. Ions in the solution move with the solvent molecules but may be lag somewhat behind. When the lag in cation motion differs from that in anion motion, because of different frictional coefficients, a small ultrasonic potential difference may develop at the frequency of the ultrasonic wave.

Except the results in Fig. 1, we carried out separately the stirring effect of each individual electrode. The pH glass electrode and the SCE were separated so that we could stir one electrode without stirring the other. Interestingly, stirring causes different effect on them for different mechanism.

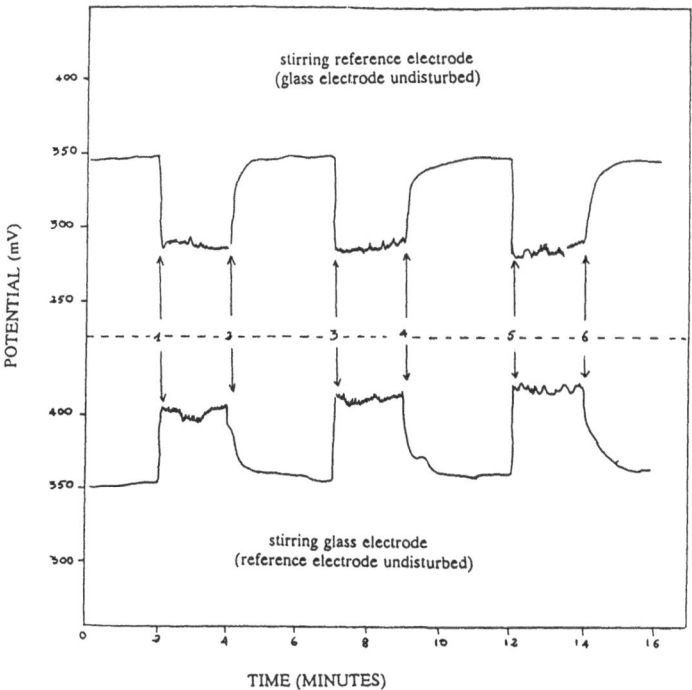

Figure 5a. Effect of ultrasonic stirring on potential of pH 1.0 solution (unbuffered)

        1: ultrasonic stirring on
        2,4,6: stirring off
        3,5: ultrasonic stirring on again

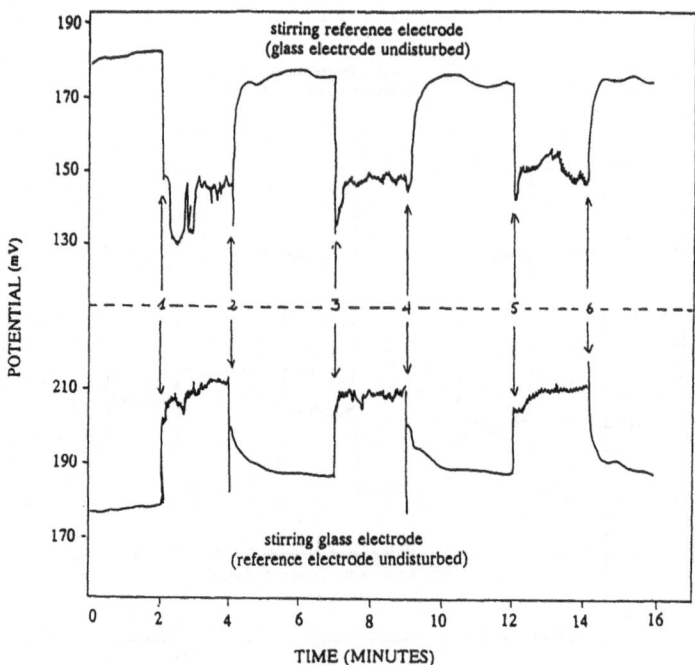

Figure 5b. Effect of ultrasonic stirring on potential of pH 4.0 solution (unbuffered, and added 1.0g NaCl)

1: ultrasonic stirring on
2,4,6: stirring off
3,5: ultrasonic stirring on again

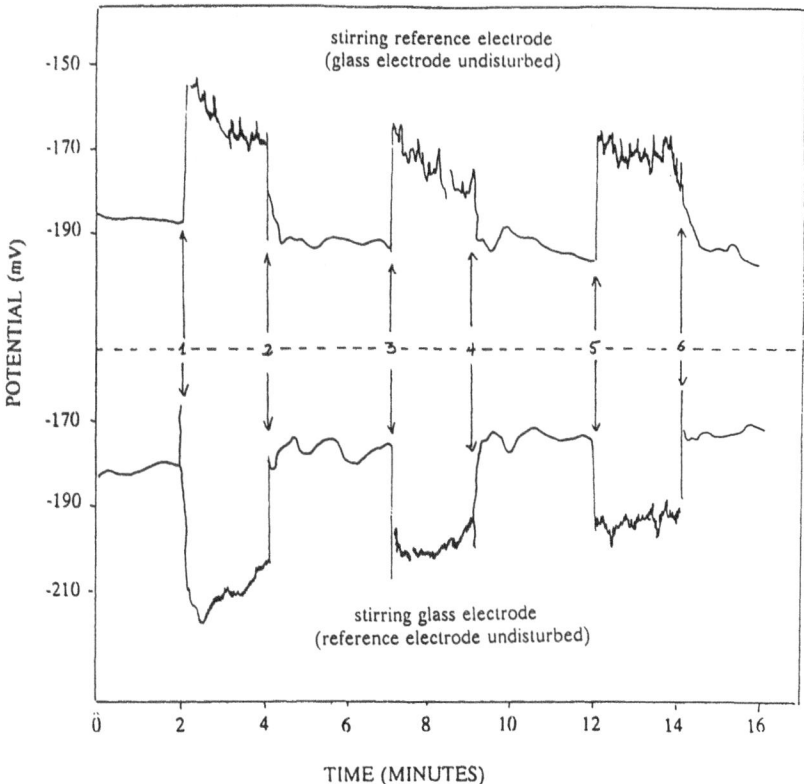

Figure 5c. Effect of ultrasonic stirring on potential of 10 solution (buffered)

1: ultrasonic stirring on
2,4,6: stirring off
3,5: ultrasonic stirring on again

Preliminary results of ultrasonic effect on soil suspensions at pH 4 are shown in Fig. 4 indicating opposite effects on pH glass electrode and SCE. The presence of soil seemed to show some additional effect on glass electrode than on SCE.

The curves in Fig. 5a, 5b, and 5c show stirring effects of reference electrode and glass electrode at opposite directions. Note marked changes of effect directions from acid to basic solution. When they are placed in the same solution they may cancel each other the stirring effect to a certain degree. The SCE may be sensitive to pH change and the dilution of its saturated KC1. The glass electrode is subject to the powerful explosive cavity formation, more H^+ ions enter into the double layer interface making the potential more positive. On the other hand, at pH 10, more OH ions enter into the double layer interface making the electrode more negative. When the ultrasonic agitation stops, the stirring potential returns to the undisturbed potential. But we noted that after the ultrasonic stirring the undisturbed potential still shows somewhat unstable motion for both SCE and glass electrode showing the residual effect.

CONCLUSIONS

The present studies demonstrate that agitation of solution with various means such as magnetic bar, manual glass rod, and ultrasound during the potential measurement using a pH glass electrode and an SCE causes considerable potential changes. Stirring is commonly practiced in the pH measurement and often recommended for vigorous stirring in order to have a uniform solution and rapid response. The presence of colloidal particles such as graphite and silver sulfide precipitates enhances the stirring effect. Further, the charged particles tend to cover up the electrode and SCE have been individually studied in separate beakers. The factors affecting the stirring effect such as stirring speed, pH, temperature, and presence of colloidal particles change the electrode potential differently. The stirring effect has been explained by the streaming potential which is based on a capillary model in literature. The term of streaming potential and its equation derived from the axillary model should not be applied to the non-capillary and free space solution which may not be in a flow motion (ultrasound). Instead of streaming potential, a term of stirring or disturbing potential is suggested. Its potential origin may be traced to the changes of double layer charge density and capacitance. It may be considered as the reversed electro-osmosis from most stirring cases. For an accurate pH measurement with a glass electrode and other potentiometric measurements using membrane electrodes should be done in a quiet solution containing no charged particles.

ACKNOWLEDGEMENT

This work was supported by the Chinese National Research Council, R.O.C.

REFERENCES

1.   J. O'M. Bockris and A. K. N. Reddy. "Modern Electrochemistry," vol.2, Plenum Press (1977).
2.   J. Koryta and J. Dvorck. "Principles of Electrochemistry," Wiley (1987).
3.   Paul C. Hiemenz. "Principles of Colloid and Surface Chemistry," Dekker (1986).
4.   D. R. Crow. "Principles and Applications of Electrochemistry," 3rd Ed, Chapman and hall (1988).
5.   "Orion Research Analytical Methods Guide," 4th Ed., October (1972).
6.   R. A. Durst, Ed. "Ion Selective Electrodes," National Bureau of Standards, Special Publ., 314, Washington D.C. (1969).
7.   S. X. Yang, K. L. Cheng, L. T. Kurtz, and T. R. Peck. Particulate Sci. and Technol. 7-139 (1989).

APPLICATION OF MEMBRANE MIMETIC CHEMISTRY TO FOSSIL FUEL CONVERSION AND ENVIRONMENTAL ENGINEERING

T.F. Yen, J. Chen and K.M. Sadeghi

Civil and Environmental Engineering Department
University of Southern California
Los Angeles, CA   90089, U.S.A.

ABSTRACT

Some recent developments of using the membrane-mimetic chemistry concept for dealing with energy conversion and environmental energy are reviewed. The heading are I. The regular framework formation in solutions containing dissolved silicates, II. In situ surfactant from petroleum and vesicles, III. Asphaltene peptization and conversions, IV. Reverse micelle multiphase biocatalysis.

INTRODUCTION

Membrane-mimetic chemistry describes the chemistry processes in simple media that mimic aspects of biomembranes. Any arrangement of amphiphilic (or surfactant) molecules forming monolayers, bilayers, multilayers, micelles, reversed micelles, unilamellar vesicles, multilamellar vesicles or polymerized vesicles can be related to biomembranes. The emerging scientific discipline concerned with development and utilization of membrane-moderated (and inspired) processes in organized surfactant assemblies has great potential engineering applications, especially in the field of environmental engineering as well as fossil fuel conversion.

In the past, as membrane-mimetic chemistry was developing, application was limited to the control of rates and stereochemistry of reactions, enhancement of solar energy conversion, micellar catalysis, and drug encapsulation and delivery systems[1,2]. In recent years, enzyme mimickers, angiogenesis[3], multiphase enzyme catalysis[4], and even control of uniform fiber growth in the textile industry[5] have yielded important technology transfers from basic chemistry studies.

Membrane-mimetic chemistry principles can be applied to both decontamination and beneficiation processes where impurities must be separated from host molecules by a micellar change. These changes are dependent on the geometrical patterns and assemblages of micelles, vesicles or large liquid crystals (gels). They can be affected by radical transfer initiated by ultrasound irradiation or other microwave sources. Reactions proceed in a submicro environment of compression, rarefaction or cavitation with localized instantaneous high pressure and temperature flux or microstreaming.

Initially, our investigations were the result of isolation of bitumens from bituminous sands using sodium silicate in aqueous systems. It was soon discovered, however, that the isolated bitumen had been upgraded by the hydrogen radicals through the water and the in-situ formed surfactant. The purpose of the in-situ surfactants is to reverse the asphaltene micelle so that polar groups can face outward and thus be accessible to interactions with agglomerations in a membrane-mimetic manner. This type of reaction is also applicable to oil shale, asphalt, and heavy oil.

Some recent examples are given below to illustrate the applications of membrane-mimetic

Advances in the Applications of Membrane-Mimetic Chemistry
Edited by T.F. Yen *et al.*, Plenum Press, New York, 1994

chemistry for the solution of problems in environmental engineering as well as fossil energy conversions:

I. THE REGULAR FRAMEWORK FORMATION IN SOLUTIONS CONTAINING DISSOLVED SILICATES

A wide range of geometric patterns and arrangements are known for silicon-containing minerals[6,7]. It is widely recognized when sodium silicate dissolved in water or the silica dissolved in aqueous solution of alkali metal hydroxides, beyond its monometric species, (othosilicate SiO_4^{4-} anion) there are other condensed silicate anions, e.g., pyrosilicate anion, $Si_2O_7^{6-}$, cyclic silicate anions such as $Si_3O_9^{6-}$ and $Si_6O_{18}^{12-}$ and infinite drain anion such as $(SiO_3^{2-})_a$ or double strand of cross linked chain $(Si_4O_{11}^{6-})_a$, etc. Furthermore, there are a number of infinite sheet anions such as $(Si_2O_5^{2-})_a$ and the framework minerals of $(SiO_2)_a$ that have been omitted for consideration. The definitive ordered structure of aqueous silicate solution was demonstrated by the spectra of $^{29}Si_8$ NMR by Harris[8,9,10]. Harris, using high field conditions (99.4 MHz, B_0 = 11.75 T) with samples enriched in ^{29}Si and applying homonuclear Si-Si decoupling technique, was able to tentatively assign singlet ^{29}Si chemical shift for a silicate solution as follows:

	$^{\Delta\delta}Si/ppm$
Monomer, Q^0	0.00
Dimer, Q_2^1	-8.62
Cyclic trimer, Q_3^2	-10.19
Cyclic tetramer, Q_4^2	-16.10
Prismatic hexamer, Q_6^3	-17.21
Cubic octamer, Q_8^3	-27.55

For the symbol Q_j^i, i is the number of attached siloxane bridges with the state of protonation is ignored and j is the number of silicon atoms.

Based on Harris studies, it is interesting to note that both the prismatic hexamer, Q_6^3 and the cubic octamer, Q_8^3 are important in solutions (see Fig. 1). Furthermore, the stepwise origins of the sequence for the formation of these framework be substantiated by the beginning and intermediate species noted in Figs. 2 and 3. Fig. 2 is adopted by the chemical shifts for the series of cyclic trimer, Q_3^2 to the final product of the prismatic hexamer, Q_6^3; or

$$Q_3^2 \rightarrow Q^1Q_2^2Q^3 \rightarrow Q_3^2Q_2^3 \rightarrow Q_2^2Q_4^3 \rightarrow Q_6^3$$

Similarly, the cyclic tetramer, Q_4^2 may lead to the cubic octamer. Q_8^3, or

$$Q_4^2 \rightarrow Q^1Q_3^2Q^3 \rightarrow Q_2^1Q_2^2Q_2^2 \rightarrow Q_2^3Q_4^2 \rightarrow Q_3^3Q_3^2Q^1 \rightarrow Q_4^3Q_3^2 \rightarrow Q_5^3Q_2^2Q^1 \rightarrow Q_6^3Q_2^2 \rightarrow Q_8^3$$

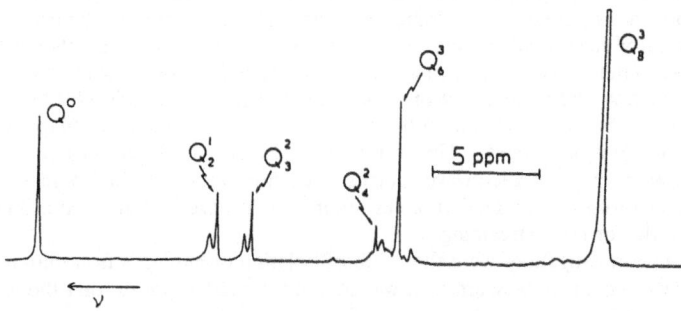

Figure 1 ^{29}Si NMR at 79.5 MHz and $5°C$ of alkaline aqueous tetramethylammonium silicate solution with atomic ratio N:Si = 1.0 and concentration ~ 2 M in Si. Sample was enriched in ^{29}Si to 95.3%. Peak assignments are indicated. Peak due to cubic octamer is not taken to its full height. Measurements show this peak is ca. 12 times as intense as that of monomer. (Based on Ref. 8)

The second series can be seen in Fig. 3. Harries also found additional species may be existing but is small amount such as $Q^1Q^2Q^2Q^1$, $Q_2^2Q_4^3$, $Q^1Q_3^2Q^3$, Q_8^3, $Q^2Q_6^3$, (see Fig. 4). Those species do not have high symmetry as a prism or a cube.

We can probably conclude that the highly, geometrically symmetric species are predominant in the silicate solution. These imply that there are ordered extended cylindrical and prismatic, channels in silicate solution. The ordered arrangements of these structures actually behave as cavities made of silicon atoms in the vertices of the framework which are surrounded by oxygen atoms. Therefore, these structures behave as double deck crown ethers in cavitands or spherands (Fig. 5). The silicate anions may have charges in the regular intervals of the cavities. In this manner, it is rather easy for the transport of the counter ions, e.g., Na, K, etc. through these cavities.

Figure 2   Identified series of species containing three-membered rings in solution. Vertices in diagrams indicate positions of silicon atoms, which are bridged by oxygens. Additional oxygens (or OH groups) are present as required for valency fulfillment. Chemical shifts are given in ppm with respect to signal due to monomer resonance on high-frequency-positive convention, as obtained for solution of alkaline aqueous potassium silicate with atomic ratio K:Si = 1.0 and concentration 0.65 M in Si. (Based on Ref. 8)

We have used the sodium silicate aqueous solution for alkaline-water improved flooding for petroleum[11-15], separation of bitumen from tar sand[16-27], upgrading of heavy oil[28-30], decontamination of oilspill in soil and land[31-32] as well as reduction of environmental pollutants[33]. In case of the destruction of halogenated compounds in water, it is a good way to remove any trace of trihalomethane (THM) in portable water. We have attempted to redefine the high concentration of chloroform in water with very attractive results. We have tried similarly with bromomethanes and polychlorobiphenyls (PCBs) and the results are very encouraging. A plausible mechanism of the catalytical activities may be due to the fixed positions of the Si atoms in solution such as the connected cubes and prisms (Fig. 6). Evidence of the existence of these fixed points in 3-d space is the commonly observed sol to gel conversion of aqueous alkali silicates. The diffusion-limited aggregation (DLA) model may be propagated from cube to cube instead of square to square.

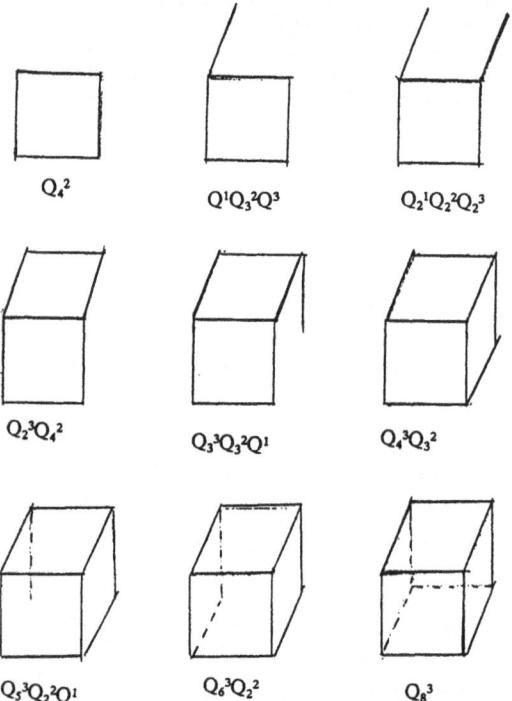

Figure 3   Series of square to cube intermediates

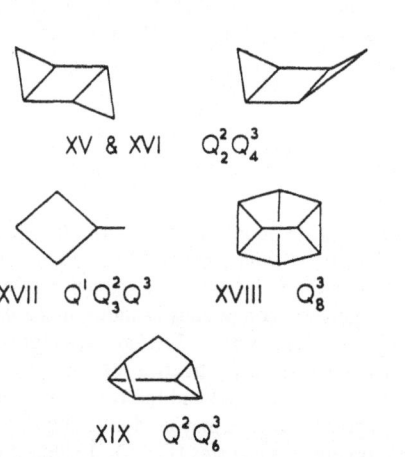

XIV linear tetramer $Q^1Q^2Q^2Q^1$

XV & XVI $Q_2^2Q_4^3$

XVII $Q^1Q_3^2Q^3$ XVIII Q_8^3

XIX $Q^2Q_6^3$

Figure 4   Additional species which may be present in solution from a spectrum. Evidence for existence of these species is not definitive. (Ref. 8)

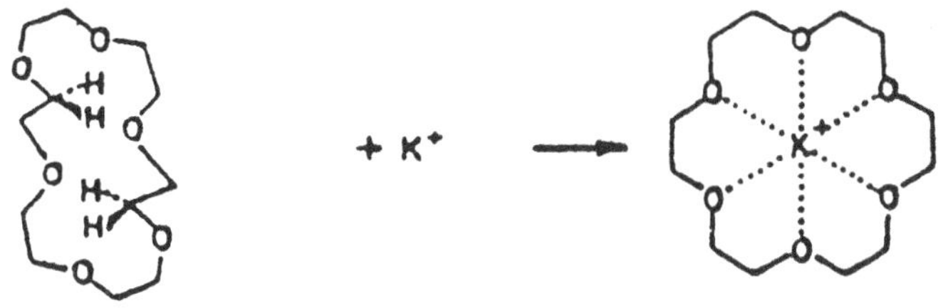

18-crown-6
(no cavity)

18-crown-6-K$^+$ complex
(cavity organized by K$^+$)

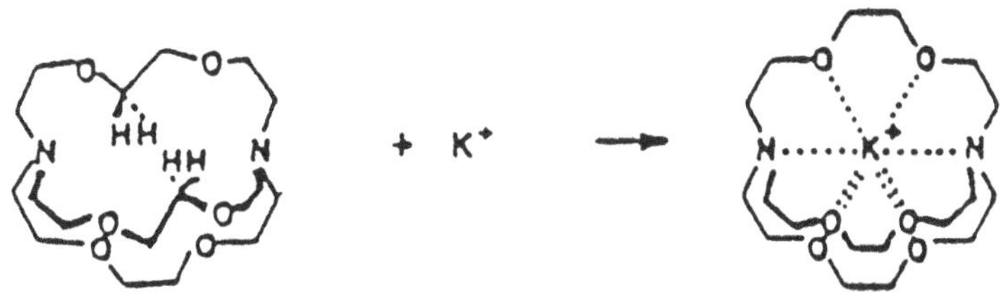

+ K$^+$ →

[2.2.2]cryptand
(no cavity)

+ K$^+$ →

[2.2.2]cryptand-K$^+$ complex
(cavity organized by K$^+$)

+ Li$^+$ →

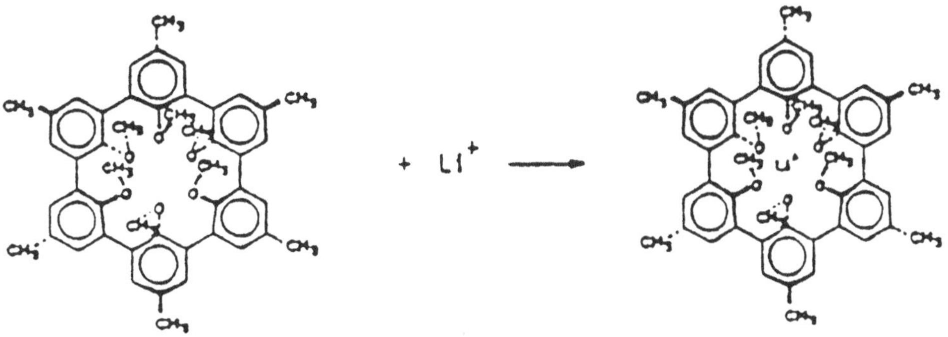

a spherand
(enforced cavity)

a lithiospnertum complex
(cavity filled by Li$^+$)

Figure 5   Comparison between crown ether, cryptand, and spherand.

Table 1

Hydrocarbons from Bacteria and Algae (percent only of total listed)

	E. coli (aerobic)	P. shermanii (anaerobic)	Chlorobium (sulfurbacteria)	Nostoc muscorum (blue-green)	Chlorogloea fritschii (blue-green)
n - C_{15}	0.5	2.1	1.5	0.35	---
n - C_{16}	1.7	2.6	0.75	0.35	0.26
Pristane	---	46.5	0.5	---	---
n - C_{17}	5.5	13.3	---	82.75	87.30
Phytane	---	1.0	50.0	---	---
n - C_{18}	27.6	3.6	0.5	0.41	0.055
n - C_{19}	12.0	3.8	1.3	---	---
n - C_{20}	10.0	3.8	1.3	---	---
n - C_{21}	5.5	4.2	1.0	---	---
n - C_{22}	6.0	4.1	1.5	---	---
n - C_{23}	8.3	3.1	3.0	---	---
n - C_{24}	7.4	1.5	4.1	---	---
n - C_{25}	6.0	1.0	6.9	---	---
n - C_{26}	3.3	0.5	10.8	---	---
n - C_{27}	3.3	0.5	13.1	---	---
n - C_{28}	0.5	---	2.1	---	---
heptadecane	---	---	---	16.10	12.29

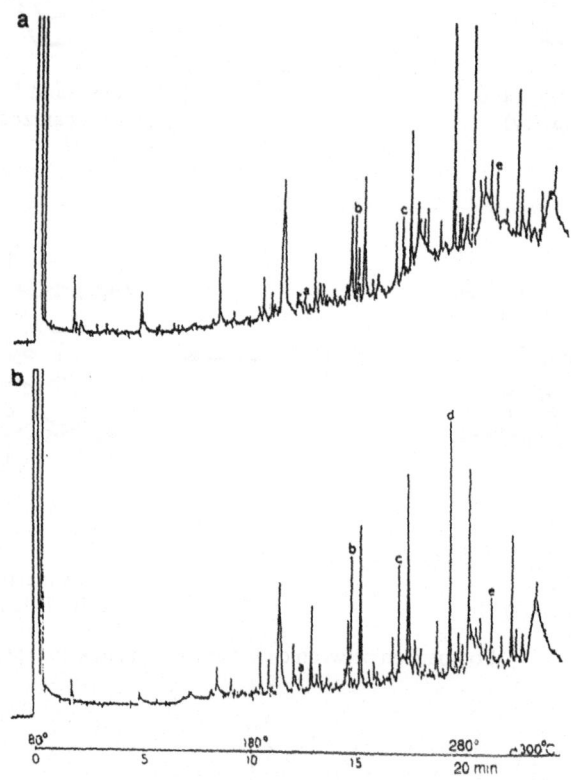

Figure 6   Gas chromatogram of methyl ester derivatives for the base-extractable carboxylic acids in (a) toluene and (b) pyridine soluble fractgions of bitumen from Athabasca tar sand. Se Experimental section for GC details. The peaks labeled are: (a) C_{14}, myristic acid; (b) C_{16}, palmitic acid; (c) C_{18}, oleic acid; (d) C_{20}, arachidonic acid; (e) C_{22}, decoranoic acid. (Based on Ref. 21)

II. IN SITU SURFACTANT FROM PETROLEUM AND VESICLES

It is known that petroleum are originated largely from marine biomass. One major component of the present day petroleum composition is what derived from lipids. The remains of the lipids classes of fatty acids, triglycerols, glycerophospholipids, sphingolipids and cholesterols are widely distributed in petroleum. The distribution of long-chain hydrocarbons correspond to those produced by microorganisms (Table 1). As a matter of fact many of the terpenoids including steranes, hiopanes and other carotenoids are recognized as molecular fossils or geological biomarkers[34].

Carboxylic acids exist in petroleum as indicated by the acid values determined usually from titration. The carboxylic acid can be easily obtained by mild oxidation of reactive hydrocarbons or by hydrolysis of the glycerides. Controlled electrooxidation can yield homologous of carboxylic acids from kerogens ar asphaltenes[35-36]. The existence of carboxylic acids in petroleum forms the basis of alkaline-water flooding or caustic flooding as a tertiary chemical recovery method belonging to the enhanced oil recovery (EOR). In actual cases, the efficiency of recovery is proportional to the in situ surfactants formed as determined by the interfacial tension (IFT). The IFT values are greatly influenced by the polar components or proton-replaceable compounds within the petroleum[12]. We have experienced that the preinjection of mineral acid followed by caustic injection can greatly improve the recovery[14]. This fact indicates that there are existing lipids requiring acid hydrolysis before the free carboxylic acids are released.

A typical GC-MS pattern from the reaction of alkali hydroxides with heavy functions of petroleum (Fig. 6). It ought to be noticed that there are many N- and S- containing carboxylic acids in petroleum. These long-chain carboxylic acids salts are the in situ surfactants. Their arrangements invariably will form the liquid crystal vesicles which will exist for years. These vesicles can be observed through transmittance and reflectance light microscopy. Similarly, if we use the spent solution of tar sand recovery by sodium silicate, these vesicles are observed[37].

Recently, we have isolated the crude biosurfactant from a thermophilic bacterium in Hot Springs of Yellowstone National Park[38]. Upon interaction with heavy crude oil vesicles are formed which are distinguishable from ordinary synthetic surfactant. The biosurfactant usually containing amino acid sequence in helical conformation (see Fig. 7) which may be of hydrodynamical advantages during aggregation process.

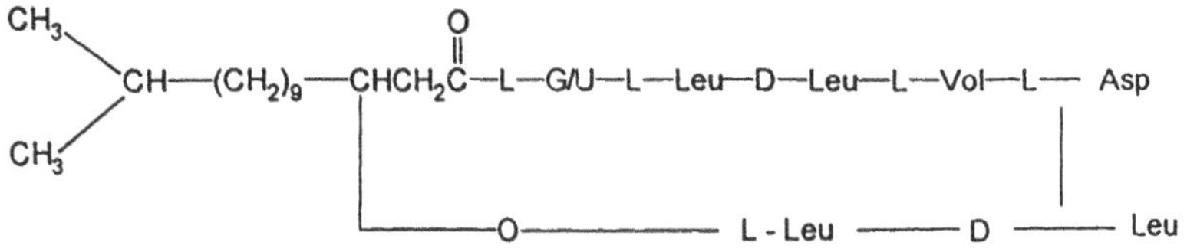

The structure of surfactin or subtilysin, a lipopeptide isolated from *Bacillus subtilis*

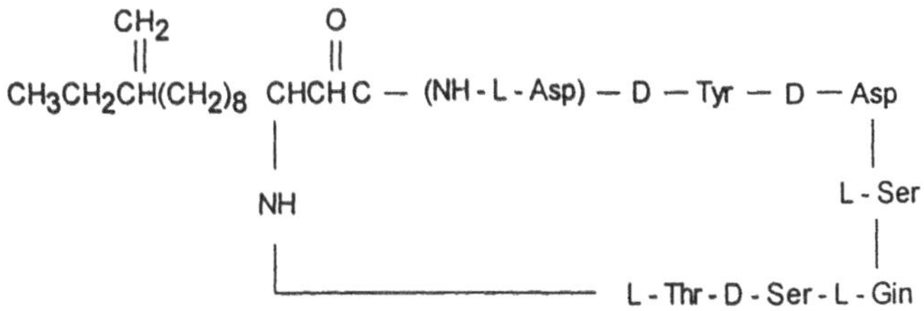

One of the surfactin analogs isolated from *Bacillus subtilis* containing a β - amino acid instead of a β - hydroxy fatty acid.

Figure 7   Examples of biosurfactants.

III. ASPHALTENE PEPTIZATION AND CONVERSIONS

During petroleum generation, micellation must play an important role just as phospholipids form liposomes under biostratinomy and thanatocenosis. During primary or secondary migrations, the polar ends of these lipids or their van der Waal's envelopes must be adsorpted to clay surfaces. In this manner, the protective colloid formed has to be a reversed micelle differing from the Hartley micelle. Often, the crude oil from wellhead assumes this form and can be reversed back to the Hartley form (water external) by the interaction with surfactant (Fig. 8). This process is peptization where the gate of solubility parameter spectrum is enlarged[39]. The most efficient surfactant is from the resin fractions isolated from crude oil. The practical applications can be adopted for EOR as well as the fouling of pipeline.

We have used ultrasound for the enhancement of heavy oil refining[30]. The primary purpose, of course, is the cracking of large molecules into smaller ones by the free radicals derived at the cavitation centers. At the water-oil interface, the feature of ultrasound is to convert the multicompartment vesicles into single-compartment vesicles, which is more efficient for the asphaltene conversion. This ultrasound reaction was emphasized by Fendler[40].

In the case of tar sand recovery, the bitumen is always found to be upgraded, e.g., the API gravity degrees constantly increased from eight to fourteen[27]. An explanation of this ease of conversion may come from catalysis within the instantly high temperature[41] and pressure[42] of the cavitation centers. Referring to section I, the Si atoms in silicate solution may have fixed location in a network of connected polygons. Also, during the tar sand recovery, Al atoms may be

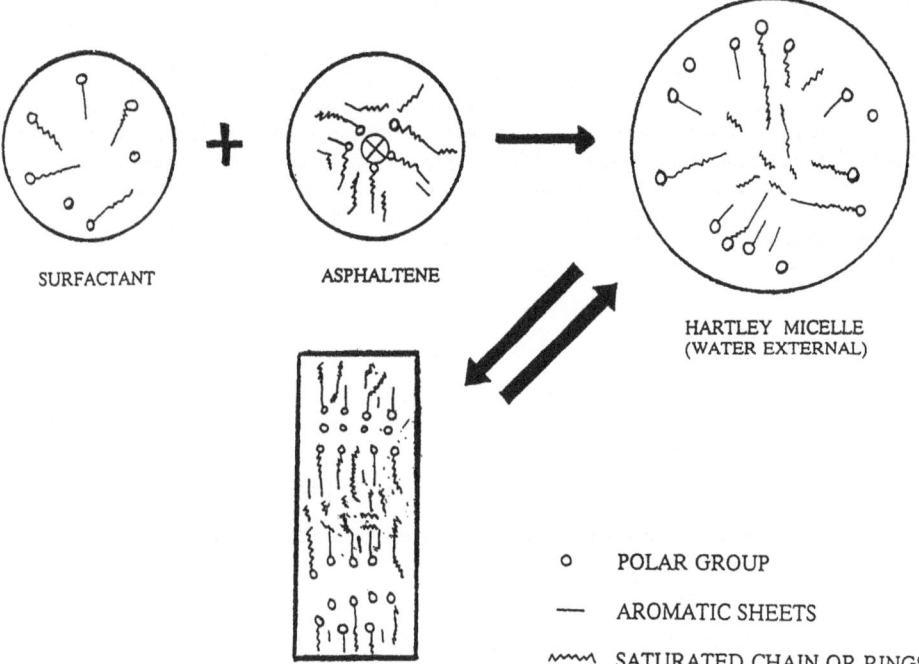

SURFACTANT   ASPHALTENE

HARTLEY MICELLE
(WATER EXTERNAL)

o   POLAR GROUP

—   AROMATIC SHEETS

⌇⌇⌇   SATURATED CHAIN OR RINGS

Figure 8   Interaction of surfactant with asphaltene to form Hartey micelle and the consequent formation of liquid crystals.

released from the inorganic portion of the tar sand. In this manner, the presence of a series of Si - Al species with different repeating distances of Si - Al is anticipated to exhibit catalytic activities as the conventional cracking catalysts (Fig. 9). Similarly, the effectiveness of the cleaning up of oil spills in soils can also be explained[32]. Furthermore, the transport of hydrated electrons or ions will be facilitated in these channels as gust-host complexation.

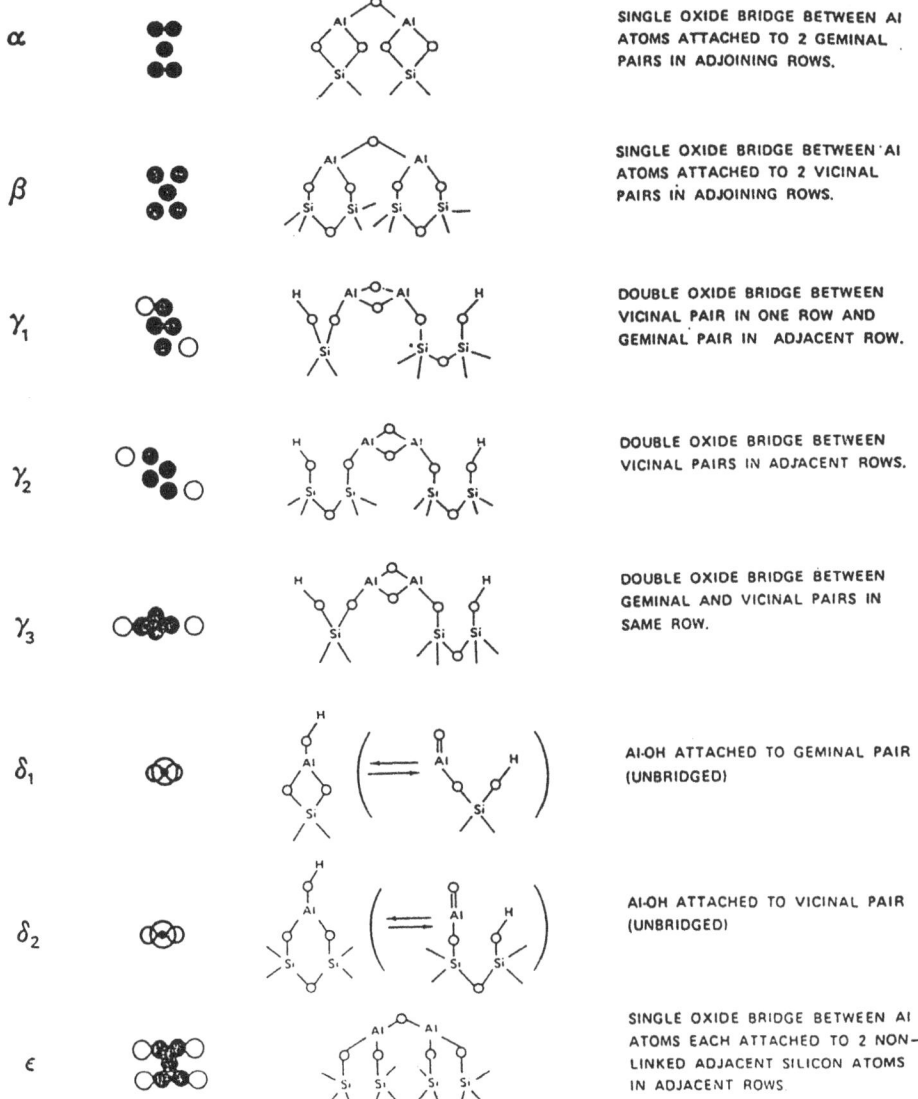

α			SINGLE OXIDE BRIDGE BETWEEN Al ATOMS ATTACHED TO 2 GEMINAL PAIRS IN ADJOINING ROWS.
β			SINGLE OXIDE BRIDGE BETWEEN Al ATOMS ATTACHED TO 2 VICINAL PAIRS IN ADJOINING ROWS.
γ_1			DOUBLE OXIDE BRIDGE BETWEEN VICINAL PAIR IN ONE ROW AND GEMINAL PAIR IN ADJACENT ROW.
γ_2			DOUBLE OXIDE BRIDGE BETWEEN VICINAL PAIRS IN ADJACENT ROWS.
γ_3			DOUBLE OXIDE BRIDGE BETWEEN GEMINAL AND VICINAL PAIRS IN SAME ROW.
δ_1			Al-OH ATTACHED TO GEMINAL PAIR (UNBRIDGED)
δ_2			Al-OH ATTACHED TO VICINAL PAIR (UNBRIDGED)
ϵ			SINGLE OXIDE BRIDGE BETWEEN Al ATOMS EACH ATTACHED TO 2 NON-LINKED ADJACENT SILICON ATOMS IN ADJACENT ROWS

Figure 9   Possible types of sites on silica-alumina. [After J.B. Peri, J. Catal. **41**, 227 (1976)].

Reconstruction or restructuring of macromolecular systems is possible because they have tertiary or quaternary structures which can be protected or interrupted by simple amphiphiles. This stabilization by the shielding of surfactant molecules is termed peptization. If the structure (micelles) is disrupted by surfactant molecules, usually through heat, the newly formed structure can be an ordered gel or liquid crystal. Hence, during thermal treatment (e.g., devolatization, pyrolysis, etc.), the optimization for avoiding the floc or solid formation depends largely on the peptization.

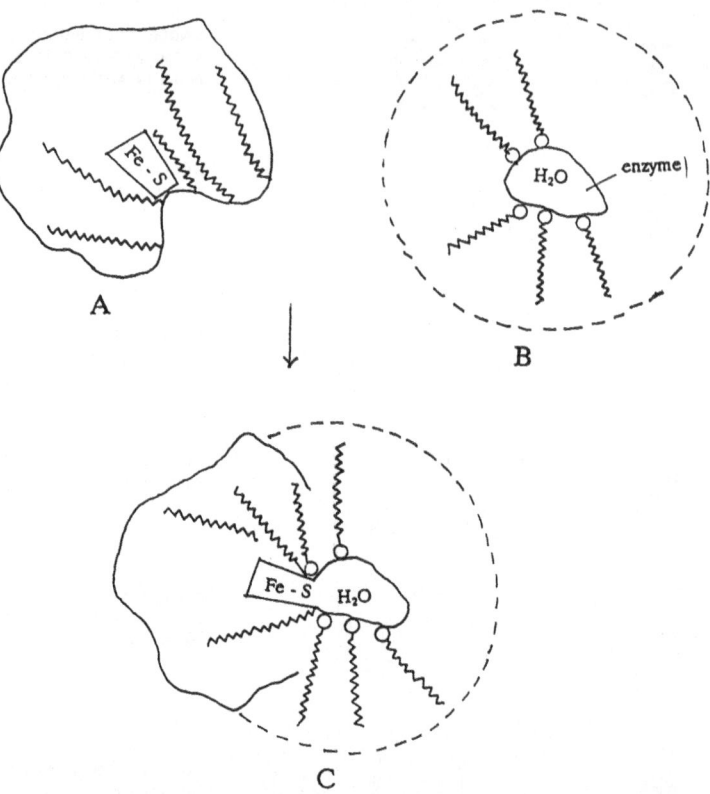

Figure 10   Coal grain, while the pyrite crystal is still imbedded into the hydrophobic chains B. reverse micelle while the water pool contains cell free enzyme extract. Notice the hydrophobic chain portions of the surfactant C. the collapse of A and B. Actually the interaction of the hydrophobic chain of A and B will be forced to squeeze the water pool to read the region of pyrite crystal.

IV. REVERSE MICELLE MULTIPHASE BIOCATALYSIS

Immobilization biocatalysis technology has been used for coal desulfurization; including surface attachment, entrapment and encapsulation. Regarding surface attachment, immobilization of cell and enzymes to a surface has been widely studied. The entrapment method entraps cells or enzymes within a porous matrix, such as glass, for subsequent multi-step bioconversion membrane-enclosed enzyme catalysis (MEEC). A wide variety of enzymatic catalysis systems has been attempted[4] with an ordinary dialysis membrane. Encapsulation has been used to envelop cells or enzymes within a set of membrane-mimetic systems, such as microcapsules formed by interfacial polymerization.

Current practice of utilizing typical sulfur-oxidizing bacteria is the microbial leaching of mineral and the desulfurization of coal requires extraordinary process time. We have developed desulfurization method[43-45] using proteins made soluble by surfactants in organic solvents. First a reverse micelle solution similar to water-in-oil microemulsion which can generate large swollen micelles (globules) is made. The water pools of those micelles created by large hydrophilic head groups of surfactants may act as microreactors. Even the solubilization of whole bacteria in this reverse micelle solution still would preserve viability. The following are the advantages of this process:

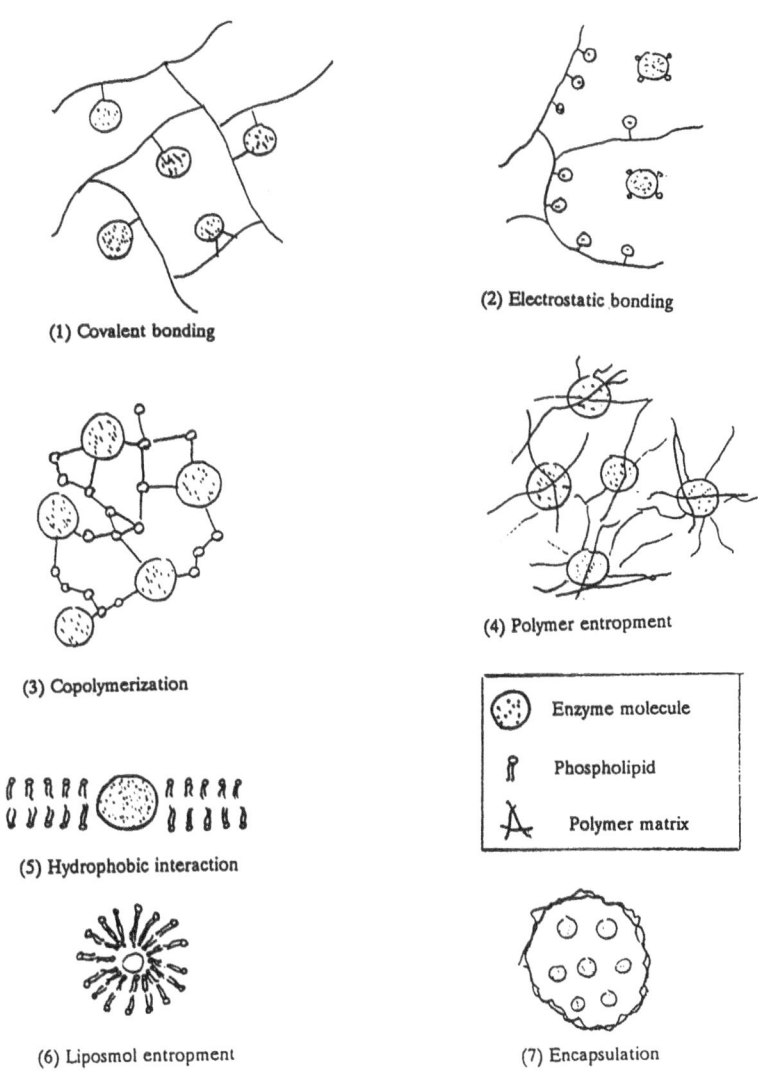

(1) Covalent bonding

(2) Electrostatic bonding

(3) Copolymerization

(4) Polymer entropment

(5) Hydrophobic interaction

	Enzyme molecule
	Phospholipid
	Polymer matrix

(6) Liposmol entropment

(7) Encapsulation

Figure 11   Possible modes of enzyme immobilization

- substrates that are poorly soluble in water are easily facilitated to react
- reaction equilibria may be shifted favorably toward the desirable products
- due to diffusion ease, both substrate and product inhibition may be reduced
- recovery or recycling of biocatalysts are possible.

A model describing how the reaction proceed is depicted in Fig. 10. Available modes of enzyme immobilization are shown in Fig. 11.

Immobilization of enzyme is widely used in industry. The immobilization can be achieved by cross-linked enzymes, or carrier-bound enzyme, or by using an encapsulated system such as semipermeable gel or membrane. There are vast amounts of examples used in the treatment of wastewaters and the degradations and reductions of environmental pollutants. For example, immobilization of the cell *Phamerochaete chrysosporium* and the derived lignin peroxidase can be used for mineralization of 3,4-dichoroaniline, dieldrin and polynuclear aromatic hydrocarbons.

REFERENCES

1. J.H. Fendler. "Membrane Mimetic Chemistry," Wiley, New York (1982).

2. J.H. Fendler and E.J. Fendler. "Catalysis in Micellar and Macromolecular Systems," Academic Press, New York (1975).

3. J.D. Barrington and D.S. Clark. "Biocatalysis and Biomimetics," ACS Symp. Ser. 392, Washington D.C. (1989).

4. M.D. Beduarski, H.K. Chenault, E.S. Simon and G.M. Whitesides, *J. Am. Chem. Soc.* 109:1283-1285 (1987).

5. Korean Institute for Industry Research, Private Communication (1986).

6. W. Eitel, Silicate structures, *in*: "Silicate Science," vol.1, Academic Press, New York (1964).

7. F. Liebau, Classification of silicates, *in*: "Reviews in Mineralogy," vol.5, P.H. Ribbe, ed., Orthosilicates Mineralogical Society of America, Chelsea, MI, pp. 1-24 (1982).

8. R.K. Harris, C.T.G. Knight and W.E. Hull, NMR Studies of the chemical structure of silicates in solution, *in*: "Soluble Silicates," J.S. Falcone, Jr., ed., ACS Symposium Ser. 194:79-94 (1982).

9. R.K. Harris, C.T.G. Knight and W.E. Hull, Nature of species present in an aqueous solution of potassium silicate, *J. Am. Chem. Soc.* 103:1577-1578 (1981).

10. R.K. Harris, J. Jones, C.T.G. Knight and D. Dawson, Silicon-29 NMR studies of aqueous silicate solutions, part II isotopic enrichment, *J. Mol. Structure.* 69:95-103 (1980).

11. P.A. Farmanian, N. Davis, J.T. Kwan, R.M. Weinbrandt and T.F. Yen, Participation of selective native, petroleum fraction in alkaline-water flooding, *in*: "Chemistry of Oil Recovery," R.T. Johansen and R.L. Berg, eds., ACS Symposium Ser. 91:103-115 (1979).

12. T.F. Yen, R.J. Hwang, M.K. Chan and P.F. Lin, Characterization of alkaline sensitive fraction of california crudes, *in*: "Proceedings, 5th Annual DOE Symposium on Enhanced Oil and Gas Recovery and Improved Drilling Technology," vol. 1, 1-14 (1979).

13. T.F. Yen, M. Chan and L.K. Jang, Some applications derived from petroleum component properties for improved waterflooding studies, *in*: "Proceedings of 1980 Heavy Oil Conference," U.S. DOE, NTIS, CONF 800350 413-426 (1980). Also T.F. Yen, M. Chan and L.K. Jang, Some useful concepts derived from petroleum component properties for improved waterflooding studies, *in*: "Enhanced Oil Recovery - Chemical Flooding," vol. 2, F-28/10, DOE/BETC/1c-80/3 (1980).

14. M. Chan, M.M. Sharma and T.F. Yen, Generation of surface active acids in crude oil for caustic flooding enhanced oil recovery, *Ind. Eng. Chem. Process Design and Dev.* 21:580-583 (1982).

15. L.K. Jang, M.M. Sharma, Y.I. Chang, M. Chan and T.F. Yen, Correlation of petroleum component properties for caustic flooding, *in*: "Interfacial Phenomena in Enhanced Oil Recovery," D.T. Wasan and A.C. Payatakes, eds., AIChE Symposium Series 212:97-104, vol. 78 (1982).

16. J.F. Kuo, K.M. Sadeghi, L.K. Jang, M.A. Sadeghi and T.F. Yen, Enhancement of bitumen separation from tar sand by radicals in ultrasonic irradiation, *Appl. Phys. Comm.* 6(2):205-212 (1986).

17. M.A. Sadeghi, L.K. Jang, J.F. Kuo, K.M. Sadeghi, R.B. Palmer and T.F. Yen, A new

extraction technology for tar sand production, *in*: "The 3rd UNITAR/UNDP International Conference on Heavy Crude and Tar Sands," AOSTRA, Edmonton, Alberta, pp. 739-747 (1988).

18. M.A. Sadeghi, K.M. Sadeghi, J.F. Kuo, L.K. Jang and T.F. Yen. "Treatment of Carbonaceous Materials," U.S. Patent 4,765,885, 22 claims, August 23, 1988.

19. K.M. Sadeghi, M.A. Sadeghi, G.V. Chilingarian and T.F. Yen, Extraction of bitumen from bituminous sands using ultrasound and sodium silicate, *Khimiya i Tekhnologiya Topliv i Masel/Chemistry and Technology of Fuels and Oils*, 8:24-28, in Russian (1988).

20. K.M. Sadeghi, M.A. Sadeghi, G.V. Chilingar and T.F. Yen, Developing a new method for bitumen recovery from bituminous sands using ultrasound and sodium silicate, *Geologiva Neft i Gaza* (Moscow), 8:53-57, in Russian (1988).

21. K.M. Sadeghi, M.A. Sadeghi, J.F. Kuo, L.K. Jang and T.F. Yen, A new tar sand recovery process: Recovery methods and characterization of products, *Energy Sources* 12(2):147-160 (1990).

22. K.M. Sadeghi, M.A. Sadeghi, M. Blazquez and T.F. Yen, Treatment of tar sand by cavitation induced sonication, *Ana. Quim.* (Madrid), 86(2):175-181 (1990).

23. M.A. Sadeghi, K.M. Sadeghi, J.F. Kuo, L.K. Jang and T.F. Yen, "Sonication Method and Reagent for Treatment of Carbonaceous Materials," U.S. Patent 4,841,131, 38 claims, Jan. 2, 1990.

24. M.A. Sadeghi, K.M. Sadeghi, J.F. Kuo, L.K. Jang and T.F. Yen, "Treatment of Carbonaceous Materials," U.S. Patent 5,017,281 (to Enersource, Inc.), 6 claims, May 21, 1991.

25. M.A. Sadeghi, K.M. Sadeghi, J.F. Kuo, L.K. Jang and T.F. Yen, "Treatment of Carbonaceous Materials," Canadian Patent 1,283,879, 47 claims, May 7, 1991.

26. K.M. Sadeghi, M.A. Sadeghi and T.F. Yen, A novel extraction of tar sands by sonication with the aid of in-situ surfactants, *Energy and Fuels* 4(5):604-608 (1990).

27. K.M. Sadeghi, M.A. Sadeghi, J.F. Kuo, L.K. Jang, J.R. Lin and T.F. Yen, A New Process for Tar Sand Recovery, *Chem. Eng. Comm.* 117:191-203 (1992).

28. A.S. Lee, X.W. Xu and T.F. Yen, Upgrading of heavy crude at low-temperature and ambient atmosphere, *in*: "Proceedings, The Fourth UNITAR/UNDP International Conference on Heavy Crude and Tar Sand, Vol. 5: Extraction, Upgrading, Transportation", 109-116, 145-146, Edmonton, Alberta (1989).

29. T. F. Yen and J. R. Lin, Upgrading of heavy oil via interfacial hydrogenation under cavitation conditions, "Heavy Crude and Tar Sands - Hydrocarbons for the 21st Century," Vol. 4, Upgrading, Government and Environment, 75-81, Vol 5, Discussions, 174-175, 213, UNITAR, New York (1991).

30. J.R. Lin and T.F. Yen, An upgrading process through cavitation and surfactant, *Energy and Fuels*, 7(1):111-118 (1993).

31. K.M. Sadeghi, M.A. Sadeghi and T.F. Yen, Uses of generated surfactants from a new tar sand process for extracting hydrocarbons from natural and man-made materials, "Proceedings, 1989 Eastern Oil Shale Symposium," pp. 18-25, University of Kentucky (1989).

32. K.M. Sadeghi, M.A. Sadeghi, L.K. Jang, G.V. Chilingarian and T.F. Yen, A new bitumen recovery technology and its potential application to remediation of oil spills, *J. Petrol. Sci. Eng.* 8(2):105-117 (1992).

33.   J. R. Chen, X. W. Xu, A. S. Lee and T. F. Yen, A feasibility study of dechlorination of chloroform in water by ultrasound in the presence of hydrogen peroxide, *Env. Technology*, 11:829-836 (1990).

34.   T. F. Yen and J. M. Moldowan, "Geochemical Biomarkers," a monograph, Harwood Academic Publishers, Chur, Switzerland (1988).

35.   C. S. Wen and T. F. Yen, Electrolytic oxidation and reduction of oil shale, *in*: "Science and Technology of Oil Shale," pp. 83-102, Ann Arbor Science Pub. (1976).

36.   T. F. Yen and C. S. Wen, J.I.S. Tang, J. T. Kwan, D. K. Young and E. Chow, Structural characterization of bitumen and kerogen from devonian shale, "1st Eastern Gas Shales Symposium," 414-430, Preprints, ERDA (1977); also "Proceedings, First Gas-Shale Symposium," 572-588, Technical Information Center, Springfield, VA (1978).

37.   M.A. Sadeghi, K.M. Sadeghi, D. Momeni and T.F. Yen, Microscopic studies of surfactant vesicles formed in tar sand recovery, "Oil Field Chemistry: Enhanced Recovery and Production Stimulation," ACS Symp. Ser. No. 396, Chapter 21 393-409 (1989).

38.   T. F. Yen, J. K. Park, K.I. Lee and Y. Q. Li, Fate of surfactant vesicles surviving from thermophilic, halotolerant, spore forming, clostridium thermohydrosulfuricum, "Proceedings of the 1990 International Conference on Microbial Enhanced Oil Recovery," 297-309, Elsevier, Amsterdam (1991).

39.   H. Lian, J.R. Lin and T.F. Yen, Peptization studies of asphaltene and solubility parameter spectra, *Fuel* (London), 73(3):423-428 (1994).

40.   J.H. Fendler, Microemulsions micelles, and vesicles for membrane mimetic photochemistry, *J. Phy. Chem.* 84:1485 (1980).

41.   E.B. Flint and K.S. Suslick, The temperature of cavitation, *Science.* 253:2397-2399 (1991).

42.   L.J. Brigg, Limiting negative pressure of water, *J. App. Phy.* 21:721-722 (1950).

43.   K.I. Lee and T.F. Yen, Coal desulfurization through reverse micelle biocatalysis process, *Am. Chem. Soc. Fuel Chem. Preprints*, 33(4):573-579 (1988).

44.   K.I. Lee, M.L. Blazquez and T.F. Yen, A new filamentous bacteria for coal desulfurization, *in*: "Bioprocessing and Biotreatment of Coal," 729-736, D. Wise, ed., Marcel Dekker, (1990).

45.   K.I. Lee and T.F. Yen, Sulfur removal from coal through multiphase media containing biocatalysts, *J. Chem. Tech. Biotech.* 48:71-79 (1990).

MESO-SCALED STRUCTURE AND MEMBRANE MIMETIC CHEMISTRY

T.F. Yen

University of Southern California
Los Angeles, CA 90089-2531, USA

ABSTRACT

A meso-scaled structure can fill the gap between the microstructure (fine structure) of atomic and chemical properties and the macrostructure (bulk structure) of mechanical and rheological properties. A meso-scaled structure is especially useful for large macromolecules. As an example, three discussions are made with the meso-scaled structure of asphaltene which is a native multipolymer from petroleum. These are I. dislocation and disclination; II. viscoelasticity and III. metal fixation, all of which are part of the meso-scaled structure.

INTRODUCTION

Meso-scaled structure is often referred to as the intermediate structure between the fine molecular level structure (micromolecular) and the bulk morphological phase structure (macromolecular). Illustrations will be given in this paper. Usually, for the solid-phase condensed matter, it is necessary to describe a given material by a given set of arrangement of the individual structural elements (could be just the repeating units or sequences), especially for the mesomorphic substances (between the crystalline and amorphous solids). For simple molecules, a clear chemical, molecular structure is sufficient to denote the substance. Yet for the macromolecules, especially when the macromolecules appear as multipolymers, a fine molecular structure cannot clearly specify the given classes of substances. For homopolymers the repeating sequence or repeating unit alone can describe a given polymer in many of its properties. But, for copolymer, terpolymer, tetrapolymer, etc., there is a need for understanding the chemical nature of the repeating units. For example, for a higher series such as proteins (multipolymers), there are many interactions within their repeating units. In this case not only the primary, secondary, tertiary, quaternary, and higher order structures become important, e.g., not only the hydrogen bonding linked to various bases is important, but also the allosteric function and the uniqueness of it will be essential to the properties of protein.

Membrane-mimetic chemistry with simple geometric arrangements and assemblages of amphiphiles will be sufficient to explain and specify the characteristics of a given class of surfactants. The same technique will be adopted to describe the petroleum and fossil fuel chemistry in the following discussion.

I. Dislocation and disclination

For linear defect there are six types of distortions according to Valterra. The first

Volterra dislocations
(dispiration)

Burgers vector, $\vec{b}$ ----- dislocation

Franks vector, $\vec{\omega}$ ----- disclination

(a)   : e is unit vector
(b), (c)   : edge dislocation
(d)   : screw dislocation
(e), (f)   : twist disclination
(g)   : wedge disclination

Figure 1   Volterra distortions in solid state

Figure 2   Brooks-Taylor spherules. (a) Alignment of mesophase layer
(b) spherules with sections

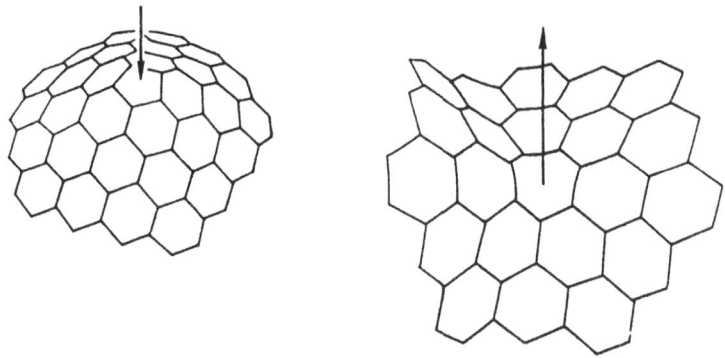

Figure 3   Wedge layer disclinations of rotation $\pi/3$ and $-\pi/3$.

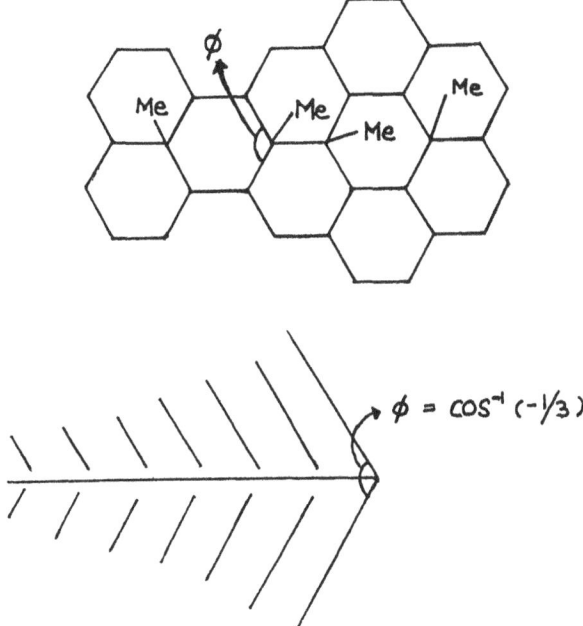

$\phi = \cos^{-1}(-1/3)$

Figure 4   Quarternary Methyl carbon undertaking a strain between 120° to 109°27′

three are translational, with Burgers vectors of $\mathfrak{b}$ as shown in Fig. 1. The initial hallow cylinder with a radical cut Γ and the $\bar{e}$ vector is shown in Fig. 1a. It follows the dislocations with Burger vector $\mathfrak{b}$ perpendicular to $\bar{e}$ vector (Fig. 1b and 1c), or the screw dislocation with $\mathfrak{b}$ opposite to $\bar{e}$ by π (Fig. 1d). The other three types are disclinations with Frank vector of $\bar{\omega}$ of rotational nature. Similar to dislocations, the first two types (Fig. 1e and 1f) are twist disclinations, and the last type (Fig. 1g) is wedge disclination[1]. In general Volterra dislocations can be expressed as

$$\vec{u} = -\vec{b} + \vec{\omega} \times (\vec{r}_w - \vec{r})$$

ū is the relative displacement of the cut surface, r̄ belongs to the surface, where r̄ᵥ indicate the position of the vector ω̄.

A clear case of disclination is that if the formation of carbonaceous mesophase from resin and asphaltene molecules. The formation of Brooks-Taylor spherule requires that all the polynuclear aromatic hydrocarbons which are 2-dimensional planar sheet to be able to bend. (Fig. 2) The bending of polynuclear aromatic hydrocarbons is actually wedge layer disclination, e.g., a case of strength s = p/6, with p being any integer, is shown in Fig. 3. the center of disclination can be pentagon or heptagon, the former case can be understood through the hemispheroid formation of fullerenes. We have observed two cases of the bending of polynuclear aromatic hydrocarbons.

1. presence of quaternary carbons in precursors in the conversion to mesophase.
2. exertion of close electron accepting metal complexes.

For Case 1, the bond angle remains between 109° 27′ and 120° which process a strain to cause the initiation of disclination (see Fig. 4). The mesophase molecules contain a greater number of methyl substitutents within the aromatics nuclei. As the tetrahedral sp^3 bonding requires smaller angle, disclination begins. Experimental evidence has been able to support Case 2, where the aromatic π-system is in close approximity to the vanadyl functional group of a porphin system, the nitrogen superhyperfine splitting can be observed[2] whereas the pyramidal distance is compressed to the nitrogen base plane of 0.3 Å. (See Fig. 5)[3]. This in- and out-of-plane movement form the basis of disclination. It is of interest to compare the biochemical pyramidal doming of the hemoglobin molecule. (Fig. 6) The T → R transition is 0.6 Å for the vertical distance of nitrogen base plane caused by the donor atoms in histidine[4]. In case of layer-stacking twist clination as defined by Zimmer and White[5], the formation of spheroid is a consequence.

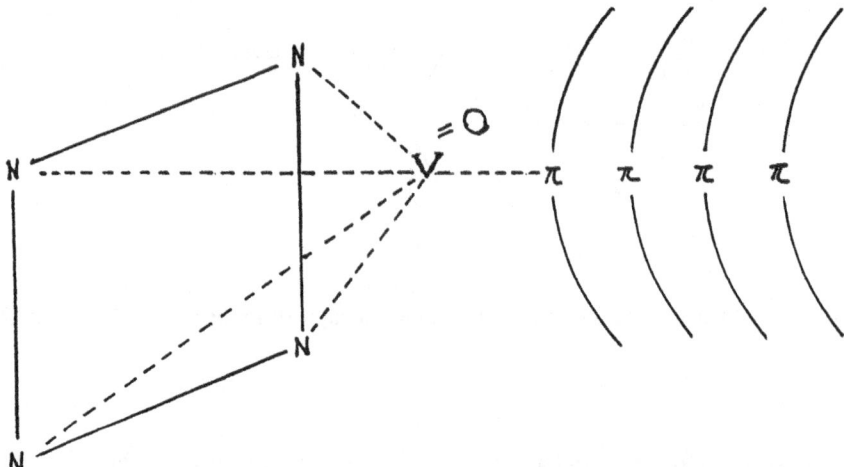

Figure 5   Influence of π-system to the vanadyl porphyrin. Due to interaction, the V to N base is reduced by 0.3Å $\Delta d_{v \to p} = 0.3$Å. Consequently, the π-system is bent.

The coarse lamelliforn morphology of freshly coalesced mesophase is observed by optical microscopy. In Fig. 7, the layer stacking defects are denoted by their polarized light response. Using Honda's nomenclature systems[6] of the Möbius morphology, the following nodes and crosses can be obtained:

Figure 6   The heme groups and its environments. T → R is 0.6Å from the nitrogen base.
             After Gelin, B.R., Lee, A. W. -N., and Karplus. M., *J. Mol.* Vol **171**,
             542(1983).]

Figure 7   The Mobius morphology of bulk mesophase with a coarse structure (a). The
             structural sketch was mapped from a series of micrographs under conditions
             of crossed polarizers. Polarized light response of a lamelliform structure
             with four common types of layer stacking defects (disclinations). The spacing
             of extinctions contours offers a measure of the bend, twist, splay and
             disclination density of a given microconstituent. The nature of each
             disclination may be determined from the rotation direction of the extinction
             contours when the plane of polarization of the incident light is rotated.

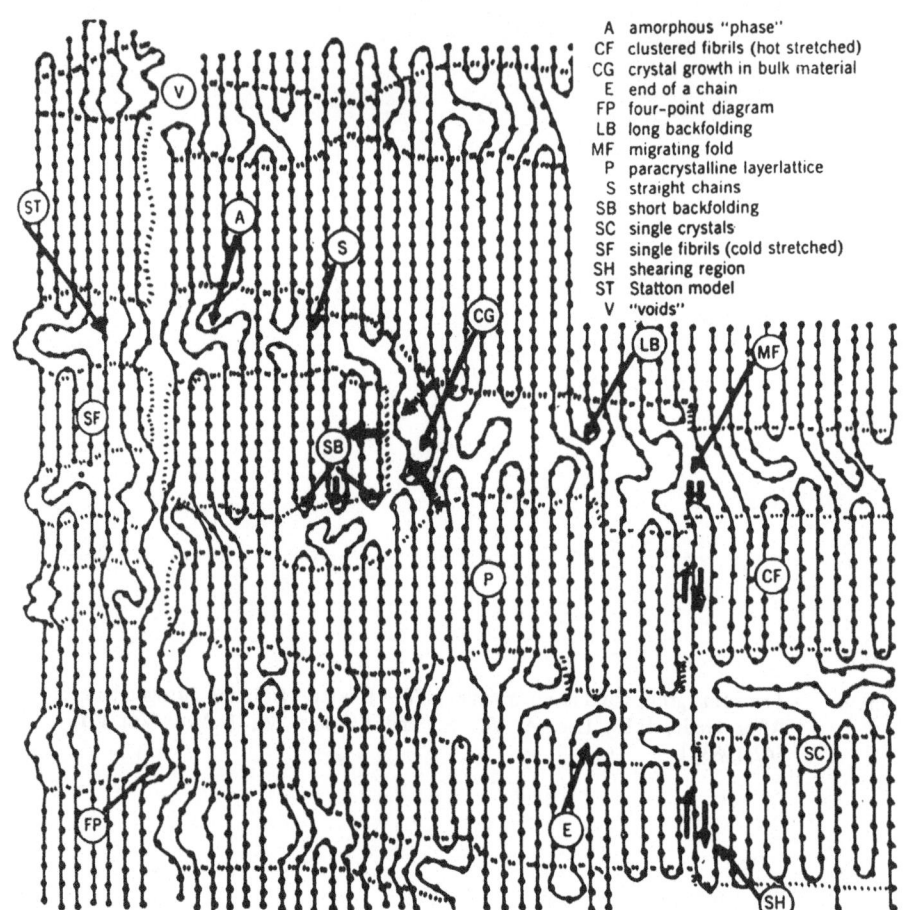

A   amorphous "phase"
CF clustered fibrils (hot stretched)
CG crystal growth in bulk material
E   end of a chain
FP four-point diagram
LB long backfolding
MF migrating fold
P   paracrystalline layerlattice
S   straight chains
SB short backfolding
SC single crystals
SF single fibrils (cold stretched)
SH shearing region
ST Statton model
V   "voids"

Figure 8   Imperfect folded chain model of a single polymer crystal, which makes up apart of spherulite. [Reprinted from Hosemann, Polymer, 3, 349 (1962).]

u   co-rotating node
y   counter-rotating node
o   co-rotating cross
x   counter-rotating cross

A structure sketch is shown in Fig. 7a and the polarized light response is in Fig. 7b. These disclinations or dispirations (defined as dislocations plus disclinations) during carbonization maybe used to characterize the precursors materials[9] or to relate the fracture process or even cracks of the finished products[10]. As emphasized by Nabarro[8] every Peierls forces play an important role in stationary dislocations. The morphology of straight chain polyethylene shows all features of kinks, twistings, loopings and thickenings with all the short and long backfoldings and cluster fibrils. (Fig. 8). The development of craze cracks and voids in polymers are related to dispirations (Fig. 9).

In general the development of dislocations, disclinations and dispirations in macromolecular materials is due to the carbonization or pyrolysis during thermal decomposition. As the saturated hydrocarbons aromatize into larger aromatic sheets, certain linkages would become defect centers during the transformation. The formation of carbonaceous mesophase is such a case and the formed product including graphite bodies would bear the mechanical properties.

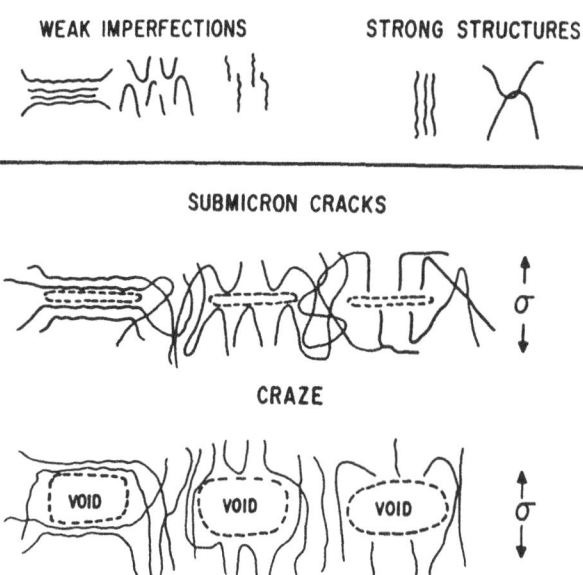

WEAK IMPERFECTIONS　　STRONG STRUCTURES

SUBMICRON CRACKS

σ

CRAZE

VOID　　VOID　　VOID

σ

Figure 9　　Cracks and craze.
Top: Regions of weakness and strength on a molecular scale in a polymer which appear to be important in the development of craze cracks. Bottom: Sequential steps in the development of voids, oriented polymer, and craze cracks as the result of a tensile stress applied in the vertical direction.

II. Viscoelasticity

The flow and mechanical properties of asphalt (defined as a mixture of asphaltene, resin and aromatic-saturated hydrocarbons), behave both Newtonian (sol asphalt) and non-Newtonian (gel asphalt)[14]. Most asphalts are sol-gel classes and exhibit pseudoplastic as well as shear-thickening behaviors (Fig. 10)[11] The time dependence of the mechanical properties is shown by the variation of stress (σ) and strain (ϵ) under different types of loading as indicated by Fig. 11[16]. Often the elements of mechanical model are spring (energy stored potentially) and dashpot (or damper, energy dissipated as heat) as shown by Fig. 12. The use of the elements, e.g., the Maxwell model is the spring and dashpot in series, whereas the Voigt model has the spring and dashpot in parallel (Fig. 13). It is very easy to compute the response to verify the experimental data shown in Fig. 11. The assumption is made that linear viscoelastic material consists of elastic component (obeys Hook's law) and viscous component (obeys Newton's law)[17]. Thus,

$$\sigma = E\epsilon$$
$$and \quad \frac{d\sigma}{dt} = E\frac{d\epsilon}{dt}$$

where E is elastic modulus. Also,

$$\sigma = \eta \frac{d\epsilon}{dt}$$

261

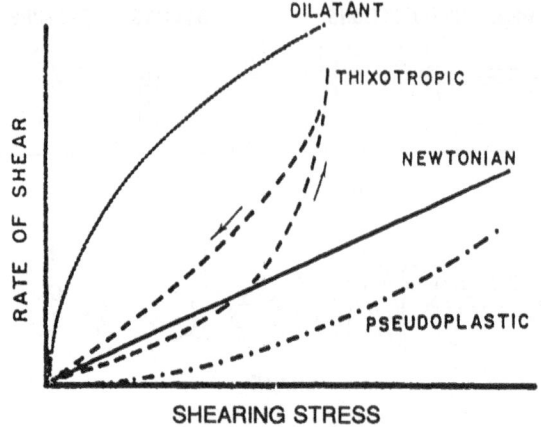

Figure 10a   Rate of shear vs shearing stress for Newtonian and three types of non-Newtonian flow.

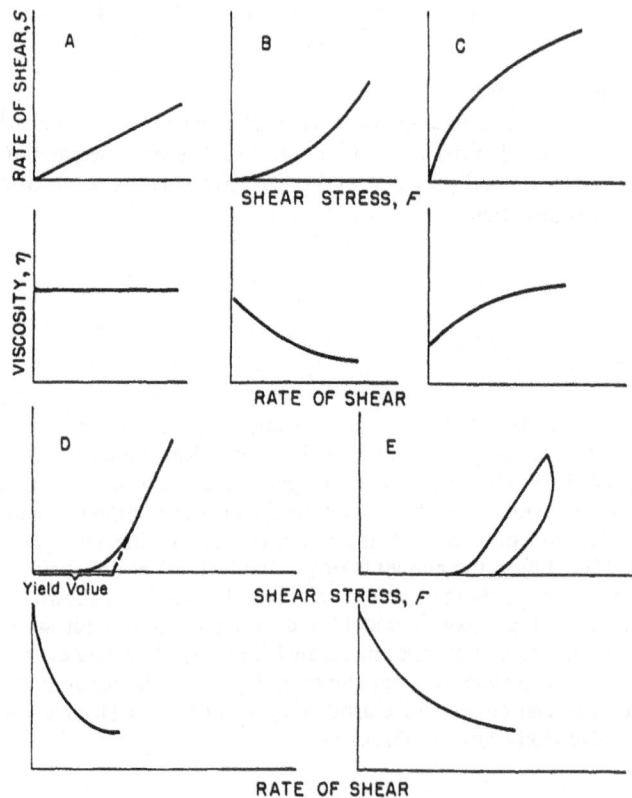

Figure 10b   Rheological types of flow commonly found in asphalts: (A) Newtonian flow; (B) pseudoplastic flow (viscous, but with elastic effect); (C) dilatant flow; (D) plastic flow (oxidized asphalts); (E) thixotropic flow.

where η is viscosity. Now Maxwell model (see Fig. 13a)

$$\sigma_1 = E\epsilon_1 \quad (spring)$$
$$\sigma_2 = \eta \frac{d\epsilon_2}{dt} \quad (dashpot)$$

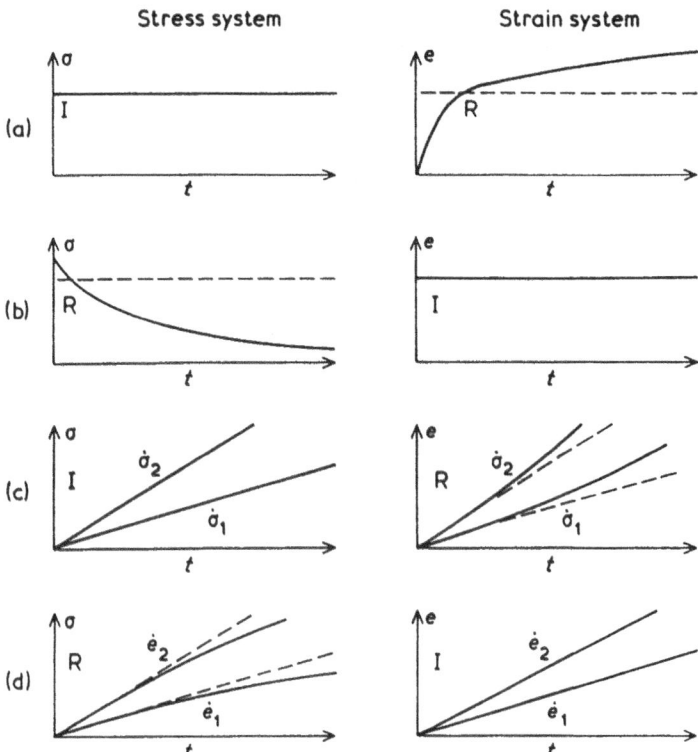

Stress system **Strain system**

(a)

(b)

(c)

(d)

Figure 11   Schematic representation of the variation of stress and strain with time indicating the input (I) and responses (R) for different types of loading. (a) Creep, (b) Relaxation, (c) Constant stressing-rate, (d) constant straining-rate (After J.G. Willaims, Stress Analysis of Polymers, Longman, London (1973).

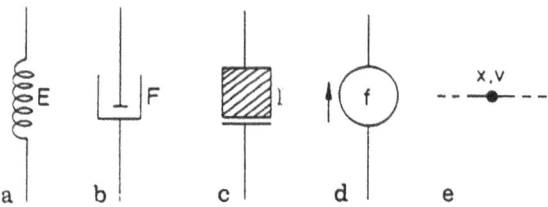

Figure 12. Conventionalized symbols for mechanical model diagrams: (a) spring, (b) dashpot, (c) mass, (d) force, (e) displacement or velocity.

$$\epsilon = \epsilon_1 + \epsilon_2$$
$$\sigma_1 = \sigma_2 = \sigma$$
$$\frac{d\epsilon}{dt} = \frac{d\epsilon_1}{dt} + \frac{d\epsilon_2}{dt} = \frac{1}{E}\frac{d\sigma}{dt} + \frac{\sigma}{\eta}$$

Two experiments are tested. For the creep test the stress is held constant $\sigma = \sigma_0$ and $d\sigma/dt = 0$, so

$$\frac{d\epsilon}{dt} = \frac{\sigma_0}{\eta}$$

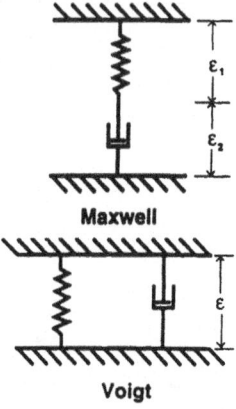

Figure 13   Maxwell and Voigt models.

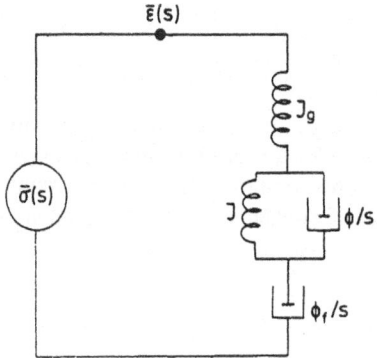

Figure 14a
Standard 4-parameter Voigt model

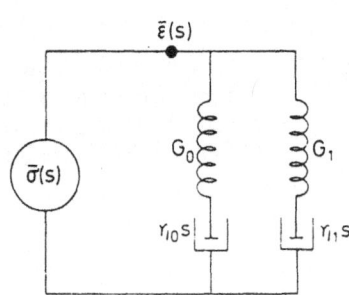

Figure 14b
Standard 4-parameter Maxwell model

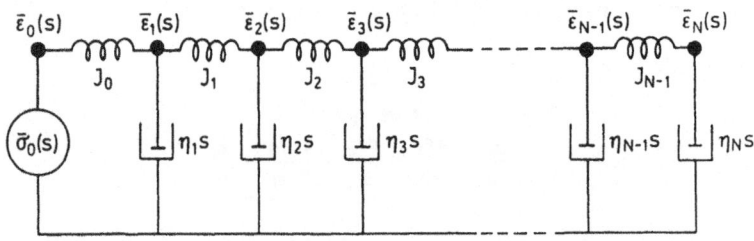

Figure 14c Obverse standard rheodictic ladder model

Thus, Maxwell model predicts Newtonian flow but not for a viscoelastic polymer. This Maxwell model is better for predicting the stress relaxation. In the case of constant strain $\epsilon = \epsilon_0$, so that $d\epsilon/dt = 0$. Thus,

$$0 = \frac{1}{E}\frac{d\sigma}{dt} + \frac{\sigma}{\eta}$$

and

$$\frac{d\sigma}{\sigma} = -\frac{E}{\eta}dt$$

or

$$\sigma = \sigma_0 \exp(-Et/\eta)$$

since

$$\frac{\eta}{E} = \tau_0 \quad (\tau_0 = constant\ relaxation\ time)$$

$$or \quad \sigma = \sigma_0 \exp\left(-\frac{t}{\tau_0}\right)$$

which predict quite well for Fig. 11b.

On the other hand for Voigt model (or Kelvin model), in parallel

$$\epsilon = \epsilon_1 = \epsilon_2$$
$$and \quad \sigma = \sigma_1 + \sigma_2$$

as previously,

$$\frac{d\epsilon}{dt} = \frac{\sigma}{\eta} - \frac{E\epsilon}{\eta}$$

or

$$\frac{d\epsilon}{dt} + \frac{E\epsilon}{\eta} = \frac{\sigma_0}{\eta}$$

$$\epsilon = \frac{\sigma_0}{E}\left[1 - \exp\left(-\frac{Et}{\eta}\right)\right]$$

as before

$$\epsilon = \frac{\sigma_0}{E}\left[1 - \exp\left(\frac{-t}{\tau_0}\right)\right]$$

which is correct way that viscoelastic polymer undergoing creep. But the Voigt model is unsuccessful in predicting stress relaxation since

$$\frac{\sigma}{\eta} = \frac{E\epsilon_0}{\eta}$$
$$or \quad \sigma = E\epsilon_0$$

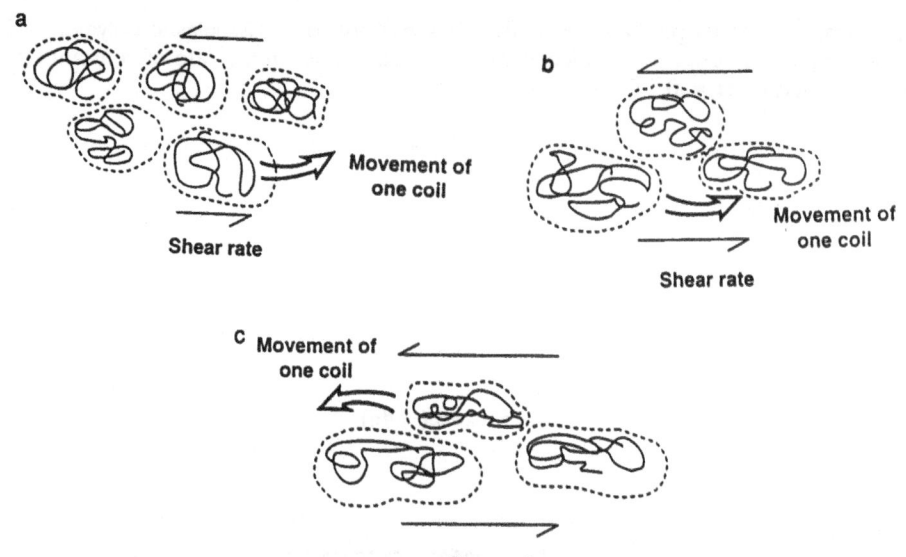

Figure 15 Movement of random coils in shear fields: (a) low shear rate, $\mu = \mu_o$, fluid motion does not distort the coils; (b) intermediate shear rate, $\mu_o > \mu > \mu_\infty$, velocity gradient elongates the coils in the shear direction; (c) high shear rate, $\mu = \mu_\infty$, coil distortion has reached a dynamic equilibrium between the elongations caused by the shear and the recoil of the molecule. (After reference 17)

which is the linear elastic case (Fig. 11b). Regardless Maxwell or Voigt model, the combination of spring and dashpot elements will exhibit the correct experimental results, e.g., a 4-element model (Fig. 14a) is much improved model[15]. Tschoegl[7] has discussed models with large numbers of elements and their solutions such as procedure X, the allocation method, the multidata method, etc. Viscoelastic behaviors can be described by some finite number of elements, e.g., the Gross-Marvin model.

In carbonaceous asphaltic material the dashpot can be considered analogous to the molecular slippage, e.g., the movement of π-system. Since the π-system are lined-up as stacks, slippage and even insertion are quite common. The insertion of proper π-system solvents can lead to the successful hydrogenation or upgrading of the residuum. The spring can be considered analogous to the elastic straining of bond angles especially for the long chain alkane substitutents or conformational changes of the naphthenics in the asphaltenes. Entanglements of the zig-zag portions of the asphaltene as well as the rearrangements of the inter- and intra-clusters within the asphaltic system will contribute greatly to the spring elements. Many of these are reversible therefore some equilibrium conditions are established automatically for the best arrangement of springs and dashpots.

Some asphalts exhibit Ostwald fluid flow, i.e., they exhibits shear-thinning behavior at moderate shear rate; at low or high shear rate they show Newtonian behavior. These phenomena can be explained by the deformation of coils at intermediate shear rate. (Fig. 15). Shear-thickening asphalts are also known. In this case, the dilatant fluids can be explained by particle aggregation under shear as shown by Fig. 16. For asphalt, the steric hardening or structural hardening is not due to chemical reactions, but due to restructuring the elements which they process. Yen has used the colloidal chemical approach to solve this practical problem. As shown in Fig. 17, the cross-linked asphalt (gel-form) can be formed at different rates depending on the meso-scaled structures of the source materials.

A word has to be mentioned here that natural polymers such as the asphaltene and its related system such as asphalt can behave as thermal plastic elastomers. They do not need chemical crosslinking; merely physical change such as the formation of soft and hard segments (similar to those of polyurethanes with prepolymer of polyols) (Fig. 18) or the domains of blocks of rigid and soft moieties (similar to ABA triblock copolymers) (Fig. 19)

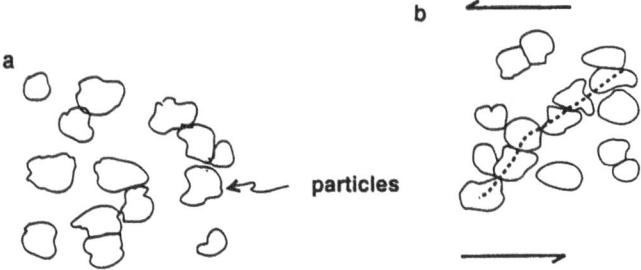

Figure 16   Dilatant fluid: **(a)** fluid at rest, **(b)** fluid under shear. Velocity gradient causes particles to pack into larger agglomerates that have increased resistance to flow. Dashed line suggests the effective diameter of the particle unit.

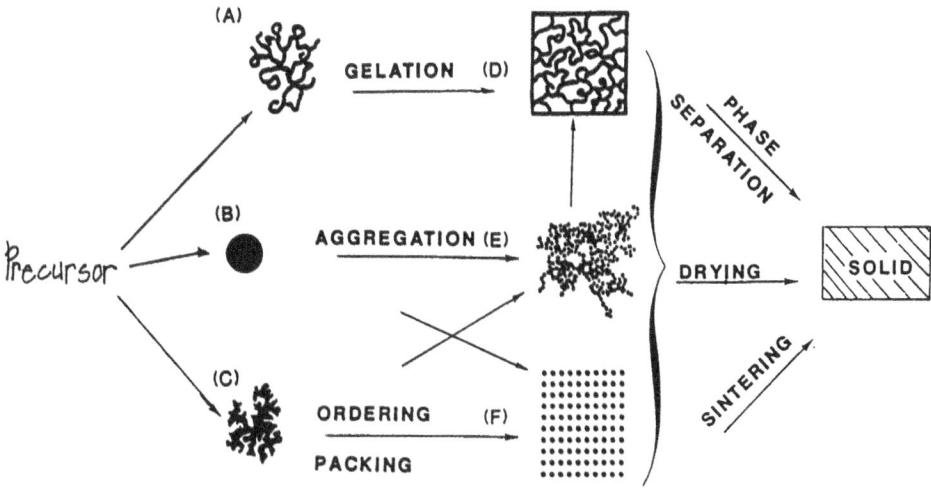

Figure 17   Conversion of a mesomorphic materials which include silicates, asphaltics and other polymers. Crosslinking is refered as gelation.

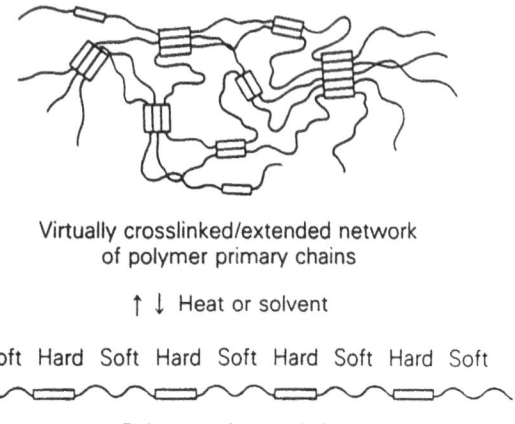

Virtually crosslinked/extended network
of polymer primary chains

↑ ↓ Heat or solvent

Soft Hard Soft Hard Soft Hard Soft Hard Soft

Polymer primary chains

Figure 18   Schematic illustration of the formation of hard segments in a thermoplastic polyurethane elastomer. (After Schollenberger and Dinbergs (1975) J. Elast. Plast. **1**, 65).

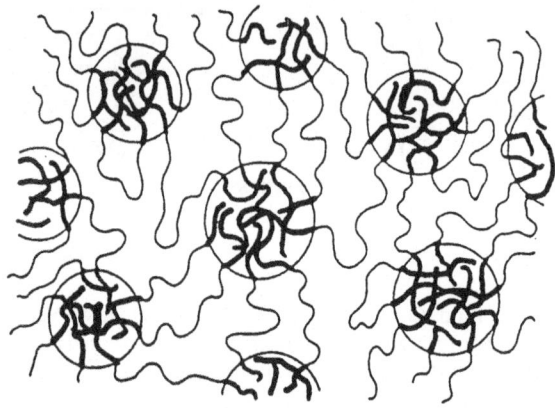

Figure 19   Schematic representation of the structure of an ABA tri-block copolymer of the polystyrene-polydiene-polystyrene type. The thicker lines represent the polystyrene blocks and the thinner lines the polydiene blocks.

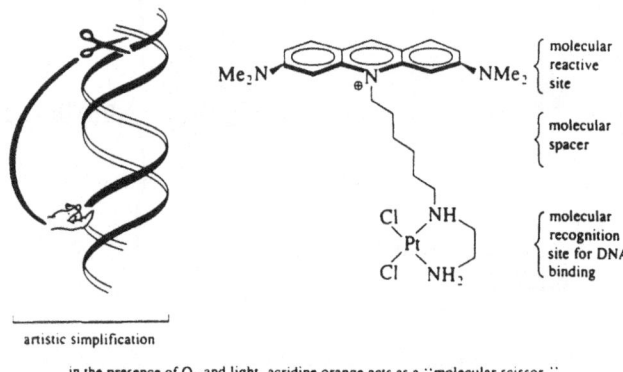

artistic simplification

in the presence of O₂ and light, acridine orange acts as a "molecular scissor."
snapping the segment of DNA to which it is attached

Figure 20   Mechanism of an Antitumor drug. In the presence of O_2 and light, acridine orange acts as a "molecular scissor, "snapping the segment of DNA to which it is attached. (After reference 20)

will be sufficient. Afterall, the asphaltene is a multipolymer by definition; it can be modified with ease to any desirable properties.

III. Metal fixation

The site of metal within the environment of other molecular functional groups is unique that the geometry of the metal will determine many important chemical and biochemical effects, such as oxygen transport, photosynthesis, nitrogen fixation, drug design, enzyme catalysis, and other chemical catalysis in general. First example we would like to discuss is bis-intercalant. For example, the antitumor drug cis-diaminedichloroplatinum (II) linked to acridine orange[12] is acting like a molecular scissor by snapping the segment of DNA duplex to which it binds, e.g., N-7 of adjacent guanidine bases as shown in Fig. 20. In case of fossil fuel chemistry and chemical engineering intercalasnts are used for catalysis, e.g., Lewis acid like salts such as ferric chloride and aluminum chloride are used as catalysts before or during air blowing (an oxidation reaction involving free radicals). The reason is

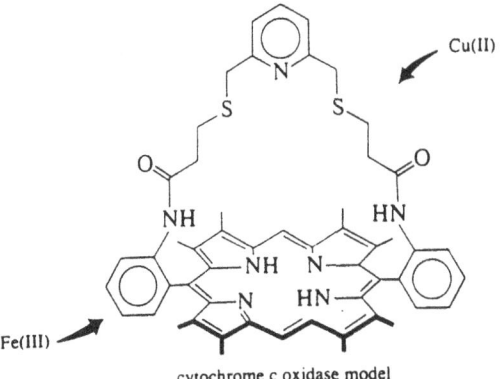

cytochrome c oxidase model

Figure 21   Model for Cyctochrane c oxidase. Both Cu(II) and Fe(III) can coordinate to this host molecule.

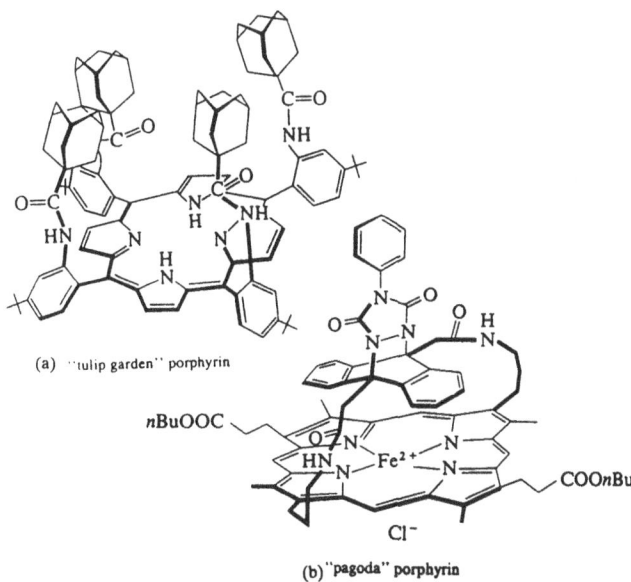

(a) "tulip garden" porphyrin

(b) "pagoda" porphyrin

Figure 22 a,b   model porphyrins synthesized and studied

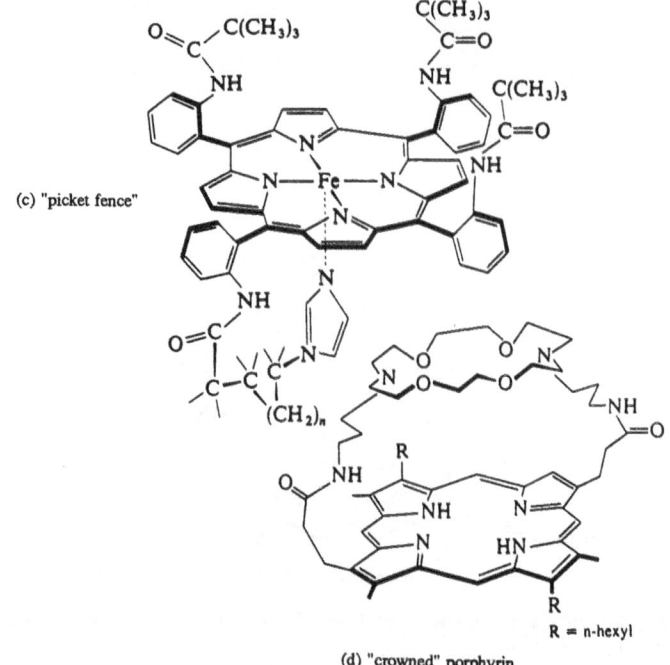

(c) "picket fence"

R = n-hexyl

(d) "crowned" porphyrin

Figure 22 (cont.) (c) " picket fence " porphyrin (d) "crowned" porphyrin.

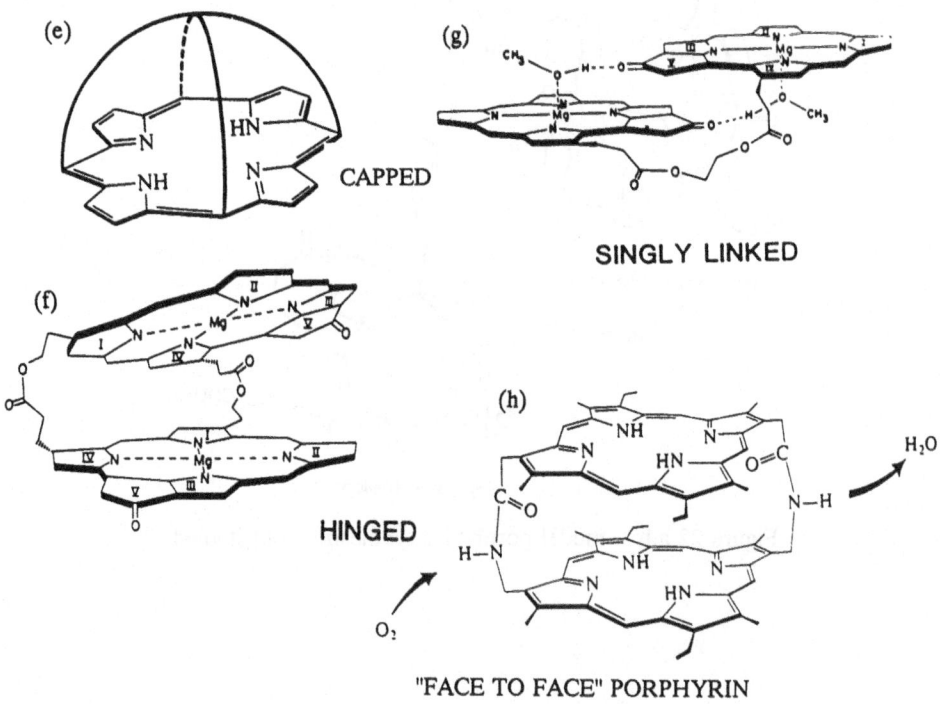

(e) CAPPED

(g) SINGLY LINKED

(f) HINGED

(h) "FACE TO FACE" PORPHYRIN

Figure 22 (cont.) (e to h)

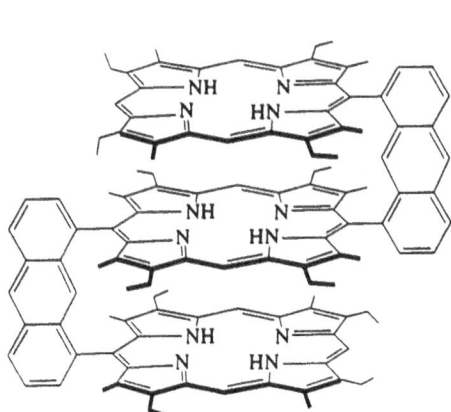

(i) "gable" porphyrin

(j) tetrameric cyclo-porphyrin

(l) triple deckered triporphyrin

(k) hexameric cyclo-porphyrin

Figure 22. (cont.) (i-l)

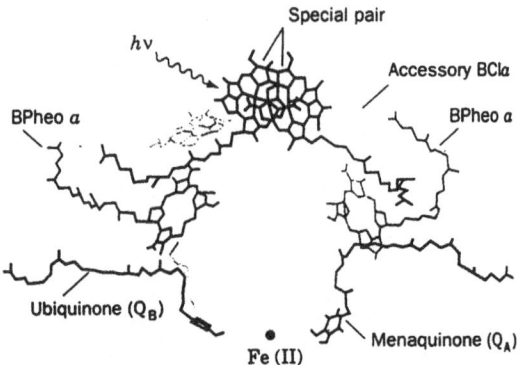

Figure 23. Excitation of the bacterial photosynthetic reaction center with chromophores (After reference 19).

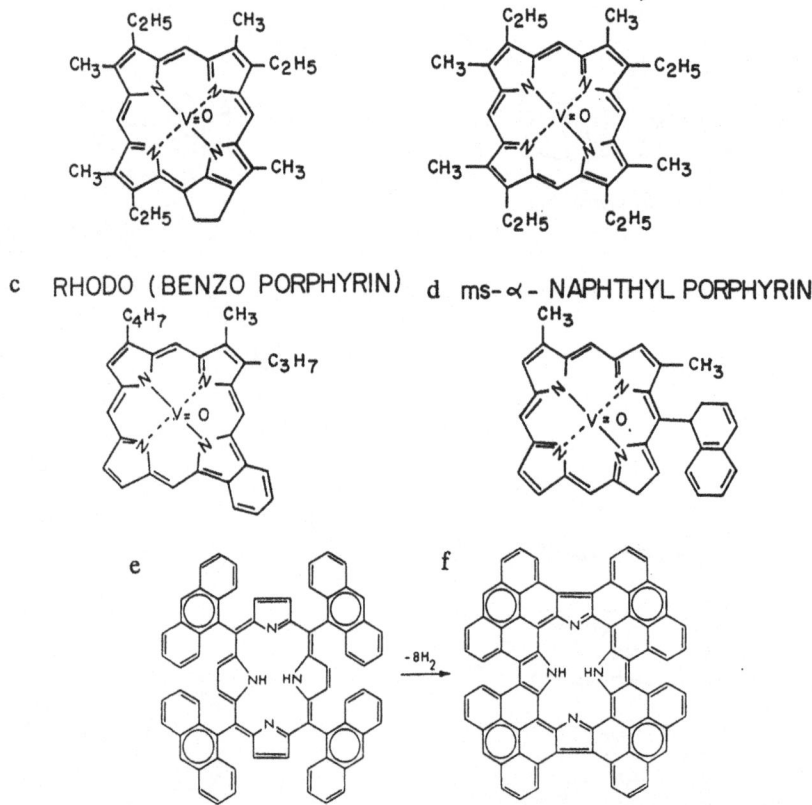

Figure 24   Fossil porphyrin from a to f depending on aging and deposition environments.

that these metals can form ferrosene-like π-complexes in the aromatic stacks of asphaltene, thus reducing the free radical nature of asphaltene. The same principle is at work doing hydrodesulfurization (HDS) of heavy residuum, the transition metals such as cobalt and nickel can behave as intercalants. Free radial hydrogenation reaction can proceed with ease at the molybdenum support centers. For the Aurobon process the vanadium tetrasulfide behaves similarly.

The second point I like to mention is that since porphyrin is widely distributed in biomass and this conjugated system, diaza-18-annulene, is very stable. Porphyrin remainants, whether from hemoglobin, chlorophyll, myoglobin, or cytochrome, can be found in petroleum, and is concentrated in asphaltic fractions. The following are some facts in this connection for the geoporphyrins[18].

1. The magnesium and iron atoms of the complexes have been replaced with mere stable vanadyl and nickel centers.
2. The reducing forms of bacteriochlorophyll (Ring II and Ring IV) and the reducing form (Ring IV) of chlorophyll are aromatized.
3. The phytyl or geranylgeranyl side chains are cleaved.
4. The adjacent isocyclic 5-membered ring (on III) of chlorophyll is opened.

These are the classical viewpoints. The following items are also added through some recent observations.

5. For bioporphyrins the cytochrome model is represented by heterodinuclear complex of porphyrin Fe (III) and heteropyrdium units (with 2 sulfur atoms) of Cu (II)[20]. In geoporphyrin both porphyrins and non-porphyrins can be located in one sample.
6. For hemoglobin and myoglobin, one side of which is covered with hydrophobic pocket and the back side, histidine base. The special pair of
   bacteriochlorophyll is lined through its pheophytin and passed to chromophores. For geoporphyrin the consequence is etioporphyrin, rhodoporphyrin and highly aromatic meso substituted etioporphyrin.
7. In heme system the N...Fe-O...N system has been established. In geoporphyrin system only the concentrated porphyrin has N...V-O...N relation; the dilute one is largely due to π...V-O...π system.

Some discussions here are made to the above statements. For Viewpoint No. 5, M.J. Gunter[13] and coworkers have studied the dinuclear receptors for cytochrome c oxidase model. The system is a capped porphyrin using a pyridino bridging units consisting of two sulfur atoms as shown in Fig. 21. Both Fe (III) and Cu (II) can be employed and utilized in this dinuclear receptor. During geochemical alteration, the labile ligands, especially those polydentate ligands with donor atoms, will be degraded, but still survived as portions of heterocyclics in resin fractions. The ligand types of N_3O have been observed so far[14].

Concerning Viewpoint No. 6, a great number of model porphyrins have been prepared, most of them having restriction of one side being tied up with hydrophobic pockets. These include the following porphyrin:

Turlip garden porphyrin	(Takagi[21])
Pagoda porphyrin	(Traylor[22])
Picket fence porphyrin	(Collman[23])
Capped porphyrin	(Baldwin[24])
Crowned porphyrin	(Chang[25])
Gable porphyrin	(Tabushi[26])
Hinged porphyrin	(Bucks and Boxer[27])
Singlely-linked porphyrin	(Bucks and Boxer[27])
Face-to-face porphyrin	(Colleman[28])

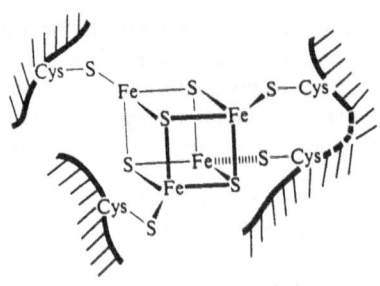

Figure 25. Cagelike cluster proposed as active site of ferredoxin (After Reference 20).

Figures 22a to 22e demonstrate that the metal in the host molecule environment is restricted or is carcerated in a bowl-shaped manner. The dinuclear and polynuclear centers are illustrated from Figs. 22f to 22k with Figs. 22f to 22i of special pair centers. The tetrameric and hexameric cyclo porphyrins are given by Figs. 22j and 22k respectively. For photosynthesis the special pair are fixed in space by their subsequent linkages with ubiquinone and menoquinone to Fe (II). (Fig. 23) The two-decked (face-to-face) porphyrin (Fig. 22h) and three-decked porphyrin (Fig. 22 l) are very efficient donors-acceptors e.g., O_2 is reduced to water with ease. The fossil porphyrins also exhibit a tendency of this property (Fig. 24 a to f). Comparing the 3-decked porphyrin with the Nonesuch porphyrin[29], the meso substitutents can act as the rim of a basket as showing by the condensed porphyrin systems[30] (Fig. 24f).

Fixation of metal in space, especially with back coödination of the 5th or 6th position as indicated in Viewpoint 7, can be best demonstrated by some enzyme catalysis. Ferredoxin is proposed to have a care-like cluster (Fig. 25) and this point can be clarified with a number of heterogeneous catalysis involving with zeolite. Many aluminosilicates contain polyhedron cavities can act as sites for the lock-in metals with certain spacial and topographic properties. (Fig. 26) In fossil fuel the asphaltic contain a number of gaps and holes in the large aromatic sheet. The edge sites are heterocyclic atoms consisting of N, S and O or X, which are perfect centers to accept the metal. In reality the metal atoms fit the space perfectly.

IV. Conclusive remarks

A meso-scaled structure of asphaltene is shown as Fig. 27; the saturated hydrocarbons are represented by zig-zag lines and the aromatic hydrocarbons are indicated by straight lines. This structure can explain many properties and features associated with asphaltic material. Any representation of fine structure of asphaltene can be included in the meso-scaled structure. To site a few: the Yen's structure of native petroleum asphaltene from Laguilinas crude structure (Fig. 28), Speight-Moschopedis's structure of petroleum asphaltene (Fig. 29), Strausz's structure of asphaltene from oil sands. (Fig. 30). All these fine molecular structure can fit this meso-scaled structure of Fig. 27.

So far we have considered three items which can be identified from the meso-scaled structure. Disclination can be viewed as bending of the aromatic stacks and dislocation can be viewed as slippage of aromatic sheets. Rearrangement or reorganization of the aromatic sheets may account for the viscous properties whereas the zig-zag spring is representative of the elastic properties. Metal in porphyrin behaves as an aromatic sheet which can be easily associated with a stack. Metals in the center of π-stacks can be intercalates by π-bonding. The meso-scaled structure has other features which will be discussed elsewhere.

This paper marks the first time that membrane-mimetic chemistry has helped geochemistry. Doubtlessly more applications will be available in the future.

Figure 26. Faujasite-type zeolites. (a) Building blocks, (b) idealized projection (c) and section through a solidite unit of faujasite. (After B.W. Wojcichowski and A. Corma, "Catalytic Cracking," Dekker, 1986).

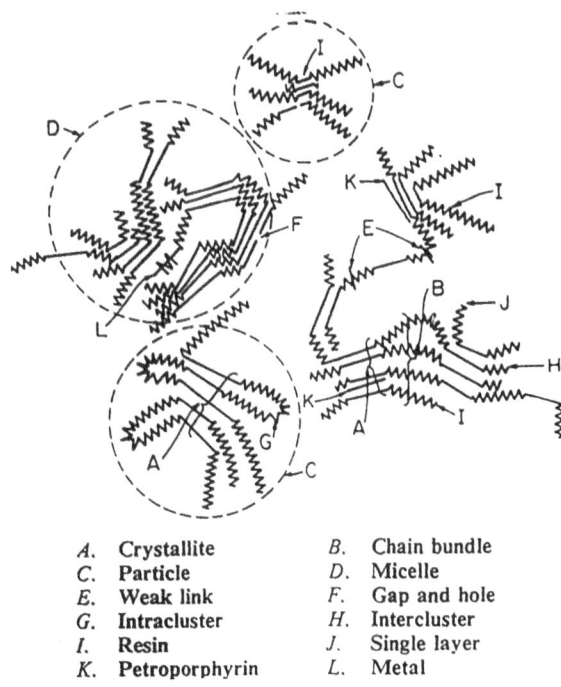

A.	Crystallite	*B.*	Chain bundle
C.	Particle	*D.*	Micelle
E.	Weak link	*F.*	Gap and hole
G.	Intracluster	*H.*	Intercluster
I.	Resin	*J.*	Single layer
K.	Petroporphyrin	*L.*	Metal

Figure 27 · A meso-scaled structure of asphaltics. (It is heavily weighted towards the microside. After J.P. Dickie and T. F. Yen, *Anal chem.* 39 1847-1852 (1967).

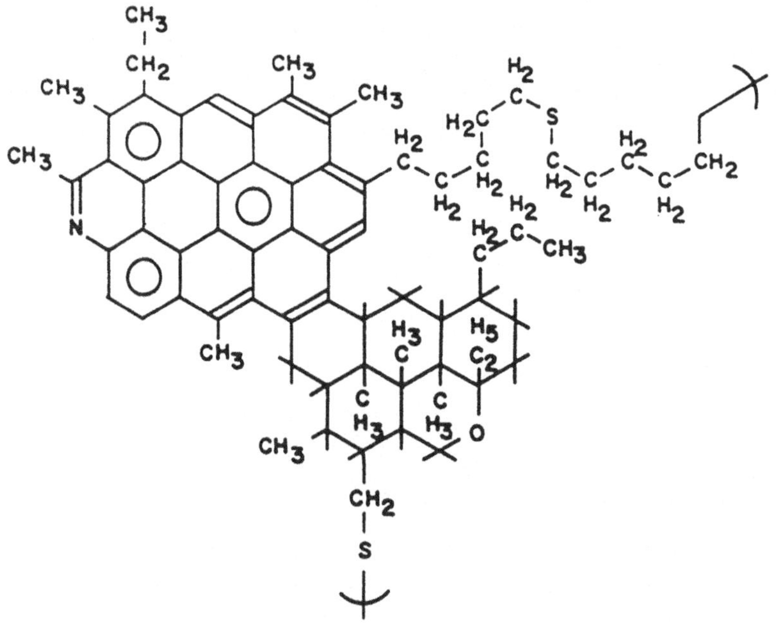

Figure 28   Hypothetic structure for a Laquinillas asphaltene. (After T.F. Yen *Adv. chem. Ser.* 195, 39-52 (1981).

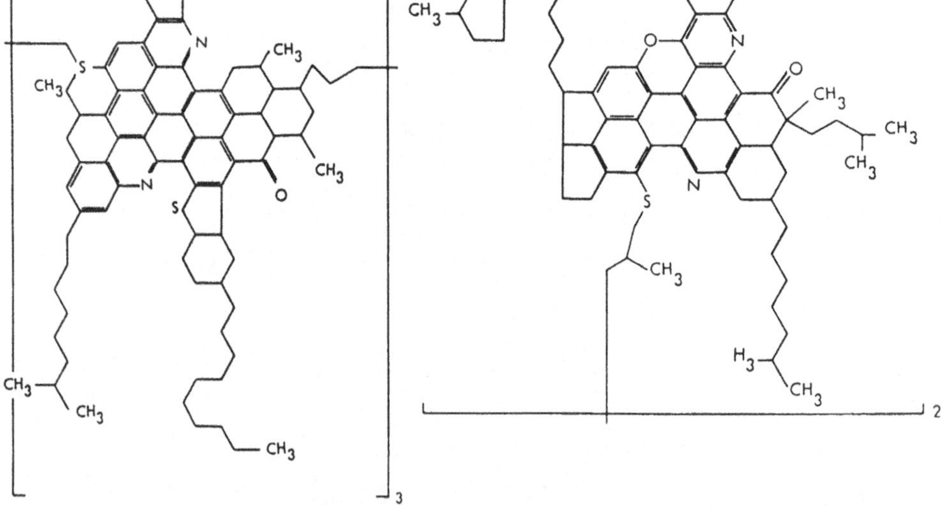

Figure 29. Hypothetical structures for asphaltenes from (a) Venezuelan crude oil and (b) Californian crude oil. (After J.G. Speight, The Chemistry and Technology of Petroleum 2nd Ed., Dekker, New York, 1991).

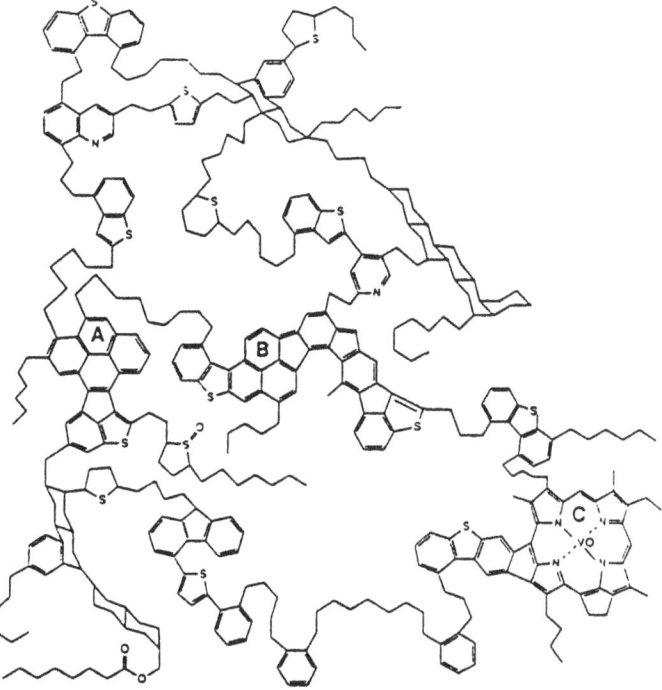

Figure 30. Hypothetical asphaltene molecule. A, B and C represent larger aromatic clusters; the rest of the structural units are based on experimental data (After Strausz et al. *Fuel* 71, 1355-63 (1992).

REFERENCES

1. A.E. Romanov and V.I. Valdimirov, Disclinations in crystalline solids, in: "Dislocations in Solids, vol. 9, of: "Dislocations and Disclinations, (F.R.N. Nabarro, ed)," 191-402 North-Holland, Amsterdam (1992).

2. T.F. Yen, Nitrogen superhyperfine splittings of vanadyl porphyrins in native asphaltenes, *Natuwissenshaften* (Berlin) 58, 267-268 (1971).

3. T.F. Yen, Vanadium and its bonding in petroleum *in*: "The Role of Trace Metal in Petroleum," 167-182, Ann Arbor Science Ann Arbor, MI (1982).

4. B.R. Gelin, A.W.N. Lee and M. Karplus, *J. Mol. Biol.* 171, 542 (1983).

5. J.E. Zimmer and J.L. White, Disclination in the carbonaceous mesophase, *Mol. Cryst. Liq. Cryst.* 38, 177-193 (1977).

6. H. Honda, H. Kimura and Y. Sanada, *Carbon* 9 695 (1971).

7. N.W. Tschoegl, "The Phenomenological Theory of Linear Viscoelastic Behavior, An Introduction," Springer-Verlag, Berlin (1989).

8. F.R.N. Nabarro, "Theory of Crystal Dislocations," Oxford University Press, London (1967).

9. J.L. White, Mesophase mechanisms in the formation of the microstructure of petroleum coke, *ACS Symp. Ser.* 21 282-314 (1975).

10. T.F. Yen, Carbonaceous mesophase formation, NSF Workshop, Morgantown, WV (1980).

11. E.J. Barth, "Asphalt Science and Technology," Gordon and Breach, New York (1962).

12. B.E. Bowler, L.S. Hollis and S.J. Lippard, Synthesis of DNA binding and photomicking properties of acridine orange linked by a polymethylene ether to (1, 2-diaminoethane) dichloroplatinum II, *JACS* 106, 6102-6104 (1984).

13. M.J. Gunter and J.M. Mander, Synthesis and atropisomer of porphyrin containing functionization at the 5, 15-meso positions application to the synthesis of binuclear ligand systems, *J. Org. Chem.* 46, 4792-5 (1982).

14. L.J. Boucher, E.C. Tynan and T.F. Yen, Spectral proerties of oxovanadium (IV) complexes IV, correlation of ESR spetra with ligand types, *in*: "Electron Spin Resonance of Metal Complexes," 111-130, Plenum Press New York (1969).

15. L.E. Nielsen, "Mechanical Properties of Polymers and Composites," Marcel Dekker, New York, Vol. 1 and 2 (1974).

16. R.J. Young and P.A. Lovell, "Introduction to Polymers, 2nd ed.," Chapman and Hall, London (1991).

17. E.A. Grulke, "Polymer Process Engineering," Prentice-Hall, Englewood Cliffs, New York (1994).

18. T.F. Yen, Chemical aspect of metals in native petroleum, *in*: "The Role of Trace Metals in Petroleum," 1-30, Ann Arbor Science Ann Arbor, MI (1982).

19. D. Voet and J.G. Voet, "Biochemistry," John Wiley, New York (1990).

20. H. Dugas, "Bioorganic Chemistry, A Chemical Approach to Enzyme Action," 2nd ed., Springer-Verlag, New York (1989).

21. S. Takagi, T.K. Miyamoto and Y. Sasaki, A new synthetic model for myoglobin: Tulip garden porphyrin, Bull. Chem. Soc. Jap. $\underline{58}$, 447-454 (1985).

22. T.G. Traylor, D. Campbell, S. Tsuchiya, Cyclophane porphyrin 2 Models for steric hinderence to CO ligation in hemoproteins, JACS, $\underline{101}$ 4748-4749 (1979).

23. J.P. Collman, J.C. Brauman, E. Rose, and K.S. Suslick, Cooperativity in O_2 binding to iron porphyrins, Proc. Nat. Acad. Sci., $\underline{75}$ 1052-1055 (1978).

24. J. Almog, J.E. Baldwin and J. Huff, Reversible oxygenation and autooxidation of a capped porphyrin iron (II) complex, JACS, $\underline{94}$, 227-228 (1975).

25. C.K. Chang, Stacked double macrocyclic ligands 1. Synthesis of crowned porphyrin, JACS, $\underline{99}$, 2819-2822 (1977).

26. I. Tabushi, S.I. Kugimiya, M.G. Kinnaid and T. Sasaki, Artificial allosteric system 2. cooperative 1-methylimidazole binding to an artificial allosteric system, zinc-gable porphyrin-dipyridylmethane complex, JACS, $\underline{107}$, 4192-4199 (1985).

27. R.R. Bucks and S.G. Boxer, Synthesis and spectroscopic properties of a novel cofacial chlorophyll—base dimer, JACS $\underline{104}$, 340-343 (1982).

28. J.P. Collman, C.S. Bencosme, C.E. Barnes, B.D. Miller, Two new members of the dimeric β-linked face-to-face porphyrin family, FTF$_4$ and FTF$_3$, JACS $\underline{105}$, 2704-2710 (1983).

29. J.H. Rho, A.J. Banman, H.G. Boetgar and T.F. Yen, A search for pophyrin in biomarkers in Nonesuch shale and extraterrestrial samples, *Space Life Sci.* $\underline{4}$, 69-77 (1973).

30. T.F. Yen, Terrestrial and extraterrestrial stable organic molecules, *in*: "Chemistry in Space Research (R. F. Landel and A. Rembaum, ed)" 105-153 Elsevier, Amsterdam (1972).

CONTRIBUTORS

N. Ashraf, Chemistry Department, University of Missouri-Kansas City, 5100 Rockhill Road, Kansas City, Missouri, 64110-2499, USA

P.A. Buchmann, Institut fur Polymere, ETH Zentrum, CH 8092 Zurich, Switzerland

Y.F. Chang, Department of Chemical Engineering, Tunghai University, Taichung, 40704 Taiwan

J. Chen, Department of Civil Engineering, University of Southern California, 3620 South Vermont Ave, Los Angeles, CA, 90089-2531 USA

Q. Chen, Institute of Photographic Chemistry, Academia Sinica, Beijing, 100101 China

K.L. Cheng, Chemistry Department, University of Missouri-Kansas City, 5100 Rockhill Road, Kansas City, Missouri 64110-2499, USA

J.H. Fendler, Department of Chemistry, Syracuse University, Syracuse, New York 13244-4100, USA

E. Friberg, Center for Advanced Material Processing and Department of Chemistry, Clarkson University, Potsdam, New York 13699-5814, USA

K. Hamdini, Department of Chemistry, University of Missouri-Kansas City, 5100 Rockhill Road, Kansas City, Missouri 64110-2499, USA

J.P. Hsu, Department of Chemical Engineering, National Taiwan University, Sec. 4 Roosevelt Rd. 1 Taipei, Taiwan

C.I. Huang, Department of Chemistry, National Sun-Yat Sen University, Kaoshiung, 80421 Taiwan

H.J. Huang, Department of Chemistry, National Sun-Yat Sen University, Kaoshiung, 80421 Taiwan

M. Hudson , Fiber and Polymer Science Program, North Carolina State University, Raleigh, North Carolina 27695-8302, USA

M.S. Lin, Department of Applied Science, Biosystems and Process Sciences Division, Brookhaven National Laboratory, Upton, New York 11973 USA

L. Jiang, Institute of Photographic Chemistry, Academia Sinica, Beijing, 100101 China

P.L. Luisi, Institut fur Polymere, ETH Zentrum, CH 8092 Zurich, Switzerland

Z. Ma, Center for Advanced Material Processing and Department of Chemistry, Clarkson University, Potsdam, New York 13699-5814, USA

D.Y. Pharr, Virginia Military Institute, Lexington, Virginia 24450, USA

E.T. Premuzic, Department of Applied Science, Biosystems and Process Sciences Division, Brookhaven National Laboratory, Upton, New York, USA

T. Rathke, Fiber and Polymer Science Program, North Carolina State University, Raleigh, North Carolina 27695-8302, USA

K.M. Sadeghi, Department of Civil Engineering, University of Southern California, 3620 South Vermont Ave., Los Angeles, CA 90089-2531, USA

P.K. Schmidli, Institut fur Polymere, ETH Zentrum, CH 8092 Zurich, Switzerland

J.M. Schnur, Center for Bio/Molecular Science and Engineering, Naval Research Laboratory, Washington, D.C. 20375-5000, USA

E. Sheu, Texaco R & D, PO Box 509 Beacon, New York 12508, USA

A.N. Singh, Center for Bio/Molecular Science and Engineering, Naval Research Laboratory, Washington, D.C. 20375-5000, USA

C.D. Tran, Chemistry Department, Marquette University, 535 North 14th Street, Milwaukee, Wisconsin 53233, USA

P. Walde, Institut fur Polymere, ETH Zentrum, CH 8092, Zurich, Switzerland

T.F. Yen, Department of Civil Engineering, University of Southern California, 3620 South Vermont Ave., Los Angeles, CA 90080-2531, USA

INDEX

Geoporphyrin, 273
Glass electrode, 209

Hamaker constant, 179, 185
Halobacterium halobium, 191
Halorhodopsin, 194
Hard sphere system, 135
Heat capacity, 60
Hemosomes, 157
Hologram recording, 201
Humicola lanuginosa, 29
Hydrodesulfurization, 273
Hydrogen, 198
    bonds, 62

Immobilization biocatalysis, 250
Immobilized enzymes, 151
In vitro, 143
Interacting surfaces, 101
Interferences, 90
Interlayer spacing, 42
Internal surface area, 211
Ionic strength, 186
Ionogenic groups, 188

Keratinocyte cultures, 41

Lamellar structure, 132
Lanthanide ions, 58
LB films, 72
    enzyme electrode, 68
Lecithin, 41
Leucocytes, 184
Ligands, 82
Lipases, 29
Lipids, 247
    lipid-heme(s), 154
Liposomes, 195
Longitudinal acoustic modes, 159
Low angle x-ray diffraction, 42, 43

Mechanical stability of vesicles, 152
Membrane-enclosed enzyme catalysis, 250
Mesophase, 19
Meso-scaled structure, 255
Metal fixation, 268
Metallic electrodes, 67
Methacrylite lipids, 145
Micro-compartmentalization, 33
Microelectronic applications, 151
Microsomes, 34
Minimal life, 37
Model, 53
    aberrant, 54
    nonlinear charge regulation model, 96
    parabolic, 53
Molecular recognition, 154
Molecules
    amphipathic, 17
    recognition, 7

Monolayers, 2
Morphological integrity, 149
Morphology modulation, 157
Multi-valent cations, 180
Multicomponent system, 126
    packing, 135
Multi-layer membranes, 195
Multiphase enzyme catalysis, 241
Multipolymers, 255

Nanofabricated quantum wires, 10
Nanometer dimension, 1
Nanosized materials, 1
Nanostructured devices, 2
Nanotechnology, 1
Nebulization efficiency, 87
    waste, 89
Newtonian fluid, 107, 261
    Newtonian region, 118
    non-Newtonian fluid, 107, 261

Oil spills, 243
Oleophilic dyes, 22
Optical microscopy, 258
Optical phase conjugation, 202
Oriented particulate, 7
Oxygen carriers, 154

Particle shape, 109
Particulate film,
    nanocrystalline, 3
    semiconductor, 4
Peierls forces, 260
Penetration, 47
Peptization, 249
Percolation phenomenon, 111
Phagocytosis reaction, 187
Phamerochaete chrysosporium, 251
Phase diagram, 120, 130
Phosphatidylcholine, 34, 143
Phospholipids, 143
    polymerizable, 144
Photoelectrical measurements, 6
Photoelectroactive membrane, 198
Photopolimerization, 146, 147
Photoreceptors, 191
Photothermal phenomena, 51
Polar head-group region, 149
Polychlorobiphenyls, 243
Polydispersity, 109, 129, 131
Polyelectrolyte mesophases, 17
Polymerized liposomes, 152
Polymerized disulfide vesicles, 152
Polymerized vesicles, 151, 157
Polypyrrole, 68
Porphyrin, 273
Potential, 98, 100
Proteovesicles, 34
Proton-pumping mechanism, 192
Purple membrane, 191

The manufacturer's authorised representative in the EU is Springer
Nature Customer Service Centre GmbH, Europaplatz 3, 69115 Heidelberg,
Germany. If you have any concerns regarding our products, please
contact ProductSafety@springernature.com

Printed and bound by CPI Group (UK) Ltd, Croydon, CR0 4YY
23/04/2026
02095607-0007